John Fryer and *The Translator's Vade-mecum*

Studies in the History of Christianity in East Asia

VOLUME 4

The titles published in this series are listed at *brill.com/hcea*

John Fryer and *The Translator's Vade-mecum*

New Perspectives on the History of Modern Chinese Scientific and Technical Lexicon

By

Gabriele Tola

BRILL

LEIDEN | BOSTON

Cover illustration: Page 40 of the "List of Botanical Terms." Courtesy of the C.V. Starr East Asian Library, University of California, Berkeley.

Library of Congress Cataloging-in-Publication Data

Names: Tola, Gabriele (Chinese studies), author.
Title: John Fryer and the translator's vade-mecum : new perspectives on the history of modern Chinese scientific and technical lexicon / by Gabriele Tola.
Description: Leiden ; Boston : Brill, [2021] | Series: Studies in the history of christianity in east Asia, 2542-3681 ; volume 4 | Includes bibliographical references and index.
Identifiers: LCCN 2020040622 (print) | LCCN 2020040623 (ebook) | ISBN 9789004442207 (hardback) | ISBN 9789004443211 (ebook)
Subjects: LCSH: Fryer, John, 1839–1928. | Science—Translations into Chinese—Handbooks, manuals, etc. | Technology—Translations into Chinese—Handbooks, manuals, etc. | China—Civilization—Western influences. | Science—China—History—19th century. | Technology—China—History—19th century.
Classification: LCC Q124 .T65 2021 (print) | LCC Q124 (ebook) | DDC 501/.4—dc23
LC record available at https://lccn.loc.gov/2020040622
LC ebook record available at https://lccn.loc.gov/2020040623

Typeface for the Latin, Greek, and Cyrillic scripts: "Brill". See and download: brill.com/brill-typeface.

ISSN 2542-3681
ISBN 978-90-04-44220-7 (hardback)
ISBN 978-90-04-44321-1 (e-book)

This book is printed on acid-free paper and produced in a sustainable manner.

In loving memory of Sofia

Contents

Acknowledgements

It is only after the completion of this book, which is the result of hours of writing and years of research, that I realise the countless number of people I should and would like to thank. Since it would not be a good idea to bore the reader with endless lists, I will simply begin by expressing my gratitude for the important guidance I received in the compilation of this work from each scholar and professor who contributed to improving it. I want to especially thank professor Federico Masini, of Sapienza University of Rome, for his guidance in the years before, during and after my doctorate; professor Zou Zhenhuan 鄒振環, of Fudan University, for his support during my research in Shanghai; and professors Uchida Keiichi 内田慶市 and Shen Guowei 沈國威, of Kansai University, for their assistance during my postdoctoral studies in Osaka. I would also like to express my gratitude to the various staff members of the libraries I have visited around the world; they were always eager to help as much as possible with my multiple, annoying, and odd requests. I would like to convey my thanks particularly to the library assistant at Bancroft Library, David Kessler, Deborah Rudolph, curator of the rare materials collection of the East Asian Library at UC Berkeley, Mark Mir, archivist and resource coordinator of the Ricci Institute for Chinese-Western Cultural History at the Center for Asia Pacific Studies of the University of San Francisco, and the whole staff of the Institute of Chinese Literature and Philosophy of Academia Sinica. I would particularly like to thank the series' editors, M. Antoni J. Ucerler, S. J. and Wu Xiaoxin, for expressing their trust in my research, and editorial assistant Stephen Ford, for his patient and helpful advice during the drafting of this text.

This work is based on my doctoral and postdoctoral research. The revision and correction of the manuscript were also made possible thanks to the suggestions of all colleagues who helped by reading this long work. I would like to particularly thank Professor Giorgio Francesco Arcodia for his precious advice on the adopted terminology. Throughout the years of research spent at home and abroad, I was able to meet many other professors and colleagues. The serious talks with them, but also the less serious ones, were all fundamental and greatly helped, from different perspectives, in the completion of the manuscript. Among them, I would like to particularly thank professors Alessandra Brezzi, Paolo De Troia, Davor Antonucci, Luisa Paternicò, Federica Casalin, Zhang Tongbing, Emanuele Raini, Serena Zuccheri, Miriam Castorina, and Chiara Romagnoli. I would also like to express my appreciation to the anonymous reviewers for the important suggestions they provided for improving this book.

Putting aside the academic world momentarily, I will never forget the support I have received from my family and friends; they were able to tolerate my draconian schedules, the thousands of kilometres separating us for many years, the countless meetings and farewells at airports, phone calls at preposterous times of the day, joys and sorrows, and celebrations and grievances only virtually spent together. I would like to particularly express my biggest thanks to my wife and colleague Yating—she always supported and encouraged me even when I could not see the end of my research—as well as Laura, Pino, Simona, Luca, Shuhui, Jingzhang, Zonghua, Ziying, and Yinxuan. I would love to also thank my grandparents; they also made this all possible, but I am unable now to tell them in person how grateful I am. I would finally like to thank my nephew and niece, Matteo and Sara, for still—hopefully—loving such an almost always absent uncle.

Lastly, I would also like to express my gratitude to all those people who, to different degrees, overtly or covertly, did not believe in what I was doing; their disapproval has been one of the greatest forms of encouragement to carry on my research even when the completion of this manuscript looked more like "catching the moon at the bottom of the sea" than an actual possibility.

Figures and Tables

Figures

Tables

Abbreviations

BCGM	*Bencao gangmu* 本草綱目 (Compendium of Materia Medica)
BTWDCL	"Botanical Terms from William's Syllabic Dictionary of the Chinese Language"
CEMMD	*Chinese and English Modern Military Dictionary*
CT	"Vocabulary of Terms in Naval Architecture"
CRMJ	*The Chinese Recorder and Missionary Journal*
DECF	*Deutsch-Englisch-Chinesisches Fachwörterbuch*
EAL	East Asian Library (University of California, Berkeley)
ECD	*English-Chinese Dictionary of the Standard Chinese Spoken Language*
ECSD	*An English and Chinese Standard Dictionary*
FP	John Fryer Papers
FRGTOC	*Fifth Report of General Terms in Organic Chemistry*
GDSE	*General Description of Shanghae and Its Environs, Extracted from Native Authorities*
HXCL	*Huaxue cailiao Zhong-Xi mingmu biao* 化學材料中西名目表 (*Vocabulary of Names of Chemical Substances*)
HXYH	*Huaxue yuhui* 化學語彙 (Chemistry glossary)
JS	*Jinshi Zhong-Xi mingmu biao* 金石中西名目表 (*Vocabulary of Mineralogical Terms*)
LBT	"List of botanical terms"
LGN	"List of geographical names"
MCST	Modern Chinese Scientific Terminologies—*Jinxiandai hanyu xueshu yongyu yanjiu* 近現代漢語學術用語研究
PACC	*The Principal Articles of Chinese Commerce*
PBXDC	*Putong baike xin da cidian* 普通百科新大詞典 (New general encyclopaedic dictionary)
QJ	*Qiji Zhong-Xi mingmu biao* 汽機中西名目表 (*Vocabulary of Terms Relating to the Steam Engine*)
RMDM	*Renming diming liang biao* 人名地名兩表 (*Vocabulary of Proper Names*)
SDCL	*A Syllabic Dictionary of the Chinese Language*
STFNC	"Syllabary for the transfer of foreign names into Chinese"
TT	*Technical Terms, English and Chinese* (Mateer)
TVM	*The Translator's Vade-mecum*
UYT	*Up the Yang-tse*
VCT	"Vocabulary of chemical terms"

VGT	"Vocabulary of geological terms"
VHCL	*A Vocabulary and Hand-Book of the Chinese Language*
VPN	"Vocabulary of proper names"
XYDCYP	*Xiyao dacheng yaopin Zhong-Xi mingmu biao* 西藥大成藥品中西名目表 (*Vocabulary of Names of Materia Medica*)

Introduction

This work is a comprehensive study of the TVM, a collection of scientific and technical glossaries with English-Chinese parallel translation, compiled by the English scholar John Fryer (1839–1928), known in China by the name Fu Lanya 傅蘭雅. More specifically, the purpose of this work is to analyse the TVM mostly from a lexicological and philological point of view, researching its role, diffusion, and contribution to the Chinese lexicon. As well as contributing to the history of language and translation in late Qing China, this work analyses the role of the TVM in the diffusion of ideas and terms conveyed between China and the West, as well as employing and integrating the approaches of cultural history, history of science, and conceptual history. Finally, with John Fryer as an epitome, it provides new insights on the topic of the connection between religious efforts by missionaries in late Qing China and their secular attitude towards translation.

I decided to study the TVM, its contents and inner evolution, the history of its publication, its diffusion, and its influence on modern Chinese lexicon for a variety of reasons.[1] From a general point of view, the choice to focus on the terminological peculiarities and translation choices adopted in the TVM derives from the assumption that researching the notions and nomenclature behind it should never be considered a mere proof of concept: "Investigating concepts and their linguistic history is as much a part of the minimal condition for recognizing history as is the definition of history as having to do with human society."[2] Furthermore, the literati during the late Qing dynasty (1644–1911) and Republican era (1911–1949) "sought to control the world, as they always had, by writing about it. Textbooks were one of the main sites of this textual production."[3] It can be said that foreign missionaries and scholars also tried to

1 Unless otherwise specified, the reference in this work is always to the standard variety of Chinese, or *putonghua*. For an excellent presentation, see Yip Po-Ching, *The Chinese Lexicon: A Comprehensive Survey* (London, New York: Routledge, 2007).

2 Reinhart Koselleck, *The Practice of Conceptual History: Timing History, Spacing Concepts*, trans. Todd Samuel Presner and others (Stanford, California: Stanford University Press, 2002), 20.

3 Robert J. Culp, "'Weak and Small Peoples' in a 'Europeanizing World': World History Textbooks and Chinese Intellectuals' Perspectives on Global Modernity," in *The Politics of Historical Production in Late Qing and Republican China*, ed. Tze-ki Hon and Robert J. Culp (Leiden, Boston: Brill, 2007), 211.

 | DOI:10.1163/9789004443211_002

accomplish the same objective through the editing of dictionaries, glossaries, and grammar textbooks.[4]

Within such a big frame, the choice of this specific text has been determined by one factor. Numerous monographs and studies that make extensive use of the approaches mentioned above, together with a history of ideas, have been already printed. They analyse the translation work of foreign missionaries and scholars in China in the bigger frame of the *Xixue Dongjian* 西學東漸, or "Transfer of Western knowledge to the East," late Qing China included, which represents the main temporal focus of this work.[5] Even so, a monograph comprehensively analysing the *TVM*, from the ideation to its birth, from the publication to its diffusion, had not been published yet.

The hypothesis that the glossaries Fryer gathered in the *TVM* could have played an important role in the history of Chinese scientific and technical lexicon, leaving a deep mark within it, derives from a basic assumption—some works have already been devoted to Fryer, most of them highlighting his contribution to the creation of a method for the nomenclature of chemical elements, also through the creation of characters, *zi* 字, illustrated in detail in section 1.1. The creation of characters, though, represents in modern Chinese an extremely low percentage of lexical innovation. The latter, instead, is carried out more frequently, if not almost exclusively, through the combination of existing characters, thus creating new words (*ci* 詞) and new meanings. This is exactly in line with the path towards polysyllabism that Chinese was already more intensively following in the temporal period considered here.[6] More specifically, one of the most important outcomes of this work is the identification of a temporal, conceptual, and lexicological line connecting Fryer's *TVM*, Protestant authors, and texts by the Jesuits, notwithstanding the widespread negative opinion held by the second on the third.[7] The *TVM* indeed adopted

---

4 Refer also to Ssu-yü Teng and John King Fairbank, *China's Response to the West: A Documentary Survey, 1839–1923* (Cambridge: Harvard University Press, 1954), 134.

5 For an overview of the history of scientific translation, focusing on the linguistic issues related to the Chinese language, refer to David Wright, *Translating Science: The Transmission of Western Chemistry into Late Imperial China, 1840–1890* (Leiden, Boston, Köln: Brill, 2000).

6 Federico Masini, *The Formation of Modern Chinese Lexicon and Its Evolution Toward a National Language: The Period from 1840 to 1898* (Berkeley: Journal of Chinese Linguistics, Monograph series number 6, 1993), 122 and following.

7 Catherine Jami, Peter Engelfriet, and Gregory Blue, ed., *Statecraft and Intellectual Renewal in Late Ming China: The Cross-Cultural Synthesis of Xu Guangqi (1562–1633)* (Leiden, Boston, Köln: Brill, 2001), 51. This was not the case, though, with Fryer; refer to John Fryer, "The Present Outlook for Chinese Scientific Nomenclature," in *Records of the Second Triennial Meeting of the Educational Association of China, Held at Shanghai, May 6–9, 1896* (Shanghai: American Presbyterian Mission Press, 1896), 155–56.

the terminology of other contemporary lexicographical materials, in turn connecting on different levels to Jesuit and native sources. This incontrovertible lexicological connection, and consequently that of the ideas behind such texts, is a difficult aspect to research, but at the same time, as confirmed in this work, can lead to major outcomes.[8]

Putting the objectives and accomplishments of this monograph aside for a moment, it should be kept in mind that when examining the birth of a text and its diffusion once published, there is an unavoidable arbitrariness in the selection of works to be considered for the research range. Even in our digital era, it would be impossible, for practical reasons, to trace back, consult, and analyse all the successive texts that cover the aspects treated in the *TVM*, and that could have used Fryer's work as a reference. The branches touched upon in the *TVM* range from chemistry to botany, from mineralogy to engineering. I had to limit the choice to some of these texts, mostly opting for those less researched in Chinese and non-Chinese secondary literature, and others that have instead been thoroughly investigated but whose employment of the *TVM* has not yet been ascertained. The use of databases can certainly help the research process and lexicographical comparison. Among the most authoritative, the *Zhongguo jiben guji ku* 中國基本古籍庫 (Database of Chinese classic ancient books), which comprises more than 10,000 texts from the Qin dynasty (221–206 BCE) to the Republican era, and the "extraordinary"[9] Modern Chinese Scientific Terminologies—*Jinxiandai hanyu xueshu yongyu yanjiu* 近現代漢語學術用語研究—A Repository of Chinese Scientific, Philosophical and Political Terms Coined in the Nineteenth and Early Twentieth Century, created by Michael Lackner, Iwo Amelung, and Joachim Kurtz.[10] Another referenced electronic resource is the repository of English-Chinese dictionaries provided by the Institute of Modern History, Academia Sinica 中央研究院近代史研究所.[11]

---

8 Benjamin A. Elman, *On Their Own Terms: Science in China, 1550–1900* (Cambridge, MA: Harvard University Press, 2005), xxxi, and Lydia He Liu, ed., *Tokens of Exchange: The Problem of Translation in Global Circulations* (Durham: Duke University Press, 1999), 138.

9 Joshua A. Fogel, ed., trans., *The Emergence of the Modern Sino-Japanese Lexicon: Seven Studies* (Leiden, Boston: Brill, 2015), 1.

10 The database includes numerous sources; in addition to dictionaries and encyclopaedias, it comprises monographs, translations of texts covering different scientific branches, teaching materials, articles, essays, travel diaries, memoirs, and diaries written by Chinese diplomats, scholars, and students. At the time of the last consultation, August 2018, the titles in the database amounted to 8,500, including monographs and articles published between 1815 and 1916, for a total of more than 125,000 terms.

11 Accessible at http://mhdb.mh.sinica.edu.tw/dictionary/index.php. Last consultation: March 2019.

Even so, I also manually analysed numerous other texts presented in this work. The choice has been in most cases directed to dictionaries and glossaries, bilingual or monolingual, composed in the late Qing dynasty and Republican era, works that by their nature could more presumably use the *TVM* as a reference. Other compilatory texts published in the same time frame, mostly encyclopaedic and collectaneous works which, due to their nature and the similarity of arguments covered in the *TVM*, could have used Fryer's text, have also been studied and illustrated in this work. Finally, I chose miscellaneous texts and different typologies of teaching materials, compiled by institutions that carried on the standardisation of Chinese scientific and technical lexicon, particularly those published in the Republican era.

The unavoidable arbitrariness in choosing the texts considered reflects the selection of Fryer's translations and is connected to their accessibility—these texts are so numerous that in most cases it was necessary to make a choice based on their availability. Those consulted and presented here for their connection with the *TVM*, in addition to texts available through databases and digital libraries, are works stored for the most part in the National Library of China in Beijing, the Shanghai Library, the section of old books of Fudan University Library, Bancroft Library, Doe Library and EAL at UC Berkeley, and the library of Ricci Institute for Chinese-Western Cultural History at the Center for Asia Pacific Studies of the University of San Francisco.

The inevitable limitation mentioned above is a starting point for further lexicological and philological research on the textual production within the considered time frame and, in wider terms, on the *TVM*. The importance of the text compiled by Fryer is duly illustrated in this work, which is an instrument for scholars and a tile in the big mosaic of research conducted on the formation of modern Chinese lexicon, and the lexicographical textual production of the late Qing dynasty and the Republican era.[12] At the same time, the lexicographical analysis conducted casts a new light on cultural interactions between China and the West that took place, with different modalities, between the sixteenth and the beginning of the twentieth century.

More specifically, one of the research paths opened in this work is closely connected to the choice of texts consulted for the analysis of the diffusion of the *TVM*. The publications that consulted and employed the *TVM* as a reference, served in turn as a platform and vector for the diffusion of the lexicological innovations devised by Fryer in the *TVM*. The diffusion these texts had

12 Consider the research conducted, among other institutions, at Sapienza University of Rome, Kansai University, Universität Heidelberg, Fudan University, Beijing Foreign Studies University, and the Ricci Institute for Chinese-Western Cultural History at the Center for Asia Pacific Studies of the University of San Francisco.

in turn is studied in this work in most cases, but the *TVM*'s influence on texts published after its first edition can be analysed more deeply when further expanding the range of research. Analogously, another research approach indicated in this work is upstream of the creation of the *TVM*, pointing out other sources Fryer employed for the compilation of his collection of glossaries. Finally, starting from the aforementioned connection between contemporary sources and those immediately preceding the *TVM*, such as those compiled by Protestant and non-Protestant missionaries, and Jesuit sources, it is possible to better clarify the line that ties traditional nomenclature and lexical innovation that is rooted much earlier than Fryer's work, for example in native Chinese literature. Both research directions, *ante* and *post quem* the publication of the *TVM*, represent an instrument and a starting point for scholars working on lexicology and lexicography. They also show the history of the diffusion of ideas and terms, and how they were conveyed between China and the West. I believe, even if only slightly, to have contributed to such an enormous task with this work, of which the contents are summarised below.

In the first chapter, I briefly introduce the main compiler of the *TVM*, John Fryer, and the most important aspects of his life. Any information connected to the birth of his lexicographical work is particularly highlighted, in addition to abundant bibliographical references. In this chapter, I especially focus on representations of Fryer, the historical significance of his work, his connections with the religious world of late Qing China, and his cooperation with Chinese translators.

In the second chapter, the structure, contents, and compositional features of the *TVM* are presented in detail. Fully drawing from sources contemporary to Fryer, I first outline the planning and publication process of the text, highlighting the history of the different glossaries that compose it, while pointing out for the first time some issues connected to the printing of a second volume of the *TVM*, which was never achieved. Different editions of the *TVM* are also compared in this chapter, indicating at the same time some discrepancies in their publication dates.

In the third chapter, I analyse Fryer's theories on the translation of scientific and technical nomenclature, stressing his commitment for its standardisation and for the publication of glossaries, for the sake of other translators and also for the development of the Chinese language itself. In this chapter, important analyses on Fryer's view on the evolution of the Chinese language, on the standardisation of Chinese relevant nomenclature and its connections with the missionary cause are also presented.

In the fourth chapter, the five glossaries published in the *TVM* in 1888 are studied separately. Starting from the English and Chinese prefaces of each glossary, I briefly present their history. In particular, the evolution of each glossary, from the manuscript versions stored in the EAL to the version included in the *TVM* is stressed, focusing on some tendencies common to each and demonstrating how they constantly reflect Fryer's theories on scientific and technical lexicon. In this chapter, some of the sources of the glossaries, as well as their source-texts, are also presented; this comparison casts a light on the meticulous editing work carried out by Fryer.

In the fifth chapter, the manuscripts of other glossaries, stored in the EAL and not printed in the *TVM*, are also presented. With each manuscript, I show the most prominent peculiarities, sources, differences in the glossaries published in the *TVM*, and the reasons why they were ultimately not printed, while referring back to the vicissitudes of the publication of the *TVM* mentioned in the second chapter. The comparison between the manuscript and the printed glossaries is particularly important to reveal the great number of lexicographical sources Fryer consulted, including Japanese and Jesuit ones, thus providing a glimpse into the amount of information representing the background of the *TVM*.

In the sixth chapter, I analyse the lexicological features of the Chinese scientific and technical nomenclature adopted in the glossaries composing the *TVM*. The translation choices adopted in the text are divided into four main categories; any connections with the lexicological peculiarities of the Chinese language and the reasons for their suitability, or lack thereof, to its lexical texture are duly illustrated for each. More specifically, some tendencies in the passage from the manuscript to the printed versions of the glossaries included in the *TVM* are shown in relation to the different translations adopted. Fryer's attempt to use existing scientific and technical nomenclature, in line with the analysis conducted in the second chapter, is also highlighted. Some of the neologisms included in the *TVM*, listed in the tables in the appendices, comprising previous dates, their meaning, possible differences in meaning with previous attestations, and *TVM*'s glossaries in which they are listed, are dated back through etymological dictionaries and lexicographical databases, and carefully analysed. In this chapter, particular attention is paid to the lexicological innovations brought forth in the *TVM*, at the same time opening up new research perspectives.

In the seventh chapter, the role, diffusion, and influence the *TVM* exercised on later texts are analysed in detail. I particularly show, with many examples, how texts such as encyclopaedias, collectanea, and periodicals used the *TVM* as a fundamental reference for a considerable length of time, extensively adopting the nomenclature employed in the *TVM*, and contributing to its diffusion and the notions behind it. I especially focus on the dictionary compiled by Yan

Huiqing 顏惠慶 (1877–1950), *Ying-Hua da cidian* 英華大辭典, a landmark in the lexicographical landscape of the Chinese language.

In the eighth chapter, the late diffusion of the *TVM* is outlined, showing how its lexicological influence and the role for the transcultural flows of knowledge faded gradually in the first half of the twentieth century, still leaving its traces even in contemporary Chinese scientific and technical lexicon and in the scholarly milieu. More specifically, in this chapter I stress how those institutions and committees that carried on the standardisation of Chinese nomenclature praised and highly evaluated Fryer's work in the *TVM*, even when their translation choices diverged from it.

In the concluding ninth chapter, different aspects pertaining to the history of scientific translation, cultural exchanges between China and the West, the modernisation process of China in the nineteenth and twentieth centuries, as well as in the history of science and linguistics presented in this work, are summarised. Finally, in the appendices I provide the transcriptions of some pages from the *TVM* edition printed in 1888, including samples from manuscript versions and unpublished glossaries, tables with the *TVM* structure and its manuscripts, and a comparison of original texts and Chinese translations adopted in the *TVM*. A detailed list of references cited in this work closes the volume.

CHAPTER 1

John Fryer and *The Translator's Vade-mecum*

1 Introduction

The life of John Fryer, author of the *TVM*, has already been duly described, but a point in which clarity is still due is his role in China; this aspect concerns the purpose of the *TVM* and consequently of its sources, origin, and diffusion.[1] Particularly, one aspect of his work is still considered somehow obscure—the relation between his original missionary purposes, which sent him to China, and subsequent abandonment.[2] This ambiguity patently reflects on the titles used to refer to Fryer, oscillating between that of missionary, translator, and the more generic "Englishman."[3] If the first one seems misleading, particularly when discussing his strenuous efforts for the Chinese translation of scientific and technical treatises, the second would exceedingly limit the range of his work background.[4] Even though John Fryer's son, Charles Fryer, personally confirmed that his father was not a missionary,[5] John Fryer's ties with the missionary milieu of the Jiangnan area were solid throughout his entire stay in China.[6] This is demonstrated by his participation in various Protestant

---

1 Unless otherwise specified, information on Fryer's life presented in the section is drawn from Adrian Arthur Bennett, *John Fryer: The Introduction of Western Science and Technology into Nineteenth-century China* (Cambridge, MA: Harvard University Press, 1967); from the FP, carton 3, folder 32, "Genealogy of Fryer Family"; and from Wright, *Translating Science*, 100–27. See also Zhongguo shehui kexueyuan jindaishi suo fanyi shi, *Jindai lai Hua waiguo renming cidian* (Beijing: Zhongguo shehui kexue chubanshe, 1981), 155.

2 Wright, *Translating Science*, 107–109. Refer also to Bennett, *John Fryer*, 7 and following.

3 Rune Svarverud, "The Formation of a Chinese Lexicon of International Law 1847–1903," in *Mapping Meanings: The Field of New Learning in Late Qing China*, ed. Michael Lackner and Natascha Vittinghoff (Leiden, Boston: Brill, 2004), 516; Patrick Hanan, *Chinese Fiction of the Nineteenth and Early Twentieth Centuries* (New York: Columbia University Press, 2004), 5 and 124; and James Reardon-Anderson, *The Study of Change: Chemistry in China, 1840–1949* (Cambridge: Cambridge University Press, 1991), 25.

4 Refer also to Sakade Yoshinobu, *Chūgoku kindai no shisō to kagaku* (Kyoto: Dōhōsha shuppan, 1983), 482–83.

5 Rosemary Levenson and Elizabeth Huff, *Elizabeth Huff: Teacher and Founding Curator of the East Asiatic Library—From Urbana to Berkeley by Way of Peking* (Berkeley: Bancroft Library, University of California, Regional Oral History Office, 1977), 136.

6 Nevertheless, Fryer is not listed in Gerald H. Anderson, ed., *Biographical Dictionary of Christian Missions* (New York: Simon & Schuster Macmillan, 1998).

 | DOI:10.1163/9789004443211_003

missionary conferences,[7] his life membership in the Educational Association of China, which he joined in 1890, and his involvement in the translation of religious texts.[8] In a late phase of his time in China, Fryer even suggested that "It would be a wise movement to extend our publishing interests to the publication of medical and theological text books."[9]

Other proof of his efforts for the missionary cause in late Qing China can be found in numerous folders of the FP. The FP represents one of the main bodies of information on Fryer's life and his translation activities, including materials personally written by Fryer for teaching Chinese, articles on the economic, political, and cultural situation of end-of-nineteenth century China, translations, notes, letters, and copious miscellaneous material.[10] In addition to providing a glimpse, through Fryer's eyes, of late Qing China, where he spent almost half of his life, there are different folders in which Fryer describes the religious zealous efforts by missionaries in Shanghai, actively providing his suggestion for the cause.[11] Notwithstanding the abundance of materials and information on the topic, the ties between his translation work, his efforts in the diffusion of science in China, and his religious views are still considered contradictory.[12] The best definition of his role is probably the one by the American missionary Alvin Pierson Parker (Pan Shenwen 潘慎文, 1850–1924), who described Fryer as "a missionary of science to the Chinese."[13]

7 John Fryer, "Scientific Terminology: Present Discrepancies and Means of Securing Uniformity," in *Records of the General Conference of the Protestant Missionaries of China, Held at Shanghai, May 7–20, 1890*, ed. W. J. Lewis, William Theodore Aquila Barber, and John R. Hykes, 531–51 (Shanghai: American Presbyterian Mission Press, 1890), and Idem, "The Present Outlook."

8 Refer to Jost Oliver Zetzsche, *The Bible in China: The History of the Union Version or the Culmination of Protestant Missionary Bible Translation in China* (Sankt Augustin: Institut Monumenta Serica, 1999), 169–70, particularly footnote no. 50.

9 *Records of the Triennial Meeting of the Educational Association of China, Held at Shanghai, May 2–4, 1893* (Shanghai: American Presbyterian Mission Press, 1893), 26.

10 For a detailed analysis, refer to Gabriele Tola, "The John Fryer Papers: State of the Art and New Perspectives," in *Italian Association for Chinese Studies. Selected Papers 2*, ed. Tommaso Pellin and Giorgio Trentin (Venice: Cafoscarina, 2018), 138–48. A partial reproduction of some contents of the FP, integrated with other sources of primary and secondary literature, can be found in Ferdinand Dagenais, *The John Fryer Papers* (Guilin: Guangxi shifan daxue chubanshe, 2010).

11 For example, "Roman Catholicism in Shanghai and the Vicinity" and "Missionary Work in China."

12 Sun Banghua, "Yu Hua yingren Fu Lanya de zongjiao kexueguan chuyi," *Xueshu yuekan*, no. 8 (1993).

13 *Records of the Triennial Meeting*, 52–53.

Putting aside the above uncertainty about the definition of John Fryer's role, other details of his life are certain. He was born on August 6, 1839 in Hythe, England, where he attended school while doing odd jobs, given the complicated financial situation of his family. Fryer's father was a religious preacher, and John moved with the family close to Bristol for reasons related to his father's work. There Fryer attended St. James' School, earning a government scholarship to Highbury Training College in London, a school specialised in the formation of teachers. Fryer was assigned to St. Paul's College, or *Sheng Baoluo shuyuan* 聖保羅書院.[14] He arrived in Hong Kong in August 1861 and worked there for about two years, taking up office as a professor of English for the *Tongwenguan* 同文館 (School of Combined Learning) in 1863.

Following the two Opium Wars that took place in 1840–1842 and 1856–1860, China had been forced to take part in the framework of international relationships. Nevertheless, the lack of language schools had a negative influence on the nation's process of strengthening and modernisation.[15] The *Tongwenguan*, a school for the study of foreign languages, was therefore established in 1862, and many of its students would later be part of the first legations sent by China abroad, acting as interpreters.[16]

In Beijing Fryer had the chance to learn the northern variety of *guanhua* 官話 (Mandarin or official speech), thus being able to lay the foundations for his future career as a translator. Fryer also became acquainted with many figures close to the Qing court, including missionaries and translators who worked in Beijing and other cities in China, who would be destined to play an important role in the *Yangwu yundong* 洋務運動.

14 As confirmed in Charles Hodge Corbett, *Shantung Christian University (Cheeloo)* (New York: United Board for Christian Colleges in China, 1955), 23.

15 For further information, refer to Zhu Youhuan, ed., *Zhongguo jindai xuezhi shiliao*, two volumes (Shanghai: Huadong shifan daxue chubanshe, 1983) I, 4–213; Masini, *The Formation of Modern Chinese*, 44 and *passim*; Melissa Mouat, "The Establishment of the Tongwen guan and the Fragile Sino-British Peace of the 1860s," *Journal of World History* 26, no. 4 (2015); and Knight Biggerstaff, "The T'ung Wen Kuan," *The Chinese Social and Political Science Review* 18 (1934).

16 Other activities were also held in the institution; refer to Gao Xi, "Jingshi Tongwenguan de yixue jiangzuo," *Zhongguo keji shiliao—China Historical Materials of Science and Technology* 11, no. 4 (1990). For a study on the competing terminologies of *Tongwenguan* and Jiangnan Arsenal, see Rune Svarverud, "Zhongguo de kexue yiming zhi zheng—Tongwenguan yu Jiangnan zhizaoju fanyiguan guanyu Zhongguo kexue shuyu fanyi de zhengyi," in *Yinshua chuban yu zhishi huanliu—Shiliu shiji yihou de Dongya*, ed. Guanxi daxue wenhua jiaoshexue jiaoyu yanjiu zhongxin, Chuban bowuguan (Shanghai: Shanghai renmin chubanshe, 2011).

The *Yangwu yundong*, sometimes translated as "Western affairs movement" or "Foreign affairs movement," also known in Chinese as *Ziqiang yundong* 自強運動, "Self-strengthening movement," refers to the political, economic, and social atmosphere that developed after China's defeat in the two Opium Wars and the resulting unequal treaties. It is exactly in this context that, in 1865, Fryer left Beijing and headed to Shanghai to establish the Anglo-Chinese School, *Ying-Hua shuguan* 英華書館. A factor that greatly contributed to the decision, as well as the loss of his status as a missionary, was the rumours of an affair between his fiancée Anna Roleston (1838–1879) and the captain of the ship she was aboard heading to China.[17] Fryer worked in Shanghai as an English teacher, and then as editor of the *Shanghai xinbao* 上海新報 (Shanghai Gazette). After growing tired of his work as a teacher, when Fryer was later offered work with the Translation Department, *Fanyiguan* 翻譯館, of the Jiangnan Arsenal, *Jiangnan jiqi zhizao zongju* 江南機器製造總局, often abbreviated as *Jiangnan zhizaoju* 江南製造局, he accepted without hesitation.[18] His cooperation with the Arsenal started in 1868; early signs of his collaboration with the Jiangnan Arsenal can be found in different FP materials, in addition to the copy of his working contract.[19] For example, in a letter to his brother George dated February 6, 1868, he mentioned, "I may perhaps get a post that has been offered to me in general terms to translate scientific books into Chinese."[20] A letter addressed to Smith Elder & Co., in which seventy books to be bought to launch the translation work at the Arsenal are listed, is dated March 18, 1868.[21] In the letter, Fryer specifically states that the "officials in charge of the Chinese Arsenal here commissioned me to procure for them the enclosed list of scientific works and apparatus [...] This being a sort of experiment on the part of the Chinese Officials, it is possible it may lead to orders on a more extensive scale."

---

17 As indicated in Wright, *Translating Science*, 107–109.

18 The establishment of the *Fanyiguan* and the educational activities of the Jiangnan Arsenal, as well as other institutions for studying languages established after the *Tongwenguan* in Beijing, are described in Knight Biggerstaff, *The Earliest Modern Government Schools in China* (Ithaca, NY: Cornell University Press, 1961), 154–99. On the *Fanyiguan* predecessor, refer to Xiong Yuezhi, "Shanghai Guang fangyanguan shilüe," in *Shanghai shi yanjiu*, ed. Tang Zhenchang and Shen Hengchun (Shanghai: Xuelin chubanshe, 1989).

19 FP, Mss C-B 968, carton 3, folder 43, "Chinese Materials." Courtesy of The Bancroft Library, University of California, Berkeley.

20 Dagenais, *The John Fryer Papers*, I, 342.

21 Ivi, 348. For the policy governing the selection of works at the Guang fangyanguan, incorporated in the Jiangnan Arsenal in 1870, see also the text *Guang fangyanguan quan'an* 廣方言館全案 [Complete documents of the Guang fangyanguan], as reprinted in Yang Yi et al., *Haishang molin—Guang fangyanguan quan'an—Fen mo congtan* (Shanghai: Shanghai guji chubanshe, 1989), 99–163, particularly 119–34.

Fryer indeed cannot be considered a scientist in the strictest sense of the term; this was something he was aware of.[22] His linguistic abilities and the fact that he realised that Chinese officials were taking an interest in scientific objects, demonstrations, and consequently translations, were probably the factors that gained him a position at the Jiangnan Arsenal. He was firmly convinced that China needed to adopt "the same curriculum of study that is found in Western lands [...], in order to satisfy his demand for new education and learning."[23] It is possible therefore to suppose that in the first phase of Fryer's translation activities, the texts translated at the Jiangnan Arsenal were selected by the Chinese officials and that later, also thanks to the trust gained, both as a translator and as a scientific divulgator, his personal curiosity also played a more relevant role. In turn, Fryer's interest for science and scientific objects can be considered born at the intersection of personal motivation and practical convenience: the fact that Chinese officials showed a penchant exactly for those objects and subjects they planned to include in their translation projects was an example of interests that tallied perfectly with those of Fryer.

In any case, there is no agreement on the number of books composed and translated by Fryer for the Jiangnan Arsenal; there is even more debate on the topic of the cooperation between the foreign translator and his Chinese colleagues, and their respective achievements.[24]

22 Wright, *Translating Science*, 117–18 and passim.

23 FP, Mss C-B 968, carton 1, folder 3, "The Recent Educational Reform." Courtesy of The Bancroft Library, University of California, Berkeley.

24 The number differs according to different lists. Among the complete ones, the following should be mentioned: Bennett, *John Fryer*; Wang Yangzong, *Fu Lanya yu jindai Zhongguo de kexue qimeng* (Beijing: Kexue chubanshe, 2000); Dagenais, *The John Fryer Papers*; John Fryer, *An account of the Department for the Translation of Foreign Books at the Kiangnan Arsenal, Shanghai: With Various Lists of Publications in the Chinese Language* (Shanghai: American Presbyterian Press, 1880); Liang Qichao, "Xixue shumu biao," in *Zhixue congshu chuji*, ed. Zhixuehui (Wuchang: Zhixuehui, 1896); Huang Qingcheng, *Zhong-Xi putong shumu biao* (Shanghai: Suanxue baoguan, 1898); Xu Weize, *Dong Xi xue shulu* (1899); Jiangnan zhizaoju, *Jiangnan jiqi zhizao zongju shumu* (Shanghai: 1902); Wei Yungong, ed., *Jiangnan zhizaoju ji* (Shanghai: Wenbao shuju, 1905), II, 15A–23A; Chen Zhu, ed., *Jiangnan zhizaoju yishu tiyao* (Shanghai: Jiangnan zhizaoju, 1909); Jiangnan zhizaoju, *Shanghai zhizaoju yiyin tushu mulu* (Shanghai: 1911); Wang Yangzong, "Jiangnan zhizaoju fanyi shumu xinkao," *Zhongguo keji shiliao—China Historical Materials of Science and Technology* 16, no. 2 (1995); and John Fryer, *Catalogue of Educational Books, Works of General Knowledge, Scientific and Technical Treatises &c., &c., &c., in the Chinese Language: Translated or Written by John Fryer* (Shanghai: Presbyterian Mission Press, 1894), also mentioned in Henry Cordier, *Bibliotheca sinica: Dictionnaire bibliographique des ouvrages relatifs à l'Empire chinois. Supplément* (Paris: Ernest Leroux, 1895), 2134. See also the collection *Jiangnan zhizaoju yishu huike* 江南製造局譯書彙刻 [Collected prints of the books translated at the Jiangnan Arsenal], particularly the index volume, available in the library of Kyoto

The technique used at the Jiangnan Arsenal, already employed for the rendering of Buddhist texts centuries before, is known as *kouyi bishu* 口譯筆述 (oral translation and annotation with the brush), or *Xi yi Zhong shu* 西譯中述 (translation by the Westerner and annotation by the Chinese). The foreign expert, able to understand and speak Chinese, provided the approximate Chinese translation of a given work, then his colleague tried to polish the language and render it in the most suitable style. Credit for the translations produced, including the method of creation of characters for chemical elements, is sometimes only attributed to Fryer, at times only to his Chinese colleagues, with his role completely neglected. This also happened in publications for which Fryer was directly responsible, such as with "徐雪村新譯化學諸書 [...]" (Xu Xuecun—i.e., Xu Shou—recently translated many books on chemistry [...]).[25] In the catalogues of the time, his role was clearly stated as *fanyi* "繙譯" (translator) and that of his colleagues as *bishu* "筆述" (annotator).[26] Considering that it is almost impossible to establish where the area of expertise of the foreign translator finishes and that of his Chinese colleagues starts, it should be remembered that Fryer's work has been unanimously recognised.[27] The favourable opinion he enjoyed among his Chinese colleagues for his mastery of the language and the conscientiousness with which he used to gather information before starting a translation, once more confirm this hypothesis. We may conclude that this translation process should be best described as a more or less balanced contribution by both parts.[28]

University, the *Tōyō bunko* 東洋文庫 (Oriental Library), and the *Kokuritsu kokkai toshokan* 国立国会図書館 (National Diet Library), and Zhou Changshou, "Yikan kexue shuji kaolüe," in *Zhang Jusheng xiansheng qishi shengri jinian lunwenji*, ed. Hu Shi, Cai Yuanpei, and Wang Yunwu (Shanghai: Shangwu yinshuguan, 1937), 422–34.

25 Zhao Ziqin, "Zhongguo yiduo pin xiren cha kuang shuo lüe," *Gezhi huibian* 8 (September 1877), also republished in Chen Pingyuan and Xia Xiaohong, ed., *Tuxiang wan Qing* (Tianjin: Baihua wenyi chubanshe, 2001), 240. Refer also to Masini, *The Formation of Modern Chinese*, 64 and footnote no. 62, and Gabriele Tola, "John Fryer and *The Translator's Vade-mecum*: The Identification of the 'Vocabulary of Terms in Naval Architecture'," *Monumenta Serica* 66, no. 1 (2018): 105. The first monograph in Chinese to cast a light on John Fryer is Wang Yangzong, *Fu Lanya*, which fully recognises his role in the translations, including the creation of the nomenclature for chemical elements in Chinese.

26 As expressed in the *Tushu fu* 圖書附 [Appendix of books], included in Wei, *Jiangnan zhizaoju ji*, II, 15A–23A.

27 See, among others, William Alexander Parsons Martin, *The Awakening of China* (New York: Doubleday, Page & Company, 1907), 288.

28 There is a difference of opinion expressed in Wright, *Translating Science*, 232–33, in which the author states that Chinese colleagues of foreign translators were considered responsible "or even blamed for any inadequacies," and in Hill Michael Gibbs, *Lin Shu,*

Notwithstanding such dispute, Chinese translators who worked with Fryer were all representative personalities of late Qing China, and the influence of their translations and their personal work for the transmission of new knowledge has already been highlighted.[29] Among the most notable, we should mention Zhao Yuanyi 趙元益 (1840–1902), particularly active in the translation of medical texts[30] and member of the delegation that accompanied the minister Xue Fucheng 薛福成 (1838–1894) in Europe,[31] the scientist Xu Jianyin 徐建寅 (1845–1901),[32] and the father of the latter, Xu Shou 徐壽 (1818–1884).[33]

Xu Shou was born in Wuxi, a city in the Jiangnan area not far from Shanghai, and was one of the most important representatives among Chinese who took

Inc.: Translation and the Making of Modern Chinese Culture (New York: Oxford University Press, 2013), 33–34. Refer also to Fryer, "Scientific Terminology," in which he emphasises how, if he had been considered personally responsible for mistakes in translations, these flaws should be ascribed to "[...] ignorance or hurry rather than wilfulness."

29 Wright, *Translating Science*, 401. A detailed list of Chinese translators working with John Fryer can be found in Richard Gregg Irwin and Ferdinand Dagenais, *John Fryer's Legacy of Chinese Writings—Edited, with Supplementary Material Provided by Ferdinand Dagenais* (Berkeley: Center for Chinese Studies, University of California, 1999), section v.

30 Zhao Pushan, "Zhao Yuanyi he ta de bishu yishu," *Zhongguo keji shiliao—China Historical Materials of Science and Technology* 12, no. 1 (1991).

31 Bridie Andrews, *The Making of Modern Chinese Medicine, 1850–1960* (Vancouver: University of British Columbia Press, 2014), 124.

32 Refer to Wright 2000, 54–61, and the primary and secondary sources indicated in the relevant footnote no. 72; Wang Gengtang and Wang Xi'nan, *Xu Shou, Xu Jianyin zhuanlun* (Beijing: Zuojia chubanshe, 2001) and Wang Guangren, "Xu Jianyin nianpu," *Qinghua daxue xuebao (Zhexue shehui kexue ban)—Journal of Tsinghua University* 12, no. 3 (1997) can also be consulted.

33 For other information on Xu Shou, refer to Yuan Hanqing, "Xu Shou—Wo guo jindai huaxue de qimengzhe," *Huaxue tongbao* (February 1955); Zhang Qinglian, "Xu Shou yu *Huaxue jianyuan*," *Zhongguo keji shiliao—China Historical Materials of Science and Technology* 6, no. 4 (1985); Zhang Zigao and Yang Gen, "Xu Shou fuzi nianpu," *Zhongguo keji shiliao—China Historical Materials of Science and Technology*, no. 4 (1981); Wang Guangren, ed., *Zhongguo jindai kexue xianqu Xu Shou fuzi yanjiu* (Beijing: Qinghua daxue chubanshe, 1998); Wang Guangren and Xu Zhenya, *Haiguo xiezhu de Xu Shou fuzi* (Beijing: Kexue chubanshe, 2000); Yang Gen, "Wo guo jindai huaxue xianquzhe Xu Shou de shengping ji zhuyao gongxian," *Huaxue tongbao*, no. 4 (1984); Idem, *Xu Shou he Zhongguo jindai huaxue shi* (Beijing: Kexue jishu wenxian chubanshe, 1986); Wang Yangzong, *Guodu shidai de qiren. Xu Shou de gushi* (Changchun: Jilin kexue jishu chubanshe, 2012); and Du Shiran, ed., *Zhongguo gudai kexuejia zhuanji* (Beijing: Kexue chubanshe, 1993), 1230 and passim. Both Xu Shou and Xu Jianyin are mentioned in Arthur William Hummel, ed., *Eminent Chinese of the Ch'ing Period, 1644–1912* (Washington: United States Government Printing Office, 1943. Reprinted in Taipei: Ch'eng Wen Publishing Company, 1970), 540; see also Xu Hong, "Xu Shou, Xu Jianyin yu Zhongguo jindai kexue yinjin," *Chuban shiliao*, no. 1 (2008) and Xu Zhenya and Ruan Shenkang, "Xu Shou fuzi, zusun yizhu jianjie," *Zhongguo keji shiliao—China Historical Materials of Science and Technology* 7, no. 1 (1986).

an interest in Western science during the period after the Opium Wars. In 1871, Fryer and Xu translated, under the title *Huaxue jianyuan* 化學鑑原 (Mirror of chemistry),[34] David Ames Wells' (1828–1898) book, *Principles and Applications of Chemistry*.[35] In chapter 29 of the first juan of the *Huaxue jianyuan*, entitled *Huazi mingming* 華字命名 (Nomenclature with Chinese characters), Fryer and Xu list their translations in Chinese of the sixty-four chemical elements known at the time the original text was printed. The list is preceded by a theoretical presentation of how they rendered them. Fryer and Xu Shou used only one character for each chemical element, with the exception of five gaseous substances: oxygen, hydrogen, nitrogen, chlorine, and fluorine.[36] Fryer and Xu put together the three radicals *jin* 金 (gold) for metals, *shi* 石 (stone) for solid non-metals, and *shui* 水 (water) for those elements that tend to be liquid in their natural state, in the left part of the character, with a phonetic marker in the right part; Fryer and Xu did not use *qi* 氣 (gas) as a radical, but as the second character in disyllabic terms for the five elements mentioned above. The *xingsheng zi* 形聲字 (semantic-phonetic compound), according to the traditional classification of *liu shu* 六書 (six categories of Chinese characters) in the *Shuowen jiezi* 說文解字 (Explaining simple characters and analysing compound characters),[37] thus created had to be composed of as few strokes as possible. This is with the exception of most of the characters already in common usage, which Fryer and Xu Shou insisted should be preserved. Some of these existing characters were chosen among those in the *Kangxi zidian* 康熙字典 (Kangxi dictionary) already fallen into disuse, giving them a new meaning; others were created by Fryer and Xu Shou only when they could not find one that fit their needs.

The principles they applied were employed for chemical elements discovered at a later stage and are still used today for newly discovered elements; the characters they suggested imposed themselves over other concurrent

34 John Fryer and Xu Shou, trans., *Huaxue jianyuan* (Shanghai: Jiangnan zhizaoju, 1871).

35 David Ames Wells, *Principles and Applications of Chemistry; For the Use of Academies, High-Schools, and Colleges: Introducing the Latest Results of Scientific Discovery and Research, and Arranged with Special Reference to the Practical Application of Chemistry to the Arts and Employments of Common Life*, 5th ed. (New York: Ivison, Phinney and Co., 1862).

36 For later developments, see Zhang Hao, "Yang qing dan de fanyi: 1896–1944 nian," *Ziran kexue shi yanjiu—Studies in the History of Natural Sciences* 21, no. 2 (2002).

37 See also Jerry Norman, *Chinese* (Cambridge: Cambridge University Press, 1988), 67; Françoise Bottéro, "La vision de l'écriture de Xu Shen à partir de sa présentation des liushu," *Cahiers de linguistique. Asie orientale* 27, no. 2 (1998); and William G. Boltz, *The Origin and Early Development of the Chinese Writing System* (New Haven, CT: American Oriental Society, 1994), 143–46.

nomenclatures.[38] Even the five disyllabic renderings for the gaseous elements mentioned above, respectively *yangqi* 養氣 (oxygen), *qingqi* 輕氣 (hydrogen), *danqi* 淡氣 (nitrogen), *lüqi* 綠氣 (chlorine), and *fuqi* 弗氣 (fluorine), which were later destined to be discarded, played an important role and were in use for dozens of years after their proposal, to be replaced only in the first part of the twentieth century. As stated by Zheng Zhenwen 鄭貞文 (1891–1969), who was responsible for the standardisation of chemical terms for the Guoli bianyiguan 國立編譯館 (National Institute for Compilation and Translation),[39] "Hydrogen, oxygen, nitrogen, chlorine 此四元素之氣態單質，譯為輕氣，養氣，淡氣，綠氣，垂數十年" (The four gaseous elementary substances hydrogen, oxygen, nitrogen, chlorine were translated as *qingqi*, *yangqi*, *danqi*, and *lüqi*; this nomenclature was employed for dozens of years).[40] The one described above is the only example of lexical innovation in Chinese in modern times via the creation of new characters to permanently enter the lexicon.[41]

Fryer's collaboration with the Jiangnan Arsenal continued steadily until 1896—that year he moved to the United States to become the first Agassiz Professor of Oriental Languages and Literature at the University of California in Berkeley. Personal reasons and the sense that his work would not be as necessary as it used to be certainly played a role in the decision. Fryer was already aware that Chinese science, and consequently the terminology to convey its notions, would be autonomously carried on by Chinese people. The context in which this decision took place is also worth noting: the professorship was established by the regent Edward Tompkins (1815–1872) at the University of California as far back as 1872. In anticipation of growing commercial relations between California and East Asia, Tompkins realised that people involved

38 For a presentation of chemical lexicon in Chinese, refer to Viviane Alleton and Jean-Claude Alleton, *Terminologie de la chimie en chinois moderne* (Paris: Mouton, La Haye, 1966).

39 For more specific references to the Guoli bianyiguan, consult Tsien Tsuen-hsuin, "Western Impact on China through Translation," *The Far Eastern Quarterly* 13, no. 3 (1954): 320–21; Wen Changbin "Minguo shiqi Guoli bianyiguan ji qi kexue mingci shending gongzuo," *Keji shuyu yanjiu* 6, no. 3 (2004); and Viviane Alleton, "Chinese Terminologies: On Preconceptions," in *New Terms for New Ideas: Western Knowledge and Lexical Change in Late Imperial China*, ed. Michael Lackner, Iwo Amelung, and Joachim Kurtz (Leiden, Boston, Köln: Brill, 2001), 16.

40 Zheng Zhenwen, *Huaxue mingming fa cao'an chugao* (Shanghai: Zhonghua xueyi she, 1933), 9–10. The same text can be read in Guoli bianyiguan, *Huaxue mingming yuanze* (Chongqing: Guoli bianyiguan, 1933), 6. On Zheng Zhenwen, refer to Qian Yimin, "Zheng Zhenwen yu wo guo huaxue mingci tongyi gongzuo," *Keji shuyu yanjiu* 4, no. 3 (2002).

41 Masini, *The Formation of Modern Chinese*, iv. For a recent study on the topic, refer to Li Li, *Jindai huaxue yizhu zhong de huaxue yuansuci yanjiu* (Beijing: Zhongyang minzu daxue chubanshe, 2012).

in such an activity needed to master the language and the culture of that region.[42] Fryer was the first to fill the position, even though at first he was not interested in applying. During his appointment, he edited his teaching materials and asked for the assistance of other eminent professors, such as the sinologist James Legge (Li Yage 理雅各, 1815–1897), always putting the teaching of linguistic knowledge in a cultural and historical context.[43] Fryer also brought his personal library with him, bequeathing it to the University on his death in 1928.[44] The far-sightedness of Tompkins' decision, coupled with Fryer's zeal and knowledge, created a perfect cradle in which a combination of scholarly progress and the study of different cultures steadily grew.

Despite his commitments in Berkeley, Fryer would still go to China every summer between 1897 and 1903 to supervise the progress of ongoing works at the Arsenal.[45] During his years at UC Berkeley, Fryer taught Chinese language, history, and culture. He also gave numerous talks on sinology in the broader sense of the term, at the same time lavishing his care on helping Chinese students studying abroad. He ended his career as a teacher in 1913 and died at the age of 89 on July 2, 1928.

2 John Fryer: Between Translation and Scientific Divulgation

Fryer's long-term activity in China cannot be easily confined to a single sector, such as translation or missionary duties. His work in China was born at the intersection of personal and historical circumstances and covered countless fields, from his career as an English teacher at the *Ying-Hua shuguan* to the scientific activities and experiments he carried out at the *Gezhi shuyuan* 格致

42 Doris Sze Chun, "John Fryer, the First Agassiz Professor of Oriental Languages and Literature, Berkeley," *Chronicle of the University of California* (Fall 2005): 1.

43 Ivi, 10.

44 Peter X. Zhou, "At the Crossroads Between East and West: The C.V. Starr East Asian Library at the University of California, Berkeley," in *Collecting Asia: East Asian Libraries in North America, 1868–2008*, ed. Peter X. Zhou (Ann Arbor: Association for Asian Studies 2010), 68–69.

45 A report of one of the travels from San Francisco can be read in the Edmund Rensselaer Lyman papers, volume four (May–September 1898), Bancroft Library, UC Berkeley. Refer also to Irwin and Dagenais, *John Fryer's Legacy*, 6, and Biggerstaff, *The Earliest Modern Government Schools*, 190, footnote no. 73. Fryer is noted as absent in *Records of the Fifth Triennial Meeting of the Educational Association of China, Held at Shanghai, May 17–20, 1905* (Shanghai: American Presbyterian Mission Press, 1906), 11.

書院 (Polytechnic institution),[46] and from his contribution to the standardisation of scientific and technical nomenclature to his work in the diffusion of scientific knowledge through the "first periodical publication with the sole purpose of introducing Western science and technology to modern China,"[47] the *Gezhi huibian* 格致彙編 (The Chinese Scientific Magazine). One of the aspects that is probably less well known is the important role Fryer played in the birth of Chinese new fiction, *xin xiaoshuo* 新小說, and his ties with leading personalities of modern Chinese history.[48] Numerous texts translated by Fryer and his Chinese colleagues were extensively read and influenced leading figures, such as the Chinese reformist Tan Sitong 譚嗣同 (1865–1898).[49]

Fryer, though, is principally famous for his translation work. The translations completed by Fryer and his Chinese colleagues, and the weight they had on the success of *Yangwu yundong* as well, can hardly be overestimated. The modality through which the ideas and theories conveyed through these texts spread in China, and the role they played in the modernisation of the country, is an aspect of Fryer's work that has only been partially analysed. Positive opinions about them were expressed by numerous prominent Chinese personalities of the time, such as the scholar Sun Baoxuan 孫寶瑄 (1874–1924)[50] and the famous scholar and reformist Liang Qichao 梁啟超 (1873–1929).[51] A glimpse into the importance attributed to the works Fryer compiled or translated can also be seen from the adoption of the *Xixue gezhi daquan* 西學格致

46 Information can be derived from the numerous reports published on the institution's activities, such as John Fryer, *Fourth Report of the Chinese Polytechnic Institution and Reading Rooms, Shanghai: From March, 1883, to June, 1885* (North China Herald, July 10. Shanghai: North-China Herald Office, 1885).

47 Chen Jiang, "Western Learning and Social Transmutation in the Late Qing," in *China's Quest for Modernization: A Historical Perspective*, ed. Frederic Wakeman Jr. and Wang Xi (Berkeley: Institute of East Asian Studies, University of California, 1997), 157.

48 On the first topic, see John Fryer, "Qiuzhu shixin xiaoshuo qi," *Wanguo gongbao* 7, no. 77 (1895) and Idem, "Shixin xiaoshuo chu'an," *Wanguo gongbao* 8, no. 86 (1896). Refer also to Zhou Xinping, ed., *Qing mo shixin xiaoshuo ji* (Shanghai: Shanghai guji chubanshe, 2011) and Li Zipeng, *Fuyin yanyi: wan Qing hanyu jidujiao xiaoshuo de shuxie* (Taipei: Taida chuban zhongxin, 2017), 161–65.

49 Refer to Luke S. K. Kwong, "Tan Sitong he Fu Lanya de yi ci huijian," *Jindaishi yanjiu—Modern Chinese History Studies*, no. 6 (1994); Idem, *T'an Ssu-t'ung, 1865–1898: Life and Thought of a Reformer* (Leiden, New York, Köln: E.J. Brill, 1996), 120–23; and particularly to Cai Shangsi and Fang Xing, ed., *Tan Sitong quanji* (Beijing: Zhonghua shuju, 1981), 458–68.

50 Sun Baoxuan also directly adopted neologisms listed in the TVM in his diary, *Wang shan lu riji* 忘山廬日記 [Diary of the cabin of the mountain of oblivion], such as *qingyang*$_{er}$ 輕養$_{二}$; the term, included in the HXCL, indicates "Binoxide of hydrogen" on page 6, and "Hydrogen binoxide" and "Hydrogen peroxide," on page 18.

51 Wang Hongxia, "Fu Lanya de Xishu zhongyi shiye," (PhD diss., Fudan daxue, 2006).

大全 (Comprehensive collection of the investigation of things and extension of knowledge of Western learning), published in Hong Kong by Xianggang shuju 香港書局 in 1897 as one of the texts used in teaching materials by Chinese students to prepare for examinations, based on the educational reforms of the *Xinzheng* 新政 (new policies).[52] The *Xixue gezhi daquan* actually includes translations Fryer carried out and employed for his *Science Outline Series*, the *Gezhi xuzhi* 格致須知,[53] published by the *Guangxue hui* 廣學會, or Society for the Diffusion of Christian and General Knowledge among the Chinese.[54] This can give a glimpse of the role of the texts which Fryer compiled during his engagement at the Jiangnan Arsenal as a vector for terms and ideas.

Nevertheless, the most renowned accomplishment by Fryer and his Chinese colleagues is the methodology devised for the nomenclature of chemical elements presented above.[55] The strong points of Fryer and Xu Shou's translations were their ability to defeat concurrent terminologies[56] and leave a clear and durable mark in the Chinese lexicon. Such influence is seen present-day and has been acknowledged by prestigious institutions in the first decades of the twentieth century: "化學之入中國，垂五十年，譯名時有變更。董其事者，在清季前後有江南製造局及京師譯學館" (Chemistry appeared in China around fifty years ago; at times, translated terms were changed. Among the experts in the sector, around the end of the Qing dynasty there were the Jiangnan Arsenal and the Beijing School of Translation).[57]

3 *The Translator's Vade-mecum*: An Authoritative Work

Although different aspects of John Fryer's work in China have been analysed, others have been neglected, particularly pertaining to its outcomes. Numerous texts rendered into Chinese by Fryer and his collaborators can be considered

52 Elman, *On Their Own Terms*, 430–31.

53 Starting from John Fryer, ed., *Gezhi xuzhi* (Shanghai: Gezhi shushi, 1882–1883). An in-depth analysis of the series can be read in Elman, *On Their Own Terms*, 321–23.

54 Chen, "Western Learning," 156.

55 More first-hand information on the topic can be found in John Fryer, "Chemical Terminology," *The Chinese Recorder and Missionary Journal, Volume XXVI*, no. 6 (Shanghai: American Presbyterian Mission Press, 1895); FP, carton 3, folder 34, "Miscellaneous Materials—Material re Chinese Scientific Terminology"; and Fryer, "The Present Outlook."

56 Xu Zhenya, "Fu Lanya yu Zhongguo jindai huaxue," *Beijing huagong daxue xuebao (Shehui kexue ban)—Journal of Beijing University of Chemical Technology*, no. 2 (2001) and Wang Yangzong, "Guanyu '*Huaxue jianyuan*' he '*Huaxue chujie*'," *Zhongguo keji shiliao—China Historical Materials of Science and Technology* 11, no. 1 (1990).

57 Guoli bianyiguan, *Huaxue mingming yuanze*, preface, iii.

among the hidden sources to take into consideration when studying his translation efforts. These translations help better highlight the *TVM* and the historical context and textual production behind Fryer's work. Many can be consulted today in different libraries around the world and have been printed as various monographs and republished in anthologies, while others remain uncompleted and were never published.[58]

The most important piece of information, though, apart from the already mentioned FP, can be drawn from the connection between the *TVM* and the manuscripts of the bilingual glossaries edited by Fryer, stored in the EAL of UC Berkeley.[59] These represent a real treasure for scholars interested in the formation process of Chinese scientific and technical lexicon, and of the historical and cultural background behind it. After 1974, these volumes were moved to the EAL, known as the East Asiatic Library until 1991: "The assiduity and the methodical manner with which Fryer compiled *The Translator's Vade-Mecum* can be observed in the several huge volumes of original manuscripts preserved and shelved in the John Fryer Collection in the Richmond Annex, University of California, Berkeley."[60]

If we start from the assumption that the modern Chinese lexicon is the complex result of interactions between linguistic but also social, political, and economic factors, deeply rooted before the May Fourth Movement—as traditionally thought—a similar conclusion can also be drawn for the modern Chinese scientific and technical lexicon.[61] Even though the influence of the Japanese language played an extremely important role, a consistent part of this specific section of the lexicon also developed from the interaction between traditional Chinese texts, works translated by Chinese and foreign missionaries and translators, and finally, veritable lexical innovation.[62]

---

58 Among the anthologies, Yuan Junde, ed. *Fuqiangzhai congshu xuji* (Xiaocang shanfang, 1901) and, more recently, Wang Deyi, ed., *Congshu jicheng xubian*, vol. 81, *Ziran kexue lei* (Taipei: Xinwenfeng chuban gongsi, 1989) can be mentioned.

59 On the history of the bequest of Fryer's collection, Levenson and Huff, *Elizabeth Huff*, 135–37, 144 and passim, and Irwin and Dagenais, *John Fryer's Legacy*, preface, can be consulted.

60 Li San-pao, "Letters to the Editor in John Fryer's *Chinese Scientific Magazine*, 1876–1892: An Analysis," *Bulletin of the Institute of Modern History* 4 (1974): 737, footnote no. 20.

61 Masini, *The Formation of Modern Chinese*, preface, iii–iv.

62 A concise presentation of the approach of Chinese and Japanese in a comparative perspective can be read in Nakayama Shigeru, "Translation of modern scientific terms into Chinese characters—The Chinese and Japanese behavior in comparison," in *Jindai Zhongguo keji shi lunji*, ed. Yang Cuihua and Huang Yinong (Taipei: Zhongyang yanjiuyuan jindaishi yanjiusuo, Guoli Qinghua daxue lishi yanjiusuo, 1990). See also Koos Kuiper, "Dutch Loan-words and Loan-translations in Modern China: An Example of Successful Sinification by Way of Japan," in *Words from the West: Western Texts in Chinese Library*

It is exactly in this context that the work of Fryer and his Chinese colleagues took place. The translation of texts into Chinese was necessarily based on existing books and therefore terms, part of the immense Chinese written tradition. On the other hand, when Fryer and his colleagues had to deal with unknown concepts, the only solution was lexical creation and innovation, all the more so pertaining to the scientific branches covered by their translations.

A thorough analysis of all the texts Fryer translated in cooperation with his Chinese colleagues would be almost impossible. Nevertheless, the TVM, as endpoint and, indirectly, the starting point of Fryer's work as a translator, includes numerous translated terms in a single publication. The terms have been gathered from other sources and frequently created from scratch, employed in various works, and related to different scientific branches. The text, initially devised as a handbook for the translator and at the same time as an instrument for the Chinese reader, represents an almost unexplored gateway to the state of the art of Chinese scientific and technical lexicon in the late Qing dynasty.[63]

The TVM is a fundamental text in several respects. On the one hand, it represents the first systematic collection of bilingual English-Chinese glossaries exclusively devoted to the scientific and technical lexicon.[64] While specific sections were already included in preceding bilingual glossaries, none can be considered as methodical as the TVM, gathering in one volume a technical nomenclature as complete as possible. On the other hand, the prestige of, and consequently the influence of the TVM on, modern scientific and technical lexicon was increased by the importance and authorship of the institution responsible for its publication, the Presbyterian Mission Press.[65]

Context; Essay to Honor Erik Zürcher on his Sixty-Fifth Birthday, ed. Lloyd Haft (Leiden: Centre of Non-Western Studies, 1993).

63 For an overview, refer to Shen Guowei, "The Creation of Technical Terms in English-Chinese Dictionaries from the Nineteenth Century," in *New Terms for New Ideas: Western Knowledge and Lexical Change in Late Imperial China*, ed. Michael Lackner, Iwo Amelung, and Joachim Kurtz (Leiden, Boston, Köln: Brill, 2001).

64 Nevertheless, Paul Perny, *Appendice du dictionnaire français-latin-chinois de la langue mandarine parlée* (Paris: Maisonneuve et cie., Ernest Leroux, 1872) should be mentioned here, particularly the appendix "Histoire naturelle," comprising terms in botany, zoology (further divided in mammals, ornithology, reptiles, ichthyology, crustaceans and molluscs, and entomology), and mineralogy. Consult also Li Nanqiu, "Shuangyu keji cidian shi lüeshu (shang)," *Shanghai keji fanyi—Shanghai Journal of Translators for Science and Technology*, no. 2 (1993) and Idem, "Shuangyu keji cidian shi lüeshu (xia)," *Shanghai keji fanyi—Shanghai Journal of Translators for Science and Technology*, no. 3 (1993).

65 Refer to Zhang Xiantao, *The Origins of the Modern Chinese Press: The Influence of the Protestant Missionary Press in Late Qing China* (London: Routledge, 2007), 108 and following. For a history of the Presbyterian Mission Press, refer to Gilbert McIntosh, *The Mission Press in China, Being a Jubilee Retrospect of the American Presbyterian Mission Press, with*

Moreover, it should not be forgotten that Fryer's translation work, which gave birth to the *TVM*, was conducted under the patronage of the Jiangnan Arsenal. The institution had been set up in the context of the Tongzhi restoration and was sponsored by authoritative political figures of the Qing dynasty: "[…] the publications of the Kiangnan Arsenal […] were still considerable […] All these translations formed an important bridge for conveying Western learning. That these translations were widely disseminated in official circles is beyond doubt."[66] As demonstrated in secondary literature and in section 8.4 of this work, the working methods set forth at the Jiangnan Arsenal were highly appreciated and adopted by institutions that would continue the work of standardisation of Chinese scientific and technical lexicon, such as the Guoli bianyiguan.[67] The historical role and prestige attributed to the Jiangnan Arsenal certainly influenced the circulation of the *TVM* and its terms, the ideas behind them, and the choice to use it as a reference by later institutions.

Furthermore, it is important to remember the importance that linguistic materials, broadly defined, that were compiled by late Qing and earlier foreign missionaries or translators have had on the interpretation and study of the history and peculiarities of the Chinese language, as well as on relevant cultural interactions according to the *shūen kara no apurōchi* 周縁からのアプローチ or "peripheral approach."[68] In light of this theory, the study of a certain phenomenon is more precise and thorough when carried out not starting

---

Sketches of Other Mission Presses in China, as well as Accounts of the Bible and Tract Societies at Work in China (Shanghai: American Presbyterian Mission Press, 1895), 6–34. The sales of the press were considerable, as shown in Milton Theobald Stauffer, Tsinforn C. Wong, Malcolm Gardner Tewksbury, and China Continuation Committee—Special Committee on Survey and Occupation, *The Christian Occupation of China: A General Survey of the Numerical Strength and Geographical Distribution of the Christian Forces in China Made by the Special Committee on Survey and Occupation, China Continuation Committee, 1918–1921* (Shanghai: China Continuation Committee, 1922), 449.

66 John King Fairbank and Liu Kwang-Ching, ed., *The Cambridge History of China, Vol. 11, Late Ch'ing 1800–1911, Part II* (Cambridge: Cambridge University Press, 1980), 169. Also, further on, 170: "The Kiangnan Arsenal's translation centre made the greatest contribution of the period." Refer also to Biggerstaff, *The Earliest Modern Government Schools*, 194–99, in particular to the statement on page 199: "There can be little doubt of the great contribution made by the translation department of the Kiangnan Arsenal to the introduction of Western knowledge into China—particularly of scientific and technical knowledge."

67 As, among others, explained in Reardon-Anderson, *The Study of Change*, 190–98.

68 Put forward in Uchida Keiichi, *Bunka kōshōgaku to gengo sesshoku: Chūgoku gengogaku ni okeru shūen kara no apurōchi* (Osaka: Kansai daigaku shuppanbu, 2010). Refer also to its English translation, Idem, *A Study of Cultural Interaction and Linguistic Contact: Approaching Chinese Linguistics from the Periphery* (Goettingen: V&R unipress, 2017), 7–21 and 154–56.

from its centre, but from its periphery. When applied to the Chinese language, this means that linguistic materials compiled from the "periphery" by Western missionaries and scholars for the study, in the broadest sense of the term, of the Chinese language, can cast a brighter light on its features than if one were to start out with materials from its core, that is, materials compiled exclusively by native speakers. Bilingual glossaries are among the most relevant documentary materials considered by this approach. The significance of this theory and the academic results of the research conducted according to it are one of the grounds for choosing the *TVM* as the subject of this monograph.

Finally, the last factor to consider is the lack of monographs and articles exclusively dedicated to the *TVM*. This path has already been indicated by scholars who have conducted research on the topic: "One possible approach, which is beyond the scope of Bennett's monograph, would be a comparison of the technical vocabulary used by Fryer and that in later Chinese-language scientific and technical works. Perhaps, too, some indication of the textbooks used in such early technical schools [...] might be unearthed. Or is it possible that the first Chinese generation upon whom Western science and technology made any real impact acquired its smattering of learning in 'quickie schools' in Japan or, by diligent application, directly from the English-language originals?"[69]

4 Conclusion

John Fryer represents an important and debated historical figure in late Qing China, a time when the destiny of scientific and technical translation was taking shape. The significance of his work has been referenced, as well as his cooperation with Chinese translators. However, even though the importance of texts compiled by missionaries and scholars has been duly highlighted in secondary literature, the *TVM* remains one of Fryer's less studied texts.

---

69 Review by Albert Feuerwerker of Bennett, *John Fryer*, in *The American Historical Review* 74, no. 2 (1968): 688–89. Not taking any allusions in general reference works into account, some texts have only analysed single aspects of the *TVM*; the author has considered them the starting point for this research. Among them, Wang Yangzong, "A New Inquiry into the Translation of Chemical Terms by John Fryer and Xu Shou," in *New Terms for New Ideas: Western Knowledge and Lexical Change in Late Imperial China*, ed. Michael Lackner, Iwo Amelung, and Joachim Kurtz (Leiden, Boston, Köln: Brill, 2001), and Wright, *Translating Science*. Specific information on the *TVM* can be found in Li Nanqiu, *Zhongguo kexue fanyi shiliao* (Hefei: Zhongguo kexue jishu daxue chubanshe, 2006).

The reasons for focusing on it are evident: the history of scientific translation in late Qing China and the knowledge transfer through translations can be analysed through the investigation of key people and their works, such as the *TVM*. By studying the *TVM*, it is possible to obtain a comprehensive overview of John Fryer's translation theories, the lexicological features of Chinese scientific and technical nomenclature, and the lexicological innovations brought forth in the *TVM*, as well as the role, diffusion, and influence the *TVM* exerted on later texts and the Chinese scientific and technical lexicon. Moreover, according to the theoretical framework of the "peripheral approach," such study has implications not only for specialists of lexicology and the history of lexicons, but also of the history of scientific translation, Eastern and Western cultural exchanges, and China's modernisation process in the nineteenth and twentieth centuries, as well as of the history of science, including chemistry, mineralogy, pharmacology, mechanics, naval architecture, and, generally, linguistics. The great number of encyclopaedias, Japanese sources, Jesuit texts, archival materials, journals, dictionaries, and miscellaneous texts that reflect and are reflected in the *TVM* can provide a different glimpse of late Qing China to the broader scholarly readership of historians and Asianists.

CHAPTER 2

Planning, Publication, and Setbacks: *The Translator's Vade-mecum*

1 Introduction

When analysing the TVM, we should first consider that there are different texts and versions ascribable to it, and that most of the glossaries composing it were first published individually. The first edition of the TVM presented in this chapter, published in 1888, can be considered a sort of master copy. The choice to focus on it arises from the fact that it is the earliest publication in a single volume of the various glossaries making up the TVM—ensuing editions that can be related to the work do not have a precise date of publication. For these other editions, only *post quem* dates are available. A specific description will also be dedicated to them, including an introduction to the glossary that represents the main difference between the edition of the TVM printed in 1888 and ensuing ones.

The TVM contents presented here can offer an outlook of the branches of science covered in the collection of glossaries. Some of the most relevant compositional features of the TVM, as well as the history of the planning and its publication process, are essential in order to better grasp the historical and linguistic contexts in which the publication of the text took place. Fully drawing from primary sources contemporary to Fryer, the issue of printing a second volume of the TVM, never achieved, is analysed; this question reflects John Fryer's influence on missionaries' translations, and the power dynamic between John Fryer and other missionary societies in terms of the standardisation of scientific terminology.

It is a difficult task to find copies of the TVM: "The papers of Fryer (1, 2) and his 'translator's vade mecum' (3) are rare, but worth reading if they can be got hold of."[1] A volume is housed in the EAL, with the full title *The Translator's Vade-mecum, A Collection of Vocabularies of Chinese Terms Used in the Translation of Scientific Books at the Kiangnan Imperial Government Arsenal, Shanghai, China*, published in Shanghai by Presbyterian Mission Press in 1888.

1 Joseph Needham, Ho Ping-yü, and Lu Gwei-djen, *Science and Civilization in China, vol. 5, part 3—Spagyrical Discovery and Invention: Historical Survey, from Cinnabar Elixirs to Synthetic Insulin* (Cambridge: Cambridge University Press, 1976), 254, note f.

 DOI:10.1163/9789004443211_004

As can be read on the title page of the text, the TVM was "Compiled by John Fryer"; the page shows the caption "Volume 1," suggesting somehow the presence of a second volume that will be explained later in this work. A brief presentation of the contents of the TVM, divided into four sections, follows: "Part 1.—Terms in Chemistry," "Part 2.—Terms in Materia Medica," "Part 3.—Terms in Mineralogy," and "Part 4.—Proper Names used in works in the above three subjects."

The information shown in the TVM title page provides important indications of the text. The TVM is said to have been "compiled" by Fryer, since it includes terms explicitly drawn from other authors; the authorship on the coining of new terms is thus not exclusively attributed to Fryer. The fact that the TVM is indicated as a collection of "Chinese terms used in the translations of scientific books at the Kiangnan Imperial Government Arsenal," creates uncertainty on the authorship of a certain term by Fryer or his Chinese colleagues. Keeping these premises in mind, in this work all lexical innovations that have been until now considered created by later authors, will be deemed coined by Fryer. In the appendices I will, therefore, date the first attestation of these lexicological features to the TVM, their presence not ascertained in other works published at an earlier stage; relevant tables are also provided to help the reader grasp its structures and features.

2 Various Dating Problems

Pertaining to the editions of the TVM, a controversial aspect of its history and birth is the date shown in the title page of the version printed in 1888. It is possible to state that at least three of the glossaries indicated in the 1888 version of the TVM, "Part 1.—Terms in Chemistry," "Part 2.—Terms in Materia Medica," and "Part 3.—Terms in Mineralogy," were first published as monographs.[2] The decision to gather them in a single publication derived both from the need of standardisation and the necessity to help other translators and Chinese readers themselves; this is something Fryer highlighted in different circumstances and tried to personally accomplish. As for the fourth glossary, "Part 4.—Proper Names used in works in the above three subjects," conceived and planned as

2 This is confirmed, also pertaining to the QJ, by Wei, *Jiangnan zhizaoju ji*, 11. In the latter, the XYDCYP, the JS, the HXCL, and the QJ are included respectively on pages 18A, 21 A, 21B, and 22B; for all of them, the text does not provide any information about the author or the publication year, presenting instead only indications on the number of books, *benshu* "本數," "一" (one), the number of juan, *juanshu* "卷數," also "一" (one), and the price, *jiazhi* "價值," "一角五分" (one *jiao* and five *fen*).

an appendix to the others, there are no certain conclusions pertaining to its publication as a monograph.

On the title page of the edition considered here, "Volume I" can be read; this is especially interesting for the compilation history of the *TVM*. I could not identify a second volume, but a plan for the execution of the *TVM* was at least conceived by Fryer. Such a plan was clear in Fryer's mind, or at least in that of the managers of the Jiangnan Arsenal, and of the committees formed by missionaries and translators working on the standardisation of Chinese scientific and technical nomenclature. This conclusion can be inferred from two factors: the first is the information pertaining to "Volume I" mentioned above; the second is the existence of various manuscripts, for the most part compiled by Fryer and stored in the EAL. In many of these manuscripts, the year of completion annotated by Fryer is a few years later than the publication date of the first edition of the *TVM*—1888. It is therefore possible to affirm that a concrete plan for a follow-up of the first volume existed. Moreover, the accurate use of legends for the different sources Fryer used for the compilation of these manuscripts would have not made sense had the volumes themselves been created for Fryer's personal use.

Going back to the 1888 edition of the *TVM*, the volume presents a Western hollow back binding, but the reading order of pages is Chinese, from right to left. On the front cover "The Translator's Vade-Mecum—Vol. I" can be read. The *TVM* edition considered here starts with the preface to the *HXCL*. The English preface to the glossary of chemical terms can be considered the first page of the volume; this page and the opposite page of the Chinese preface are unnumbered. Also unnumbered is the page where the first terms are listed; the heading in this page reads *Vocabulary of names of substances occurring in various works on chemistry; Chiefly in "Bloxam's Chemistry," first edition.*

The page count starts from page 2 and continues until page 36; the subsequent one, the last of the glossary, is unnumbered and its opposite page is blank. Pages are to be leafed through from right to left; the same order applies to the two columns in which the pages are divided. In the columns, the entries are arranged in alphabetical order according to the English or scientific Latin lemmata, which precede their corresponding Chinese translations.[3] It should finally be noted how, for original lemmata, there is no graphic separation when a term with a new initial letter begins; this will be the case, on the contrary, with the *JS*.

3 In the *TVM* it is frequently possible to find the same lemma listed first in Latin and then in English or vice versa.

After the HXCL it is possible to read the preface to the section designated *Vocabulary of names of Materia Medica occurring in the translation of "Royle's Manual of Materia Medica, and Therapeutics." With lists of names of persons and places occuring in the same work, and in various treatise on Chemistry and allied subjects*. Notwithstanding the information provided in the title, the list of anthroponyms and toponyms can be found only after the glossary of mineralogical terms, the second to last one. Opposite to the page with the English preface is the Chinese preface; both pages are unnumbered. After the two prefaces, the body text of the XYDCYP begins. The numeration, as in the HXCL, starts from page 2 and continues until page 68, the last page. The structure of the entries is the same as that of the HXCL.

Without any separation, after the XYDCYP the preface to the *Vocabulary of Mineralogical Terms Occurring in the Manual by J. D. Dana A. M.* follows. The entries in the JS are divided into three columns: the first one corresponds to the original terms, i.e. "Terms used in J. D. Dana's manual";[4] the second, to the "Terms used in Dr. Macgowan's translation";[5] and the third, to the "Terms in general use and chief elements." In the JS, when terms with a new initial letter are listed, the letter is written with a bigger font in the first column. The numeration of the JS starts from page 2, the first of the glossary, since the prefaces are shown on page 1, and continues until page 37. The glossary ends with a page of "Errata," where typographical errors are listed.

Without any separation after the JS, the *Vocabulary of proper names occurring in Royle and Headland's Materia Medica and Therapeutics, and in various works on medicine and chemistry—Part I. Names of persons* follows. The RMDM page structure is divided into three columns; the original term is followed by the Chinese translation. The first page is unnumbered, with the numeration starting from page 2 and continuing until page 13, the last of this section. On the opposite page, "Part II. Names of places" begins. The graphical structure is the same, with the numeration continuing from the previous part; page 14 is the first of this section. The glossary continues until page 22, the last one. After a blank page, another rigid page, corresponding to the one where the label for borrowing is glued, follows. The volume ends with the inside back cover and the outside back cover. In this TVM edition, the total number of pages is 176, not considering the rigid pages.

4 The reference is to the American scientist James Dwight Dana (Dan Na 丹納 or Dai Na 代那, 1813–1895).

5 The American medical missionary Daniel Jerome Macgowan (Ma Gaowen 瑪高溫, 1814–1893). See Wright, *Translating Science*, 86, footnote no. 34.

It is important to highlight one of the features of the volume presented above. The total numeration of the pages starts again with every single glossary. Since page numbering methods were complicated in science translations in late Qing China, this may not serve as evidence of the fact that the glossaries included in the TVM were first published independently as monographs. However, combined with other proof presented later on in this work, this can contribute to reinforcing the hypothesis that they were gathered in the TVM at a later phase.

3 *Translator's Vade-mecum* and *Chinese and English Technical Vocabularies*

Other than the 1888 edition, there are at least two other complete versions available of the TVM. The first has a different title, namely *Chinese and English Technical Vocabularies* and it includes, along with the four aforementioned glossaries, the QJ. The order of the glossaries in *Chinese and English Technical Vocabularies* is slightly different: first is the XYDCYP, second the RMDM, third the QJ, fourth the HXCL, and lastly the JS. A copy of this volume is housed in Doe Library, UC Berkeley; no indications on the publication year are available. Nevertheless, the volume certainly has a *post quem* date; since it includes the glossary of engineering terms, the QJ, whose preface was signed by Fryer in October 1889, the publication of the *Chinese and English Technical Vocabularies* was surely completed after that year.

Another version of the TVM is also stored in the Doe Library, UC Berkeley. In this volume, with the title *Translator's Vade-mecum*, the order of the glossaries is different.[6] First is the QJ, second the JS, third the HXCL, and finally the XYDCYP; the RMDM is not included. In the same way as the *Chinese and English Technical Vocabularies*, it does not provide any information on the publishing year. Once more, information on the *Translator's Vade-mecum* can be drawn from the preface to the glossary of engineering terms; since the volume comprises the QJ, as did the *Chinese and English Technical Vocabularies*, the publication was definitely completed after 1889.

Even though, as indicated above, there are only *post quem* dates for these two editions, published after the first edition of the TVM in 1888, the two volumes

6 The title is actually only written on the spine, but there is no title page. Below the title, in the lower part of the spine, four characters can be read: *Gezhi Zhong Xi* 格致中西 (Investigation of things, extension of knowledge, and China and the West). A label glued onto it, though, prevents the reader from seeing the other characters that probably follow.

provide the reader with other important clues. It is possible to suppose a later publication of the *Translator's Vade-mecum* compared with *Chinese and English Technical Vocabularies*—discrepancies among some of the entries in the two volumes, the only among the almost 20,000 total lemmata included in the glossaries, can be found. On pages 18 and 19 of the *XYDCYP*, "Cubeb," "Cubeb pepper," "Cubebs," "Cubebs camphor of," "Cubebs oil of," and "Cubebs tincture of" are translated in the *Translator's Vade-mecum* as *bichengqie* 蓽澄茄, *bichengqie jiao* 蓽澄茄椒, *bichengqie* 蓽澄茄, *bichengqie zhangnao* 蓽澄茄樟腦, *bichengqie you* 蓽澄茄油, and *bichengqie jiu* 蓽澄茄酒. In the *Chinese and English Technical Vocabularies*, on the contrary, they are rendered as *gebaiba* 嗝拜把, *gebaiba jiao* 嗝拜把椒, *gebaiba* 嗝拜把, *gebaiba zhangnao* 嗝拜把樟腦, *gebaiba you* 嗝拜把油, and *gebaiba jiu* 嗝拜把酒. One of Fryer's consistent corrections in the passages from the manuscripts to the printed version of the glossaries in the *TVM* is the transformation of phonemic loans into hybrids or other categories, in which the phonetic aspects of the original term could be ignored, and the transformation of phonemic indications into semantic indications, but never the opposite. It is therefore possible to infer that the printing of *Chinese and English Technical Vocabularies*, in which the aforesaid lemmata are translated with terms bearing more phonemic indications than those of the *Translator's Vade-mecum*, precedes that of the latter.

The glossaries included in the 1888 edition of the *TVM* do not show differences of content, compared with the *Chinese and English Technical Vocabularies* and the *Translator's Vade-mecum* versions. The only discrepancies are related to the printing process, the most evident of them being the order of the glossaries in the three volumes, and at times the reading order of the pages inside the single glossaries. Some of them, in the two aforesaid volumes, are to be leafed from left to right. A peculiarity, though, differentiates the *HXCL* inside the *Translator's Vade-mecum*. In this volume, even though the pages are to be leafed from left to right, the two columns in which terms are divided should be read from the left column to the right one only in the first two pages of the glossary, not the opposite, as is the case with the other thirty-five pages.

The version of the *TVM* published in 1888 and the two following copies mentioned above are the only three that can be considered complete monographic publications; partial collections of the glossaries can also be located. The first among these can be traced to a book containing miscellaneous material written by or related to John Fryer, stored in the Bancroft Library, UC Berkeley. The volume, composed of different heterogeneous works, includes a smaller booklet, without a cover. This booklet gathers the *RMDM* and the *XYDCYP*; the glossary of chemical terms and engineering terms are not included. As for the *Chinese and English Technical Vocabularies* and the *Translator's Vade-mecum*, indications on the publishing year of this volume are also not provided; the hypothesis proposed in which the glossaries included in the *TVM* were first published as monographs is once more confirmed by this volume. The first page of the *XYDCYP* gathered within it presents the Chinese title, *Xiyao dacheng yaopin Zhong-Xi mingmu biao* 西藥大成藥品中西名目表, together with the English one, *Vocabulary of Names of Materia Medica, in English and Chinese, Together with Lists of Persons and Places Mentioned in Various Works on Medicine, Chemistry and Allied Sciences, Compiled at the Kiangnan Arsenal, Shanghai, by John Fryer*, indicated as published in Shanghai by Presbyterian Mission Press in 1887. Therefore, the publishing year reported for this glossary, correspondent to the one in which the *XYDCYP* preface was signed, is 1887, one year before the first edition of the *TVM* was published.

The second of these partial reproductions is included in the second volume of this book. The volume comprises the *JS*, the *RMDM*, and the *XYDCYP*; the *XYDCYP* title page is identical to the one reported above.[7]

4 Planning and Birth of *The Translator's Vade-mecum*

Before analysing the glossaries composing the *TVM* in detail, the history of its planning, compilation, and publication is outlined in this section. Delineating the conception of the *TVM* is important from at least three perspectives. Firstly, from its references in contemporary sources, it is possible to get a better glimpse of its complicated interactions with other lexicographical materials. Secondly, these references provide additional information on the relations between the different actors in the translation and missionary world and of the intersections between them. Lastly, from such a context it is possible to derive a better understanding of the transcultural flows of knowledge between the places represented by those actors and the regions, particularly China, they were active in.

The most relevant information from this point of view can be derived from sources such as *The Chinese Recorder and Missionary Journal*, and the conference proceedings of missionaries residing in China during the late Qing dynasty. These sources better reflect the power relations between different missionaries and their societies for both the religious enlightenment of China

7 Other reproduction of single glossaries can be found in numerous collectaneous and encyclopaedic Chinese publications printed in the late Qing dynasty, some of which will be analysed in this work. Contemporary reproductions can also be located, for example, a reprinting of the *HXCL* is traceable in Wang Deyi, *Congshu jicheng xubian*.

and the related question of the standardisation of Chinese scientific and technical lexicon.

What can be considered the first *TVM* mention can be read in the fourth number of the tenth volume of the *CRMJ*, within the article "Uniformity of Terms in The School and Text Book Series," printed nine years before the *TVM*.[8] In the text, composed by the Protestant missionary Alexander Williamson (Wei Lianchen 韋廉臣, 1829–1890),[9] the author notes how he repeatedly asked for a report concerning the progress in the field of the standardisation of the translations by the "School and Text Book Series Committee," the "predecessor" of the Educational Association of China.[10] Williamson also adds that "only one list has as yet been sent in," even though the committee requested the dispatch of these lists of terms with the aforesaid purpose numerous times.[11] In the text, Williamson then spurs the other authors of the school and textbook series to "[…] send in their catagories [sic] of names and terms as soon as possible even though incomplete, or merely tentative. […] The Committee hope to print the lists when finally arranged, and so provide a sort of translator's '*vade-mecum*', as an excellent correspondent has suggested. I hope the 'writers' […] say whether, and how far they have adopted the 'lists' in Doolittles 2nd Vol."[12]

The text quoted above contains the first reference to the title of the work that Fryer will publish nine years later. First of all, Williamson demands other translators send lists of terms pertaining to different branches of knowledge, in order to facilitate the work of standardisation of Chinese scientific and technical nomenclature in the committee publications; he regrets that only one list had been provided at the time of publication of the article in *CRMJ*. The purpose of the work promoted by Williamson was to supply a sort of "translator's '*vade-mecum*'," as suggested by an "excellent correspondent"; this is most likely a reference to Fryer himself. Other important information conveyed by the aforementioned passage is Williamson's request to other translators to communicate the possible adoption of the lists furnished by the missionary of

8 Alexander Williamson, "Uniformity of Terms in the School and Text Book Series," *The Chinese Recorder and Missionary Journal* 10, no. 4 (1879).

9 Refer to Wright, *Translating Science*, 79 and particularly footnote no. 18.

10 John Fryer, *Educational Directory for China* (Shanghai: Educational Association of China, 1895), preface. Consult also John Fryer, *Descriptive Catalogue and Price List of the Books, Wall Charts, Maps, &c., Published or Adopted by the Educational Association of China (Formerly the School and Text-Book Series Committee)* (Shanghai: American Presbyterian Mission Press, 1894) and Wang Yangzong, "Qing mo Yizhi shuhui tongyi keji shuyu gongzuo shuping," *Zhongguo keji shiliao—China Historical Materials of Science and Technology* 12, no. 2 (1991).

11 Williamson, "Uniformity of Terms," 304.

12 *Ibidem.*

the American Board of Commissioners for Foreign Missions, Justus Doolittle (Lu Gongming 盧公明, 1824–1880). As will be examined further on, the work by Doolittle, who was also co-editor of the CRMJ from 1869, will be one of the sources used for the TVM and for the manuscripts compiled by Fryer that are housed in the EAL.[13]

The second reference in chronological order comes from the March and April 1880 CRMJ issues.[14] In the text, the minutes of the committee meeting on the publication of "A Series of School and Text-books," previously mentioned, can be read:

> Mr. Fryer showed specimens of various lists of scietific [sic] terms and proper names which he had been accumulating for several years past from all available sources. He hoped to have some of these lists ready for publication during the present year; but as they have to be done at leisure moments, it would be impossible to make rapid progress. He was advised to write to Peking and Yeddo[15] for lists of the Chinese terms employed in the translation of foreign books at the Universities of those places, so as to accumulate materials for a *Translator's Vade Mecum*, which the Committee hoped he would place in their hands for publication. With respect to proper names a list of above 25,000 geographical terms in most frequent use, drawn up by H. E. Li Fung-pao,[16] the present Chinese Minister at Berlin, with the assistance of Dr. Kreyer and Dr. Allen, was brought before the notice of the Committee by Mr. Fryer […][17] After long discussion, the following seven resolutions were passed: […]

---

13 On Doolittle's role in the CRMJ, see Kathleen Lorraine Lodwick, *The Chinese Recorder Index: A Guide to Christian Missions in Asia, 1867–1941* (Wilmington: Scholarly Resources Inc., 1986), II, xi.

14 "Educational Works for the Chinese—Minutes of Meeting of the Committee for the Publication of 'A Series of School and Text-books', Held at the London Mission, Shanghai, 15th, 16th and 17th March, 1880," *The Chinese Recorder and Missionary Journal, Volume XI*, no. 1 (Shanghai: Shanghai, American Presbyterian Mission Press, 1880). A reproduction of the text can be found in Dagenais, *The John Fryer Papers*, II, 349–51.

15 Edo, an old variant for contemporary Tokyo.

16 Namely Li Fengbao 李鳳苞 (1834–1887), the Chinese diplomat in Berlin. Refer to Koos Kuiper, *The Early Dutch Sinologists (1854–1900): Training in Holland and China, Functions in the Netherlands Indies* (Leiden, Boston: Brill, 2017), 397 and footnote no. 28, and Wright, *Translating Science*, 16, footnote no. 36.

17 "Dr. Kreyer" refers to Carl Traugott Kreyer (Jin Kaili 金楷理, 1839–1914); "Dr. Allen" is Young John Allen (Lin Lezhi 林樂知, 1836–1907). They were two of the foreign translators hired for the Jiangnan Arsenal's Translation Department; see, respectively, Wright, *Translating Science*, 57, footnote no. 83, and Wu Xiaoxin, ed., *Christianity in China: A Scholars' Guide to*

> 4.—That Li Fung-pao's list be harmonized and printed for circulation as a portion of the proposed *Vade mecum.*
> 5.—That Mr. Muirhead and Mr. Fryer be constituted as sub-committee to agree upon a syllabary for rendering geographical names and names of persons into Chinese.[18]

The information included in the minutes provides a better outlook on the *TVM*'s compilation history and of its entangled links with other contemporary lexicographical works. Firstly, from the year of publication of the text included in the *CRMJ*, it is possible to ascertain that Fryer was planning the publication of the glossaries gathered in the *TVM* since at least 1880. The commitment required for his work with the Jiangnan Arsenal, though, relegated this task to the spare time left by his work as a translator. Moreover, the fact that Fryer already had the publication of some of these lists of terms in mind leads us to suppose that the relevant work had been in progress for years, as also confirmed by the text quoted above.

Another even more important indication is the fact that, in the text published in the *CRMJ*, a suggestion for Fryer to write to Beijing and "Yeddo" to receive more lists of terms used in the translations of texts in the respective universities is mentioned. Whether Fryer has accepted the suggestion, whether the lists have been provided to him, and whether, once received, he used them or not are questions to which there are no definite answers.

In the text it is also mentioned that Fryer had presented a list of 25,000 frequently used geographical terms to the Committee; the list was compiled by Li Fengbao with Kreyer and Allen's help. The impossibility to consult such a detailed list makes it difficult to establish whether the terms later merged into the publication by Fryer or if the text remained only in manuscript form. For example, in volume number one of the manuscripts housed in the EAL, namely the LGN, or volume ten, to wit the VPN.

Nevertheless, the resolutions adopted by the Committee, quoted in the *CRMJ*, supply fundamental indications on the history of the birth of the *TVM* and of the glossaries gathered within. More specifically, the fourth resolution establishes that the list Li Fengbao edited should have been amended in the offing of its publication within the *TVM*. The latter will indeed comprise a glossary of anthroponyms and toponyms, the *RMDM*, in which terms related to geography are also included, but on an extremely smaller scale compared

Resources in the Libraries and Archives of the United States, 2nd ed. (London and New York: Routledge, 2015), 103.

18 "Educational Works," 141–42.

with the 25,000 in the list mentioned in the *CRMJ*. In order to establish if these fewer terms included in the *RMDM* had been concretely taken from the list by Li Fengbao, it would be necessary to consult the original, which has not been located. Finally, resolution number seven envisages that Fryer himself and the Protestant missionary William Muirhead (Mu Weilian 慕維廉, 1822–1900), would have taken the responsibility of drafting a syllabary for the phonetic rendering of toponyms and anthroponyms.[19] It was exactly different phonetic renderings of a single name that Fryer harshly criticised and pointed out as one of the main causes for discrepancies in the Chinese scientific and technical lexicon, as will be seen in the next chapter. The aforementioned syllabary might be identified in volume two of the EAL manuscripts, the STFNC.

Continuing in tracing the history of the compilation of the *TVM*, another reference comes from the April 1887 issue of the *CRMJ*. On page 162, in the section "Correspondence" of the "School and Text Book Series Committee," in a letter sent by Williamson to the editor of the periodical, it can be read: "In consequence of the inconvenience arising from the non-uniformity of nomenclature it was agreed to endeavour yet more to secure a translation of a vade mecum, and that henceforth no new book be received for publication without a glossary both in Chinese and English."[20] As indicated by the date of the passage quoted, the reference to the *TVM* comes one year earlier than the publication of its first edition, printed in 1888. Williamson restates in the *CRMJ* the need to avoid the exacerbation of confusion in Chinese scientific and technical nomenclature; no other books should have been published without a glossary in the appendix.

Later in the same year, in the June 1887 issue of the *CRMJ*, another reference to the *TVM* can be found, providing more details on the text. In the section "Report of the School and Text Book Series Committee," among the "Editorial notes and missionary news," the *CRMJ* editors point out: "This pamphlet gives in a very condensed form the results of the Committee's labors from May, 1877, to December 31st, 1886. [...] In the 'Summary of Work Done', four lists are given; I. School Books already Published—II. Text Books for Students already Published—III. Books in course of preparation and IV. Books still needed." As specified further on in the same page, "In the second list are [...] a History of

19 On Muirhead, refer to Zou Zhenhuan, *Wan Qing xifang dilixue zai Zhongguo: Yi 1815 nian zhi 1911 nian xifang dilixue yizhu de chuanbo yu yingxiang wei zhongxin* (Shanghai: Shanghai guji chubanshe, 2000), 90 and passim.

20 "Correspondence—School and Textbook Series Committee," *The Chinese Recorder and Missionary Journal, Volume XVIII*, no. 4 (Shanghai: Presbyterian Mission Press, 1887), 162.

Russia and a Translator's Vade Mecum, by John Fryer, Esq."[21] It is clear that the "Translator's Vade Mecum" indicated in the *CRMJ* passage is a reference to the glossaries published before December 31, 1886, namely the *JS* and the *HXCL*, almost certainly printed individually in the same years the relevant prefaces were signed, 1883 and 1885.

The hypothesis that the glossaries were published in individual volumes before a collective volume was printed is corroborated by documents external to Fryer's work. Another reference can be found in the November 1887 issue of the *CRMJ*, in the "Abstract of Minutes" of the School and Textbook Series Committee: "The Editor [...] also laid on the table the third part of his *Vade mecum* [...]."[22] In the passage it is confirmed that in 1887 the third part of the *TVM* had already been put into print; the glossary published in that year is the *XYDCYP*, after the *JS* and the *HXCL*, referred to in the previous passage.

Another hint to the planning and printing history of the *TVM* can be found in the September 1889 issue of the *CRMJ*. In a letter by Williamson reporting the minutes of a School and Textbook Series Committee meeting, it is possible to read that "[...] the Editor [...] placed on the table proofs of his Vocabulary on Steam as another instalment of the *Vade mecum*."[23] In the passage, which represents an addition to the two quoted above, it is once more clearly indicated that the *TVM* was initially published in single volumes. The reference in this case is to the *QJ*, which went into print in 1889, with the prefaces signed exactly that year. The *QJ* was indeed the last glossary gathered in the *TVM*; successive mentions of the text do not refer to any other new glossary.

The last hypothesis is confirmed again in the *CRMJ*. In the March 1890 issue, in the "Editorial notes and missionary news" section dedicated to the "School and Text Book Series Committee," composed by Williamson, it is possible to read that "[...] the editor [...] placed on the table the fifth part of his translation—*Vade Mecum*—giving the terms in Chinese relating to the steam engine [...]."[24] Contrary to the passage quoted above, here Williamson explicitly states that the *QJ* represented the fifth section of the *TVM*, not the fourth. In this passage, he most probably considered the *RMDM* an independent section,

21 "Editorial Notes and Missionary News," *The Chinese Recorder and Missionary Journal, Volume XVIII*, no. 6 (Shanghai: Presbyterian Mission Press, 1887), 243.

22 "Abstract of Minutes—School and Textbook Series Committee," *The Chinese Recorder and Missionary Journal, Volume XVIII*, no. 11 (Shanghai: Presbyterian Mission Press, 1887), 442.

23 "Correspondence—School and Text Book Series Committee," *The Chinese Recorder and Missionary Journal, Volume XX*, no. 9 (Shanghai: Presbyterian Mission Press, 1889), 430.

24 Alexander Williamson, "Editorial Notes and Missionary News—School and Text Book Series Committee," in *The Chinese Recorder and Missionary Journal, Volume XXI*, no. 3 (Shanghai: Presbyterian Mission Press, 1890), 138; the text is signed February 17, 1890.

not an appendix to other glossaries previously published; consequently, the *QJ* was the fifth. As proved from the quotation above, the *QJ*, in any case, had been published by 1890. Nevertheless, this excerpt from the *CRMJ* does not contradict with absolute certainty the hypothesis of its publication in 1889.[25]

Other useful indications to reconstruct the history of the *TVM*, in connection with concerted efforts by foreign missionaries and scholars for the translations of scientific concepts in Chinese, can be drawn from the *Records of the General Conference*, in "Appendix E," inside the "Report of the School and Text Book Series Committee." In the text, Williamson, the honorary secretary, affirms: "A translator's *Vade Mecum*, by J. Fryer, Esq., has been in progress for several years, and some portions are already issued. When completed, as they hope it soon will be, it will be a valuable help and guide to all who may in future write on scientific subjects. They have also agreed to urge that no new MS. be printed or new editions issued without a glossary in Chinese and English."[26] Further on, in "List B," which according to the words of the "General editor" of the "School and Text Book Series Committee," namely Fryer himself, included the books published "at the editor's own expense and afterwards adopted by the committee," the "Translators' Vade Mecum:—Part I., Vol 1. Mineralogical Terms, 2. Terms in Chemistry, 3. Terms in Materia Medica, 4. Terms on the Steam Engine" is listed.[27] As can be inferred from the passage, the *TVM* had been in progress for a number of years. In addition, the plan to publish a second volume of the *TVM* is confirmed. According to Williamson, in fact, by 1890 "some portions" of Fryer's publication had already been printed. In light of the analysis conducted, though, the *QJ*, most probably published in 1889, represents the last glossary added to the *TVM*.

On the other hand, the words of Williamson, who was hoping for the completion of the *TVM*, infers that Fryer had already devised a plan to expand the *TVM*. Further on, in list B, another confirmation of this hypothesis can be found: among the books already published at the expense of the editor of the School and Textbook Series Committee, the first part of the "Translators' Vade Mecum" is included. In this first part of the *TVM*, as indicated in the *Records of the General Conference*, the aforementioned four glossaries were gathered; in this case the *RMDM* was considered an appendix to the others.

25 Wang Yangzong, "Jiangnan zhizaoju," 15, indicates the glossary as published in 1890, specifying in footnote no. 1 that in the aforementioned number of the *CRMJ*, Fryer's work is indicated as printed that year.

26 W.J. Lewis, William Theodore Aquila Barber, and John R. Hykes, ed., *Records of the General Conference of the Protestant Missionaries of China, Held at Shanghai, May 7–20, 1890* (Shanghai: American Presbyterian Mission Press, 1890), 714.

27 Ivi, 717.

The subsequent reference from a chronological point of view to Fryer's *TVM* is indirect and comes from the text "Educational Notes—The Educational Association of China," included in the January 1892 issue of the *CRMJ*.[28] Inside the section "Technical terminology," among the principles recommended by the association for the compilation of lists of terms in English and Chinese, number one suggests: "The collector is to designate by a system of signs the sources from whence the terms are derived and to mark in some way the one which he prefers."[29] In many manuscripts stored in the EAL, Fryer will clearly indicate for each term a legend with colours representing the sources from where the terms were taken; the sources will only be explicitly written in one case, though.

Even more important is what can be read further on in the same text. According to the indications provided by the association, "The various technical subjects are roughly divided among the Committee as follows: [...] Dr. Fryer—Chemistry, mineralogy, meteorology, lithography, electro-plating, chemical apparatus, philosophical apparatus, founding and moulding, gunnery, ship-building, mining, engineering."[30] The various branches of knowledge covered by the bilingual glossaries Fryer edited do not correspond exactly to these recommendations; this will be true only for naval engineering, chemistry, mineralogy, and engineering. They do overlap, however, with some of the texts mentioned in the aforesaid letter addressed to Smith Elder & Co., in which seventy books to be bought to launch the translation work at the Arsenal are listed, and also with some of Fryer's personal interests in science. Nevertheless, as indicated further on in the same issue of the *CRMJ*, this list should have been considered provisional, subject to changes whenever necessary.[31]

Further on in the same *CRMJ* issue, a direct reference to the *TVM*, or at least to some of the glossaries gathered in it and already published as individual volumes, can be found: "Copies of Dr. Fryer's Essay on 'Chinese Scientific Terminology, its present Discrepancies and Means of Securing Uniformity', read at the General Missionary Conference at Shanghai, and of his technical vocabularies already published, will be forwarded gratis to any member of the Association applying to him."[32] This "essay" represented a data point for all translators dealing with the Chinese scientific and technical lexicon; those

28 "Educational Notes—The Educational Association of China," *The Chinese Recorder and Missionary Journal, Volume XXIII*, no. 1 (Shanghai: American Presbyterian Mission Press, 1892).

29 "Educational Notes," 33.

30 *Ibidem.*

31 Ivi, 34.

32 *Ibidem.*

who wished, could have obtained a copy of this article, in addition to the glossaries already compiled by Fryer.[33] Perhaps the most interesting information on the genesis of the *TVM*, on the destiny of its manuscripts, and on the power dynamic between late Qing missionaries—and also the last from a chronological perspective in the *TVM*'s textual history—can be found in the speech Fryer delivered during the second triennial meeting of the Educational Association of China, titled "The Present Outlook for Chinese Scientific Nomenclature."[34] In the text, after a long digression by Fryer on China's request of texts pertaining to science, all the more pressing after its defeat in the First Sino-Japanese War (1894–1895), Fryer returns once more to the topic of scientific and technical nomenclature. As he reproaches, the question was too frequently ignored by the Publication Committee of the Educational Association of China itself:

> We have had a separate existence six years, and during that time, with so many members of a scientific and practical turn of mind, we ought surely ere now to have published a *Technical Vocabulary*, embracing at least all the generally used elementary terms in English and Chinese. The matter, as you know, was referred to our Publication Committee. During the first three years a good deal of work in this direction was done. The part assigned to me as a member of that committee, was carried out as far as time and circumstances permitted, and I believe my colleagues also really made considerable advances in their portions. The subjects of materia medica, chemistry, mineralogy and the steam engine had been carefully gone over by me, collecting all existing terms that could be found separate. Four vocabularies on these subjects respectively, as complete as could be made under the circumstances, were previously published by me independently of the funds of the Association, and are still on sale. Subsequently, geological, botanical, geographical, biological and other terms were also treated by me in the same way, as well as naval architecture, but were not published, because I wanted to use, if possible, only such terms as the Publication Committee approved of. A specimen of one of these collections I now show you. It is a large, thick book, full of botanical terms. I have about a dozen like this. During the second triennium, the whole question of terminology being relegated to the new Publication Committee, to which I was not re-elected as a member, all my previous work proved to have been thrown away as far as this Association is concerned. I am quite in the dark as to what the Publication Committee

33 Fryer, "Scientific Terminology."
34 Fryer, "The Present Outlook."

> has done during this interval, but I merely know that no vocabularies of scientific terms have been issued. It was a sore disappointment to me on hearing the Report of the Publication Committee read yesterday morning not to find this all-important subject of nomenclature even alluded to […].[35]

In the passage, Fryer reiterates that he took care of the compilation of glossaries in the fields of "materia medica, chemistry, mineralogy and the steam engine"; these are exactly the four branches covered by the glossaries included in the TVM edition published after 1889. The reference is extremely important since it confirms once more that, in 1896, seven years after the publication of the QJ, other glossaries had not been printed.

Another matter of importance is that Fryer points out these glossaries included existing terms; new terms were coined, as far as possible, only when suitable ones did not exist. The glossaries were still for sale at the time of the conference, in 1896, which indicates that these texts were still available and circulating in 1896, even though thirteen years had passed since the publication of the first one, namely, the JS.

The successive words in Fryer's speech are the key elements of the quoted text. As already mentioned, after the four glossaries gathered in the first and following editions of the TVM were published, Fryer devoted himself to the compilation of others. Among those explicitly indicated during the *Records of the Second Triennial Meeting* are glossaries in the fields of geology, botany, geography, biology, and more, including one in naval engineering. The works Fryer indicates in the passage quoted can be easily identified with the manuscripts housed in the EAL: volume seven, VGT, for geology; volume three, BTWDCL, and volume eight, LBT, for botany; and volume one, LGN, for geography.[36]

As Fryer already specified, these glossaries were never printed. The reason lies in Fryer's will to first receive the approval of the committee in charge of the publication; this may also supply a partial answer as to why the EAL manuscripts were not published. Another fundamental motivation is the lack of funding for printing: "Several other manuscripts will be offered us for publication soon, and unless we can get help from some source the printing of these much needed books will have to be postponed indefinitely."[37] In order to display the results of his work, Fryer showed during the conference a specimen

35 Fryer, "The Present Outlook," 157–58.

36 With respect to the volume on biology, it has still not been possible to locate the text Fryer is referring to; it may indicate a section inside volume six, the VCT.

37 *Records of the Triennial Meeting*, 54.

of the edited volumes, a very thick one in which numerous terms referring to botany were included. The reference, in this case, could be to the aforesaid volume three, the BTWDCL, or more likely, according to the description, volume eight, including the LBT. Fryer added that he was in possession of a dozen texts similar to the one shown during the conference; the EAL volumes that, in the strictest sense, include glossaries, are actually nine, even if at times a volume includes more than one glossary.

Finally, in the second part of the text quoted from the *Records of the Second Triennial Meeting*, it is possible to identify one of the reasons that could have led Fryer to abandon his work for the Jiangnan Arsenal and move to the United States, a decision already taken by the time of the conference.[38] This motive was the lack of, or at least inadequate, acknowledgement of his work as a translator. Fryer had not been re-elected as a member of the Committee and he regretted that, also due to this decision, all efforts lavished by him on the publication of scientific and technical glossaries proved to be wasted.

Notwithstanding this regret, further on in the *Records of the Second Triennial Meeting*, in which he urges the Committee not to spare any effort to make such a collection of glossaries available, Fryer restated his willingness to cooperate for this aim even in 1896: "It means, of course, a lot of downright hard work; but if the new Publication Committee, about to be elected, will divide it up fairly, so that each member of the Association, able to undertake the nomenclature of anyone or more subjects, will select his subject or have it assigned to him, this pressing want would soon be supplied. I am willing to do my full share in the matter."[39]

5 Conclusion

The analysis of the planning of the *TVM*, its publication process, and the abandonment of the idea of a second volume provides significant information for the readership. Firstly, the words expressed by Fryer and his colleagues reflect

38 Fryer accepted the Agassiz Professorship on November 2, 1895, as can be read in a letter to John Harmon Charles Bonté (1831–1896), Secretary of the Board of Regents of the University of California. Nevertheless, in a letter to his brother George, dated August 26, 1892, he had already expressed concern about his financial state in China. Fryer showed for the first time his interest in the Professorship and in moving to Berkeley, where his family was already living, in a letter to Martin Kellogg (1828–1903), president of the University of California, dated January 12, 1895. All letters are stored in the FP and have been republished in Dagenais, *The John Fryer Papers*, III, respectively 44, 17, and 20.

39 Fryer, "The Present Outlook," 158.

the dynamics between them, the societies they were affiliated with, and the ideological and theoretical struggle for the standardisation of the scientific and technical lexicon of Chinese taking place in late Qing China. Secondly, the fact that the TVM was printed after the glossaries is significant: it confirms they were already circulated among other missionaries well before the publication of the first edition of the TVM, as shown in the sections above, and that they were not only used as a personal reference by Fryer. Consequently, the translation principles behind the text, the scientific and technical notions conveyed through it, and the lexicological innovations put forward in the TVM started to circulate in a systematic way as early as the publication of the JS. Finally, the texts quoted in this chapter give a glimpse of the reasons for which a second volume of the first collection of scientific and technical English-Chinese bilingual glossaries was never published. In spite of the efforts spent for almost thirty years for the Jiangnan Arsenal and, in a broader sense, for the standardisation of Chinese scientific and technical nomenclature, as well as the promise to continue his work with the aforementioned aim, numerous glossaries Fryer composed remained manuscripts. Many years passed before the standardisation of Chinese scientific and technical lexicon would be accomplished.[40] This showcases the complex interactions at work between linguistic factors,

40 Other allusions to the TVM can also be found in different types of secondary literature sources. For example, in James Howard Gore, *American Members of Foreign Orders* (Washington: Press of W. F. Roberts Co., 1910), 40, Fryer is described, among others, as the "secretary to Imperial Chinese ambassador, Kwo-Sung-Tao" and as the author of the "Translator's Vade-Mecum, or Vocabulary of Scientific Terms in Chinese and English." Basically, the same words are used in Henry Woldmar Ruoff, *Masters of Achievement: The World's Greatest Leaders in Literature, Art, Religion, Philosophy, Science, Politics and Industry* (Buffalo, NY: Frontier Press Company, 1910), 706, and Albert Nelson Marquis, ed., *Who's Who in America: A Biographical Dictionary of Notable Living Men and Women of the United States; Vol. VI, 1910–1911* (Chicago: A. N. Marquis & Company; London: Kegan Paul, Trench, Trubner & Co., 1910), 697. Nevertheless, in the previous edition of the last text, namely John William Leonard, ed., *Who's Who in America: A Biographical Dictionary of Notable Living Men and Women of the United States; 1901–1902* (Chicago: A. N. Marquis & Company, 1901), 406, while the mention of Fryer as the author of the TVM is essentially the same, he is indicated as "second interpreter to Chinese ambassador, Kwo-Sung-Tao," not as secretary. The same can be read, almost with the same words, in *The Blue and Gold 1904* (Junior Class, University of California, 1903), 29. A mention of Fryer as the author of the "Translator's Vade-mecum, or Vocabulary of Scientific Terms in Chinese and English" can also be found in Frederick Converse Beach and George Edwin Rines, ed., *The Encyclopedia Americana*, vol. 7 (New York, Chicago: The Americana Company, 1904), and Frank Moore Colby and Talcott Williams, *The New International Encyclopædia: Volume IX*, 2nd ed. (New York: Dodd, Mead and Company, 1917), 320, but no indication in these two texts of Fryer's role in the embassy of the Chinese diplomat Guo Songtao 郭嵩燾 (1818–1891) can be located.

historical circumstances, and power dynamics for the success of the *TVM* and, analogously, of other relevant texts. It is not only the translation methods, lexicographical suitability or prestige of the translator to determine the success of a certain term, and therefore the possibility to convey the idea behind it. If the manuscripts had been published, it is plausible that other traces of the work by Fryer and his Chinese colleagues, analysed in detail later on in this work, could be found in the Chinese language, and the notions expressed by them could be spread in a quicker way in late Qing China.

After this long journey through primary sources to trace the compilation and publication history of the *TVM*, we will—before immersing ourselves in its contents—consider the theoretical and linguistic background of Fryer's translation work and, consequently, of the *TVM*. This is the task the next chapter is devoted to.

CHAPTER 3

Fryer's Theories on Translation into Chinese

1 Introduction

Analogously to the way that the *TVM* planning and publication process reflects the historical circumstances in which it took place, the examination of the theoretical principles that lie behind the *TVM*, particularly from the point of view of translation methods, is a precious source of information. It can help to understand the reasons for the translation choices adopted by late Qing translators and the personal and institutional interactions that led to them. At the same time, it can clear up the topic of the connection between religious efforts by missionaries in late Qing China and their more secular attitude, at least for some of them, towards translation.

The translation principles behind the *TVM* have been illustrated by Fryer in different circumstances; they can be considered the real background of his work as a translator and as a former missionary. In order to carry out this task, a fundamental resource is the paper "Scientific Terminology: Present Discrepancies and Means of Securing Uniformity."[1] From a broad perspective, the conference proceedings of the Protestant missionaries represent a useful instrument for scholars, since they "[…] provide good indications concerning the views held by individuals and groups within the […] Protestant missionary community in China."[2] More specifically, the aforesaid text, other than being the basis for Fryer's speech at the conference held in Shanghai in 1890, integrates many of his theories on translation and the evolution of Chinese lexicon, expressed in different circumstances.[3] As can be inferred from the year the conference was held, the paper was published two years after the first edition of the *TVM* was published in 1888. The speech, therefore, was delivered at a later phase of Fryer's time in China, and can be considered a summary of his work for the Jiangnan Arsenal and of his theories on translation into Chinese.

1 Fryer, "Scientific Terminology." The article was republished in Dagenais, *The John Fryer Papers*, II, 376–410.

2 Gary Tiedemann, *Handbook of Christianity in China—Volume Two: 1800–present* (Leiden, Boston: Brill, 2010), 53.

3 See for example, "The Advisability, or the Reverse, of Endeavouring to Convey Western Knowledge to the Chinese through the Medium of their Own Language," *Journal of the North China Branch of the Royal Asiatic Society*, New series, vol. 21, no. 1 (1886): 9–11; Fryer, *An account*, 8–11; and Idem, "The Present Outlook."

 | DOI:10.1163/9789004443211_005

2 Religious Views and Technical Nomenclature: Discrepancies and Responsibilities

A topic in which there is still much debate is the entangled interaction between Fryer's religious opinions, his relations with the missionary world of late Qing China, and the questions of nomenclature and translation. As already pointed out in chapter 1, Fryer abandoned his missionary aspirations after arriving in China, but his connection with the missionary milieu remained strong throughout his stay in Shanghai. However, during the conference, and in the relevant proceedings, other participants always refer to him simply as "Mr.," "Esq.," or "Dr."[4] At the beginning of the aforesaid speech, in the first part, entitled "Scientific terminology in its relation to the Chinese Language," Fryer emphasises the importance of the scientific terminology for the missionary cause in what he calls "[...] the present crisis in the history of China [...]."[5] By placing the two questions on the same plane, Fryer hoped that terminology and its standardisation would receive the attention they deserved. In his speech, Fryer repeatedly expressed his disappointment precisely because the scientific and technical terminology had been ignored in the time since the previous conference. In this regard, Fryer is assuming an attitude analogous to that of Jesuits during the Ming (1368–1644) and Qing dynasties; his task was as difficult as that of Jesuits during the Rites Controversy, even if from a different perspective.[6] In dealing with the question of terminology, Fryer highlights it as one of the main focuses of the conference, whether for the sake of the Chinese language or for the missionary cause, drawing the attention of the other participants.

In any case, it is evident that Fryer's attitude towards translation is tightly connected to his religious views, and it can help to understand that of the other conference participants. Many missionaries believed that religious propaganda and the translation of scientific texts should be separated.[7] Fryer can be counted among them, but even so, it is significant that he presented his theories on the linguistic aspects of his work precisely at a conference of Protestant missionaries. Even though he never rendered into Chinese texts of

4 This can also be seen in Matthew Tyson Yates, R. Nelson, and E. R. Barrett, ed., *Records of the General Conference of the Protestant Missionaries of China, Held at Shanghai, May 10–24, 1877* (Shanghai: Presbyterian Mission Press, 1878) and *Records of the Triennial Meeting*, 27.

5 Fryer, "Scientific Terminology," 531.

6 Elman, *On Their Own Terms*, xxvii. Refer also to George H. Dunne, *Generation of Giants: The Story of the Jesuits in China in the Last Decades of the Ming Dynasty* (Notre Dame: University of Notre Dame Press, 1962), 28.

7 As indicated in Tsien, "Western Impact on China," 311.

religious content, he always stressed the importance of the religious aspects of the work by foreign doctors, translators, and experts living in China.

In order to better illustrate such a connection and the power dynamic between John Fryer and other missionaries in terms of the standardisation of science terminology, the reply to his speech by the American Presbyterian missionary Calvin Wilson Mateer (Di Kaowen 狄考文, 1836–1908) is enlightening:

> I have been a member of the Committee on School and Text Books from the first, and I feel bound to say that there are two things in which the action of the committee does not represent me. First, there has been a considerable number of books published by the committee, which are not properly school books at all but tracts. I have always opposed the publication of these books by the committee. They are religious tracts and fall properly within the province of the religious tract societies, of which we have several in China [...]
>
> I agree with Mr. Fryer's paper in the main. I certainly agree with him that a dictionary of technical terms is needed. It should not, however, be a dictionary of terms manufactured by any one man, but should rather be a thesaurus of all the terms heretofore used, rejecting only such as are ridiculously unsuitable. Nor should any attempt be made by any committee or society to impose authoritatively a set of terms upon authors or teachers. A technical term has just as much authority as its intrinsic excellence and the reputation of its author will give it, and no more.[8]

Mateer is opposing Fryer on two levels here. The first is the connection between the translation of scientific and technical texts and the religious publications by the School and Textbook Series Committee. According to Fryer, the competence of the Committee should not include religious treatises. The second level is about the methods for the standardisation of the nomenclature adopted in such translations. Mateer did not want it to be the result of the commitment of only one person, referring overtly to Fryer, but instead a concerted effort of all Committee members. This shows how, even for a relatively secular figure such as Fryer, the religious attitudes and power dynamic in terms of the standardisation of science terminology were tightly interconnected and not easily detachable.

---

8 Fryer, "Scientific Terminology," 550–51. On Mateer, consult Daniel Webster Fisher, *Calvin Wilson Mateer: Forty-Five Years a Missionary in Shantung, China; A Biography* (Philadelphia: The Westminster Press, 1911).

Going back to Fryer's speech, he conducts a detailed analysis to illustrate how, in Western countries, the evolution of terms pertaining to scientific branches was already in a relatively advanced phase.[9] According to him, in the West, scientific and technical nomenclature was already in a mature stage of its development, as modern terms had already passed through different phases that tested their suitability for expressing the preciseness of scientific reality. Those terms unable to cope with it, "[…] have either to be discarded, or else retained as a useless encumbrance."[10] According to Fryer, China had the undeniable advantage of being able to draw from this body of knowledge without all the encumbrances that its evolution would have implied. The only problem was which nomenclature would express this knowledge.

It is important to note that Fryer did not consider the question of terminology purely in itself; for him, it also had implications on the "[…] progress of China […]."[11] Fryer's biggest concern lay in the fact that Chinese scientific and technical nomenclature was becoming more cumbersome, with various systems employed by different authors already following diverging paths. As he had personally ascertained, "The language is already beginning to be burdened with different sets of technical terms expressing one and the same idea."[12] This concern pertained to the educational and scientific development of China itself and was also shared by other scholars and missionaries; a few years later "Rev. John C. Ferguson," namely the American scholar John Calvin Ferguson (Fu Kaisen 福開森, 1866–1945), would confirm, "The great lack of all was a common system of terms, which would enable the student to pass from one work to another with freedom and ease […] In one work terms were translated, in another they were transliterated, and in still others native terms had been used."[13]

Fryer believed that "Nearly everything in the way of modern science is still new to China."[14] This situation would have helped Chinese and foreign translators, all the major modifications that scientific terminologies had undergone in the West, mentioned earlier, being unnecessary. Apart from Fryer's

9 This was true, for example, for chemistry. A history of the terminology in this field can be read in Maurice Pierre Crosland, *Historical Studies in the Language of Chemistry* (London: Heinemann, 1962. Reprinted Mineola, NY: Dover Publications, 2004), particularly chapters 7 and 8.

10 Fryer, "Scientific Terminology," 532.

11 Ivi, 533.

12 *Ibidem*. See also Arthur Frederick Wright, "The Chinese Language and Foreign Ideas," in *Studies in Chinese Thought*, ed. Arthur Frederick Wright (Chicago: University of Chicago Press, 1953), 295.

13 *Records of the Triennial Meeting*, 18.

14 Fryer, "Scientific Terminology," 533.

paternalistic considerations concerning the possibility of transmitting a whole body of knowledge supposedly almost unknown in China, it was true that some of the sectors covered in the translations by Fryer and his Chinese colleagues, such as chemistry, fell outside the traditional imperial examination system, *keju* 科舉.[15] For this and other sectors, trying to create *ex novo* a standardised nomenclature would have been a relatively easy task, if only there had been cooperation between every single translator, foreign or Chinese, who "[…] takes part or interest in the advancement of China."[16]

According to Fryer, one of the principal causes that led to the existence of different nomenclatures was the absence of coordinated efforts by translators, as well as the lack of process. He particularly regretted the fact that nobody really realised the importance of Chinese scientific and technical nomenclature, and devoted efforts in this direction.[17] Fryer was deeply aware of the complexity of the situation and it was exactly for this reason that he had been working for the Jiangnan Arsenal since 1868.

Fryer also had to face numerous claims that the Chinese language hindered providing a more precise definition of scientific and technical reality.[18] According to him, problems arising in the translation of scientific and technical texts could not be ignored, but instead were mainly attributed to the faults of foreign translators and the impediments advanced by their Chinese colleagues.[19] In other circumstances, Fryer would even affirm that Chinese is not only as capable as Western languages to express scientific reality, "[…] but on the direct contrary there is a special adaptability for that purpose not to be found in any other language all over the world. There is a peculiar elasticity, expressiveness and terseness in the written language of the Middle Kingdom, through which it lends itself or accommodates itself to the reception of foreign ideas generally, and to the formation of new but easily intelligible scientific terms in particular."[20]

15 Refer to Reardon-Anderson, *The Study of Change*, 77 and passim, and to Calvin Wilson Mateer, "School Books for China," *The Chinese Recorder and Missionary Journal* 8, no. 5 (1877): 428.

16 Fryer, "Scientific Terminology," 533.

17 *Ibidem.*

18 Refer also to "The Advisability," 9–11.

19 Fryer, "Scientific Terminology," 533–54.

20 Fryer, "The Present Outlook," 159.

3 Between "The Descriptive" and "The Phonetic" Methods

The analysis Fryer proposed on the various systems that different institutions, or at times even individual translators, adopted to express the same concepts, required practical solutions. The topic was not a purely lexicological concern, since it affected the pace of diffusion of those notions that he deemed necessary for the scientific development of China. The question, therefore, started from the lexicon and had other complicated and higher ramifications; that is why, after the aforesaid introduction on Chinese scientific and technical nomenclature, Fryer makes an excursus in which he explains the modalities Chinese language employed to enrich its lexicon. According to his analysis, in order to express new concepts Chinese have absorbed neologisms "[…] in three ways. The first is what may be called the descriptive method, the second the phonetic, and the third a combination of the other two."[21] In the first case, he makes the example of *yanye* 烟葉, its meaning composed of the two terms "smoke" and "leaf"; according to Fryer, *yanye* was a popular name for tobacco and could be described as a "descriptive term." The second example provided is *jinji'na* 金雞那. The term refers to "cinchona bark" and, according to Fryer, had been introduced in Chinese by the Jesuits; *jinji'na* is an instance of a "phonetic term." Finally, the third example, *jiasha* 袈裟, is according to his words a phonetic term to indicate the *kāṣāya* of Buddhist monks, and at the same time is a descriptive term; this definition is due to the presence of the radical for clothes, *yi* 衣.

Analysing these considerations in light of modern linguistic nomenclature, it is evident that the first category Fryer pointed out, "descriptive term," can be identified with calques and descriptive labelling.[22] On the other hand, the definition of "phonetic term" indicates phonemic loans. Finally, pertaining to the third group, Fryer was presumably referring to what nowadays are known as hybrids and loan blends. The example provided, though, is misleading, since *jiasha* 袈裟 has to be considered a fully-fledged phonemic loan; here Fryer is confusing the graphical aspect of the character, a *xingsheng zi*, with its phonetic value. It can be noted that the use of radicals, particularly *kou* 口 (mouth), to indicate a specific character is only being used for its phonetic value is one of the most salient features of Chinese phonemic loans. This property had been

21 Fryer, "Scientific Terminology," 534.

22 The relevant nomenclature varies according to different articles and monographs; in this work, the author uses the one adopted in Benjamin Ka-Yin T'sou, "Language Contact and Lexical Innovation," in *New Terms for New Ideas: Western Knowledge and Lexical Change in Late Imperial China*, ed. Michael Lackner, Iwo Amelung, and Joachim Kurtz (Leiden, Boston, Köln: Brill, 2001).

in existence for centuries, employed by Jesuit missionaries as well, and would be increasingly disseminated to avoid using common characters, bearers of their own specific semantic connotations—so much so that a set of characters specifically intended for this purpose would become standardised.

A certain evolution of Fryer's linguistic theories about the Chinese language can be noticed here. Pertaining to the three modalities of creating neologisms, he had previously stated that, "There are two ways by which new terms seem to be introduced into Chinese, and which may be termed the descriptive and the phonetic methods."[23] Therefore, the idea of a third method was developed by Fryer at a later stage.

Other important considerations follow. Fryer reinforces that, whenever possible, Chinese tended to prefer "descriptive terms" and only as a last option "purely phonetic terms." According to him, Chinese readers and learners despised phonemic loans, particularly those composed of more characters, since they do not provide any indication on the meaning of the referent of a certain word. The only case in which phonemic loans can be employed is "for terms that are absolutely untranslatable in any other way";[24] the fault for their excessive use should be attributed to "lazy or ignorant translators."[25]

Fryer here is highlighting one of the peculiarities of Chinese lexicon: phonemic loans are not easily integrated within its texture, especially if composed of many characters. In contemporary Chinese, phonemic loans represent a very low percentage of the whole lexicon; among these, terms composed of more than three characters are particularly rare.[26] Phonemic loans are more easily accepted when accompanied by one or more characters, usually as right-hand components, expressing semantic information on the referent, thus creating hybrids. Another modality that disguises phonemic loans in the lexical texture is the adoption of characters that, in addition to their phonetic value, bear meanings associated with the referent; the results are, in this case, loan blends.

This characteristic, successively studied in different articles and monographs,[27] was not universally recognised by other contemporaneous translators and

23 "The Advisability," 10. In the text, a series of opinions by different late Qing foreign scholars on the likelihood to transmit Western knowledge to China, through the use of the Chinese language, can also be read.

24 Fryer, "Scientific Terminology," 534.

25 *Ibidem.*

26 Among others, refer to Giorgio Francesco Arcodia, *Lexical Derivation in Mandarin Chinese* (Taipei: Crane Publishing Co., Ltd, 2012), 120.

27 Among the most important: Gao Mingkai and Liu Zhengtan, *Xiandai hanyu wailaici yanjiu* (Beijing: Wenzi gaige chubanshe, 1958); Zdenka Novotná, "Linguistic Factors of the Low Adaptability of Loan-words to the Lexical System of Modern Chinese," *Monumenta Serica* 26 (1967); Eadem, "Contribution to the Study of Loan-words and Hybrid Words in

scholars. Among the replies to Fryer's speech at the conference of Protestant missionaries, it is possible to recall one by the Chinese translator and pastor Yan Yongjing 顏永京 (1838–1898), who suggested adopting phonemic loans, rather than "having them translated."[28] According to his analysis, phonemic loans are easily accepted in Chinese, as demonstrated by the term for "*ultimatum*," "rendered ai-ti-mei-tun," and by "telephone," "known as teh-li-fung."[29] Analogous considerations were also expressed by foreign missionaries, such as Mateer—he supported Yan Yongjing's proposal, preferring to adopt phonemic loans rather than proper translations for "scientific terms."[30] Interestingly, the divergent opinions expressed echo the different attitude towards phonemic loans and other translation methods of early Jesuits and their successors. Some of them were criticised precisely because they avoided coining news terms and tried to attribute new meanings, frequently religious ones, to words already listed in classical Confucian texts.[31] In any case, Mateer apparently changed his mind a few years later, pertaining to both the nomenclature for chemical elements and other sectorial terms such as "affinity," "quantivalence" or "acid," as expressed in the letter sent "To the Members of the Committee on Terminology," signed by him and the Presbyterian missionary Watson McMillan Hayes (He Shi 赫士, 1857–1944).[32]

In this context, Fryer accused contemporaneous translators of using, and sometimes overusing, phonemic loans, either for lack of will to cooperate with other colleagues, or not having sufficient scientific and linguistic knowledge. The relevant examples provided are "granite," translated as *gelanituo*

Modern Chinese," *Archiv Orientální* 35 (1967); 36 (1968); 37 (1969); Eadem, "Morphemic Reproduction of Foreign Lexical Models in Modern Chinese," *Archiv Orientální* 43 (1975); Christopher D. Godwin, "Writing Foreign Terms in Chinese," *Journal of Chinese Linguistics* 7, no. 2 (1979); Ren Xueliang, *Hanyu zaocifa* (Beijing: Zhongguo shehui kexue chubanshe, 1981); Masini, *The Formation of Modern Chinese*; Liu Zhengtan et al., *Hanyu wailaici cidian* (Shanghai: Shanghai cishu chubanshe, 1984); Alleton, "Chinese Terminologies"; T'sou, "Language Contact"; and Thekla Wiebusch and Uri Tadmor, "Loanwords in Mandarin Chinese," in *Loanwords in the World's Languages: A Comparative Handbook*, ed. Martin Haspelmath and Uri Tadmor (Berlin: De Gruyter Mouton, 2009).

28 Fryer, "Scientific Terminology," 549.

29 *Ibidem*. For the latter, refer to the relevant entry in the appendix of neologisms in Masini, *The Formation of Modern Chinese*, 165.

30 Fryer, "Scientific Terminology," 550.

31 See Wright, "The Chinese Language and Foreign Ideas," 289–90.

32 FP, Mss C-B 968, carton 3, folder 34, "Miscellaneous Materials—Material of Chinese Scientific Terminology." Courtesy of The Bancroft Library, University of California, Berkeley.

"合拉尼脫," and "gypsum," rendered as *juebusi'en* "絕不斯恩."[33] These two translations are listed in the *JS* under the column of terms appearing in the *Manual of Mineralogy* by Dana. The theory of the inadaptability of phonemic loans was reaffirmed by Fryer in many circumstances, particularly when facing statements by other translators on the impossibility for Chinese to clearly express scientific and technical reality:[34] "It really is often a matter of little importance whether we phoneticize a new term, or whether we make the term self-descriptive or self-evident."[35]

Fryer explained that branches such as pharmacology and legal medicine rapidly developed in China, also from the point of view of nomenclature; the relevant translations could simply adopt terms already recognised and accepted.[36] According to Fryer, other branches of science, in more recent times, had swiftly expanded; the difference for these sectors, though, is that nomenclature was still not ready and there was a necessity to coin new terms, following the three methods mentioned before. For Fryer, Chinese could express, like any other language, the scientific and technical reality through the use of an adequate nomenclature. If that was not happening, translators should be blamed, particularly "foreigners," since they loosely adopted phonemic loans and coined neologisms.[37] The reasons lay in the lack of systematicity to avoid the annoyance of "[…] investigating either what nomenclature the Chinese have had in use for centuries, or what recent translators have already done."[38]

These factors all lead to the discrepancies between existing nomenclatures, also caused by the lack of publication of lists of terms. The last factor, in particular, would have greatly helped the standardisation process and the work by other translators, while in reality they simply adopted their own system and did not share it with others: "The number of vocabularies in Chinese and English that have been given for general use or comparison is extremely few."[39] The standardisation of nomenclature could be achieved only by consulting works that treated similar aspects; the purpose of this task, to be accomplished before starting a translation, was to consistently adopt the nomenclatures

33 Fryer, "Scientific Terminology," 539–40. The first was actually listed twice in juan 13 of the *Shiwu tongkao* and in juan 14 of Chen Longchang, *Zhong-Xi binglüe zhizhang* (Dongshan caotang, 1897).

34 See also Wright, "The Chinese Language and Foreign Ideas," 286–88.

35 Fryer, "The Present Outlook," 157–58.

36 Refer also to Elman, *On Their Own Terms*, 29–34.

37 Fryer, "Scientific Terminology," 535.

38 *Ibidem.*

39 *Ibidem.*

employed in those works, avoiding the creation of redundant and unnecessary terms.[40]

Leaving aside for the moment theoretical considerations on the translation of scientific and technical nomenclature in Chinese, in the speech Fryer provided important details on some of the work he personally consulted to achieve this purpose:

> [...] the works on Astronomy and Mathematics, by Mr. A. Wylie; Mechanics, by Dr. Edkins; Natural Philosophy and Medical Science, by Dr. Hobson; Political and Physical Geography, by Rev. W. Muirhead and others; and Botany, by Dr Williamson. With these should be mentioned the voluminous works of the Jesuit missionaries, which, though two or more centuries old, were often of considerable service, especially in astronomical and mathematical terms. The only useful vocabularies of scientific terms in English and Chinese at that time were: the very limited ones of Mr. Wylie, giving astronomical and mathematical terms; of Dr. Edkins, giving terms in mechanics; of Dr. Hobson, giving terms in natural philosophy and medicine; and of Dr. Bridgman, giving a long series of terms published in his Chrestomathy.[41]

The importance of this information on the origin of the *TVM* is fundamental; Fryer is mentioning texts translated by the Protestant missionaries Alexander Wylie (Weilie Yali 偉烈亞力, 1815–1887), Joseph Edkins (Ai Yuese 艾約瑟, 1823–1905), Benjamin Hobson (He Xin 合信, 1816–1873), William Muirhead, and Alexander Williamson, in addition to earlier texts by Jesuit missionaries.[42] Fryer also quoted some of the very few scientific and technical bilingual glossaries available at this time, including that by the American Protestant missionary Elijah Coleman Bridgman (Bi Zhiwen 裨治文, 1801–1861). As can be read in the passage quoted above, Fryer is particularly referring to dictionaries compiled by Jesuits, though they are not specified. Some of these texts

---

40 *Ibidem.*

41 Ivi, 536.

42 For the relevance of Joseph Edkins and Li Shanlan, trans., *Zhongxue* (Shanghai: Mohai shuguan, 1859), translation of *An Elementary Treatise on Mechanics* by William Whewell (1794–1866), refer to Deng Liang and Han Qi, "'Zhongxue' banben liuchuan ji qi yingxiang," *Wenxian*, no. 3 (2009). Li Shanlan 李善蘭 (1811–1882) was a famous mathematician and translator; for his role in the rendering of mathematical texts, see Wann-sheng Horng, "Li Shanlan: The Impact of Western Mathematics in China During the Late 19th Century" (PhD diss., City University of New York, 1991) and Benjamin A. Elman., *From Philosophy to Philology: Intellectual and Social Aspects of Change in Late Imperial China* (Cambridge, MA: Council on East Asian Studies, Harvard University, 1984), 120.

are identified and presented in this work and prove the important connection between the translations carried out during the Ming dynasty and the late Qing dynasty.

4 Advice of the Translator: Fryer's Suggestions

The analysis Fryer proposed on the causes of discrepancies in Chinese scientific and technical lexicon was not a proof of concept for him. Other than involving power relations with other missionaries and colleagues, the question of terminology also affected the possibility of China's scientific and educational development. For this reason, Fryer did not limit himself to dissecting these causes and the theoretical background of translation practice; he also provided detailed methodological and linguistic suggestions for creating a valid system for scientific nomenclature. Such a system had already been devised at a much earlier stage by his Chinese colleagues and was published for the first time in "[…] January, 1880, when it appeared in the *N.-C. Daily News* and *Herald*, and subsequently in *Nature*."[43]

His proposals are divided into three main points.[44] In the first one, Fryer suggested searching existing, or presumably existing, terms in native texts and in works compiled by Jesuit and Protestant missionaries, as well as among people who had to deal with specific scientific and technical terminologies for their work. This suggestion was later praised and recognised by other scholars and missionaries, such as John Calvin Ferguson:

> The whole subject of scientific terminology, which was so ably treated by Dr. Fryer in his paper read before the General Conference, is of the greatest importance […] Some of us must find time to search out the terms used in native books and in the Imperial Dictionary of K'ang Hsi […] The early works of the Jesuit Fathers will be of great service, as many

43 The text can also be read in Fryer, *An account*, 9–10. The reference here is instead to its original publication in the *North China Herald* (1880, January 29): 77–81, and to the version printed in "Science in China," *Nature* (1881, May 5): 9–11; (May 19): 54–57. The text was later published in its Chinese version, in John Fryer, "Jiangnan zhizao zongju fanyi xishu shilüe," *Gezhi huibian* 3, no. 5 (June 1880): 10A–12B; no. 6 (July 1880): 9A–11B; no. 7 (August 1880): 9A–11B; no. 8 (September 1880): 9A–10B. The latter version was reprinted in *Zhongguo jindai chuban shiliao chubian*, ed. Zhang Jinglu (Shanghai: Shanghai chubanshe, 1953), 1–23. The reference comes from Wang Yangzong, "A New Inquiry," footnote no. 4, and Wright, *Translating Science*, 238 and 479, footnote no. 141.

44 Fryer, "Scientific Terminology," 536.

> common terms were given special technical meanings by them and are now known by general students. Our work can also be greatly facilitated by constant communication with teachers in Japan and by reference to Japanese works. Many Western books have been translated into Japanese by natives who are specialists in the subjects of which these works treat, and as the majority of the terms used by them are expressed by Chinese characters we have an excellent guide [...]."[45]

For Fryer, only when terms could not be located should new ones be coined. The first method for coining neologisms suggested the creation of characters (*zi*), not of *ci* (words).[46] The typical example is the nomenclature for chemical elements, creating *xingsheng zi* or attributing a new meaning to a rare character. The second method recommends the adoption of a "descriptive term," already analysed; it is only as a last resort that Fryer envisages the use of phonemic loans. In the latter case, Fryer underlines the need to use the same character for the same phoneme, giving preference to characters already used by other translators. Nevertheless, phonemic loans should have been considered provisional and replaced by existing terms, in case they could be found according to the methods suggested above, or if more suitable ones were coined.[47] Finally, the third point restates the necessity of gathering such terms in lists, in order to print them.

Nevertheless, from the date of the first publication of his proposal, in 1880, the relevant efforts were not satisfactory—confusion in terminology prevailed in the works produced by the Jiangnan Arsenal itself. If translators had adhered to these proposals, they would have been able to publish various lists of terms by the time of the conference, "instead of only the four or five already published, containing about 18,000 terms, and about the same number in manuscript."[48] The reference here is important for drawing a better description of the TVM's history and diffusion; the "four or five" glossaries Fryer

45 *Records of the Triennial Meeting*, 18–19. This, as Ferguson explains, could have helped to prevent critics from blaming the texts for not being translated by natives or specialists.

46 The distinction and the nomenclature between *zi* and *ci* will be theorised at a later stage, as stated for example in Lydia He Liu, *The Clash of Empires: The Invention of China in Modern World Making* (Cambridge, MA: Harvard University Press, 2004), 207. Refer also to Uchida Keiichi, *Kindai ni okeru tōzai gengo bunka sesshoku no kenkyū* (Osaka: Kansai daigaku shuppanbu 2001), 167–99, and Idem, *Bunka kōshōgaku*, 113–24. Even so, Fryer seems here to be consistent with his definitions; refer to Joachim Kurtz, *The Discovery of Chinese Logic* (Leiden, Boston: Brill, 2011), 133, in which another example of Fryer's use of the term *zi* for "word" can be read.

47 Refer also to Zhou Dingyi, "Yinyici he yiyici de xiaozhang," *Zhongguo yuwen*, no. 119 (1962).

48 Fryer, "Scientific Terminology," 537.

mentions are exactly those included in the first edition of the *TVM* printed in 1888, considering the *RMDM* an appendix and including the *QJ*, published in 1889. As for "the same number in manuscript," the reference is to the volumes stored in the EAL: this is further confirmation that Fryer had drawn these manuscripts with the purpose of publication and not exclusively for personal reference.

Once more, Fryer praises the role played by Jesuits; the principles he listed during the conference of Protestant missionaries determined the value of the texts translated in Chinese and were "[...] evidently well understood by the Jesuit missionaries."[49] This statement confirms that Fryer read their works, appraising the perfection of the terminology adopted, and at the same time certifies the link between late Qing lexicographical materials and earlier Jesuit texts, analysed further on in this work. As these texts were few, Fryer also tried to consult glossaries with Latin and Chinese parallel translation compiled by Jesuits, but he could not find them. Fryer was willing therefore to adopt the terminology coined and employed by Jesuit missionaries; he was merely unable to consult it in the form of bilingual glossaries.

In the second part of his speech, "Some of the Essential Features for a System of Scientific Nomenclature for China," Fryer explains in detail the different translation choices mentioned earlier, as well as their pros and cons. The first suggestion envisages that "*New terms ought to be translations, where possible, and not mere transliterations.*"[50] He particularly rejects the accusations toward the Chinese language regarding its supposed scarcity of technical terms, blaming instead those scholars who believed that its lexicon could be enriched only with phonemic loans from Western sources. The voices promoting this idea simply ignored the importance that "The original meanings of the said characters, or the number that will be required to express an ordinary technical term" had in Chinese.[51] Fryer therefore stressed the importance of avoiding as much as possible the use of phonemic loans; it was not clear which source language should be used as a reference, and in different regions of China the phonetic

49 *Ibidem*.

50 Ivi, 538. Fryer always indicates phonemic loans as "transliterations" or "phoneticization," referring to other solutions as "translations." Read, for example, further on in the same text, page 539: "In the same way it no doubt often happens that even our best translators paraphrase or phoneticize where they ought to translate." See also Shen Guowei, "Yici yu jieci—Chongdu Hu Yilu 'Lun yiming' (1914)," *Wakumon*, no. 9 (2005): 107–108, and Idem, "Kindai kanji yakugo no kenkyū ni tsuite: chūgokugo kara no shiten," in *Higashi Ajia gengo sesshoku no kenkyū*, ed. Shen Guowei and Uchida Keiichi (Osaka: Kansai daigaku shuppanbu, 2016).

51 Fryer, "Scientific Terminology," 538.

aspects of the same phonemic loan could differ extremely.[52] The outcome of this situation was evident discrepancies among renderings adopted by one translator or the other, not to mention the cases of those working in different parts of China.[53] This was not a mere problem of linguistic inconsistencies, but involved instead the possibility itself of transmitting knowledge by means of the Chinese language and, consequently, hindered scientific progress at all levels.

Fryer, though, was not *a priori* against the use of phonemic loans; translators should employ them whenever required for practical reasons. For instance, when they could not trace the etymology of a given word or because the amount of work made it impossible to search another adequate term. According to Fryer, languages are able to borrow and integrate new words "[...] but the advantages are in proportion to the similarity of the language."[54] Nevertheless, adding different examples of unsuccessful translations, he explained that it would be a mistake to adhere too rigorously to the etymology of the original word.[55] Even so, the example mentioned by Fryer earlier in the text, *jinji'na*, which he attributed to Jesuits, is still listed in contemporary Chinese dictionaries to indicate plants of the genus Cinchona, with the character *na* 納 in place of the homophone *na* 那.

The ideal method for Fryer was to look for those characters fallen into disuse in dictionaries, particularly the *Kangxi zidian*, conferring them a new meaning; this was exactly the solution adopted by him and his Chinese colleagues for the nomenclature of chemical elements.[56] Nevertheless, Fryer did not realise that, with this exception, the path followed by the Chinese language, including scientific and technical terminology, was polysyllabism, through the combination of existing characters, thus creating new words and, consequently, new meanings.[57]

In any case, Fryer accused other translators of not being willing to cooperate with him; he specifically mentions two colleagues, "[...] medical missionary like Dr. Dudgeon" and "some eminent botanist like Dr. Faber [...]."[58] The

---

52 *Ibidem*.

53 In the TVM, phonemic loans based on English lemmata and others derived from scientific Latin terms are indeed listed.

54 Fryer, "Scientific Terminology," 539.

55 For an example in the field of physics, refer by way of illustration to Iwo Amelung, "Some Notes on Translations of the *Physics Primer* and Physical Terminology in Late Imperial China," *Wakumon*, no. 8 (2004).

56 For some examples belonging to the lexicon of chemical elements, see Wright, *Translating Science*, 224–25.

57 Refer to Masini, *The Formation of Modern Chinese*, 121–27.

58 Fryer, "Scientific Terminology," 539.

first is the Scottish missionary and physician John Dudgeon (De Zhen 德貞, 1837–1901), and the second is the German protestant missionary Ernst Faber (Hua Zhian 花之安, 1839–1899).[59] Fryer spurs them to find in the *Kangxi zidian* Chinese terms within their respective fields, anatomy, and botany, therefore compiling relevant bilingual glossaries with English translations.[60] Such sectors were indeed not among those Fryer chiefly dealt with; therefore, he needed the cooperation of other translators in a concerted effort for the sake of Chinese scientific and technical nomenclature.

The second point Fryer highlighted is the adoption of adequate characters in case phonemic loans had to be employed; this was particularly true for anthroponyms.[61] Another suggestion provided for a better use of phonemic loans, was to adopt "the same characters invariably for such affixes as ic, ia, ine, ite, etc., etc., when they have the same meaning." The purpose was to avoid confusion and let the reader infer when different terms, broadly speaking, belonged to the same semantic field.[62] Among the characters Fryer suggests to use for this purpose, he cites "the numerous lists from various Asiatic sources that are given in a work by Li Yu-wang known as the 李氏音鑑."[63] The latter is a reference to the *Lishi yinjian* 李氏音鑑 (Mirror of phonetics by Mister Li), composed by the renowned novelist and phonologist Li Ruzhen 李汝珍 (1763–1830) and published for the first time in 1810; the text was a renowned treatise on phonetics in the nineteenth century.[64] This statement by Fryer is evidence of the connection of late Qing lexicographical materials with earlier native sources, analysed in detail further on in this work.

Listing these characters, Fryer provides other indications on the sources adopted for the compilation of the TVM, including a "set of characters on page

59 On Dudgeon, consult Gao Xi, *De Zhen zhuan: yi ge Yingguo chuanjiaoshi yu wan Qing yixue jindaihua* (Shanghai: Fudan daxue chubanshe, 2009) and Eadem, "'Jiepouxue' zhongwen yiming de youlai yu queding—Yi De Zhen *Quanti tongkao* wei zhongxin," *Lishi yanjiu*, no. 6 (2008).

60 For Fryer, the analysis conducted by Dudgeon in John Dudgeon "Review of a New Medical Vocabulary," *The Chinese Recorder and Missionary Journal* 13, no. 1: 30–43; no. 3, 177–82; no. 4, 259–65 (Shanghai: American Presbyterian Mission Press, 1882) was clearly too sketchy and the terms provided in the text lacked the systematicity he was advocating.

61 Fryer, "Scientific Terminology," 540.

62 *Ibidem.*

63 Ivi, 541.

64 As reinforced in Elisabeth Kaske, *The Politics of Language in Chinese Education, 1895–1919* (Leiden, Boston: Brill, 2008), 67 and relevant footnote no. 244. A possible explanation for the transcription of the name of the author used is that Fryer confused him with Li Ruhuang 李汝璜 (n.d.), Li Ruzhen's brother. The author thanks Professor Zou Zhenhuan for pointing out this possibility.

408 in Doolittle's Dictionary, Vol. II., giving English syllables with Chinese equivalents, by F. H. Ewer, Esq.," another "by a committee in Peking for rendering proper names," and a "list of geographical names in English and Chinese" based on the latter and prepared by "Rev. L. D. Chapin."[65] The first is a reference to F. H. Ewer's "List of English Syllables with Chinese Equivalents."[66] Pertaining to the second, among Fryer's manuscripts housed in the EAL, volume one is exactly a "List of geographical names" signed by "L. D. Chapin," i.e. Lyman Dwight Chapin (Jiang Daide 江戴德, 1836–1894). In the preface to the volume it is possible to read that "The following list of the most common foreign names has been prepared on the basis of the accompanying syllabary," and in turn the syllabary can be identified with volume number two of the manuscripts, the STFNC. Nevertheless, it has not been signed by the same author.

The question of phonemic loans, though, was further complicated by the fact that, apart from local varieties, there could be significant pronunciation discrepancies even in the standard language among the north and the south of China. Despite this, Fryer was still confident he could find a series of characters whose phonetic dissimilarities could not be so significant across different regions.[67]

The third point stresses that "*New terms ought to accord as far as possible with the general construction of the language*."[68] This should not be interpreted as a recognition of the morphosyntactic peculiarities of Chinese language, but simply of the importance of radicals in selecting those characters already in disuse from the *Kangxi zidian*. Even so, Fryer realised that newly-created characters were not easily authorised.[69] Nevertheless, he particularly highlights the productivity of the radical *kou* 口 and the role it could play to help readers in distinguishing characters used solely for their phonetic value and those for their semantic value.[70] The relevant infelicitous example mentioned by Fryer is the "Vocabulary of Medicines" by the Reverend and doctor Joseph Clarke Thomson (Lao Tan Yuese 老譚約瑟, 1856–1926):[71] "If we use descriptive terms,

65 Consult Wright, *Translating Science*, 219, footnote no. 95.

66 Justus Doolittle, *Ying-Hua cuilin yunfu—A Vocabulary and Hand-Book of the Chinese Language, Romanized in the Mandarin Dialect* (Fuzhou: Rozario, Marcal, and Company, 1872–1873), 408–12.

67 Fryer, "Scientific Terminology," 541.

68 Ivi, 542.

69 *Ibidem.*

70 Ivi, 543.

71 Joseph Clarke Thomson, *Vocabulary of Medicines, in English and Chinese* (China Medical Association, 1889). For more information on Thomson, refer to Xu Guangqiu, *American Doctors in Canton: Modernization in China, 1835–1935* (New Brunswick, London: Transaction Publishers, 2011), 27, 32, and passim.

as I notice Dr. J. C. Thomson does, for these words, we must have the same character 廿 common to all, as he appears to do."[72]

The fourth point proposed by Fryer is "[…] *new terms should be short and terse*."[73] He thus referenced the tendency, established in contemporary Chinese, to shorten those terms composed of numerous characters; long terms are cumbersome and only those "that are short, or such as are capable of being shortened" could effectively establish themselves.[74] The Chinese lexicon was always more polysyllabic, scientific and technical nomenclature included.[75] Fryer's idea that only terse terms would survive in the scientific and technical lexicon can also be seen in the passage from the manuscripts to the glossaries published in the *TVM*.

The fifth point advocates the accuracy of the definitions of terms; Fryer exhorted his colleagues to provide one each time a term appears for the first time in a written text.[76] According to him, this was in line with the Chinese tradition of annotation and explanations, for example, to clarify archaic and obsolete terms. This device, in the form of a glossary appended to the translated scientific and technical treatises "arranged according to the radicals or any recognized Chinese system," would have proved extremely useful for the relevant standardisation work.[77]

The following point advanced by Fryer, "*New terms must bear an analogy with all others of the class they belong to*," stresses the importance, when translating terms that belong, in a broad sense, to a common semantic field, to keep a link between them, therefore helping the reader to identify their common root.[78]

The last point, "*new terms must be elastic*,"[79] finally highlights how only practice can determine whether a technical term is adequate or not within the Chinese lexicon; it is not the presence in a glossary that decides its suitability, but only its factual use.[80] That is why Fryer concluded the second part of his speech by stating that his generation would not provide the Chinese language with a scientific and technical nomenclature; the latter required instead a variable lapse of time, depending on the suitability and adaptability

72 Fryer, "Scientific Terminology," 545.

73 Ivi, 544.

74 *Ibidem*. See also Wright, "The Chinese Language and Foreign Ideas," 295.

75 Refer to Arcodia, *Lexical Derivation*, 126 and the relevant footnote no. 5.

76 Fryer, "Scientific Terminology," 544.

77 Ivi, 545.

78 *Ibidem*.

79 *Ibidem*.

80 *Ibidem*.

of the nomenclature itself.[81] The purpose of the *TVM* has to be interpreted exactly in light of the last consideration. Other than listing terms required for the translation into Chinese of scientific and technical texts, the *TVM* above all laid theoretical foundations in the field, particularly by highlighting some of the lexicological features of the Chinese language.

Within the third part of the speech, Fryer analysed the main causes of divergences in Chinese scientific and technical nomenclature, also thanks to his experience in editing the *TVM*. Among the main problems was the inadequacy of the translators' expertise, as well as their lack of knowledge on the subjects of the translations; the absence of a specific organisation for the standardisation of lexicon, particularly devoted to the publication of scientific and technical dictionaries; and the unfamiliarity with the native technical literature already published in Chinese, also due to insufficient cooperation between translators.[82]

On the last point, though, it should be noted that Fryer himself was not completely flawless—he was not so inclined to adapt and change his translation choices to conform to other standards. In a letter addressed to "To the Com. of the Educational Association on Terminology, and The Com. of the China Medical Association on Terminology" signed by Mateer and Hayes, he stated: "Your committee ought not to change my terms unless they are radically wrong and impossible to be used. Should any term of mine be shown to be erroneous, absurd, or otherwise unserviceable and another be without defects I will gladly yield to it and not otherwise."[83]

---

81 Ivi, 545–46. Refer to the words expressed on the topic many years later in Luo Jialun, "Zhongguo ruo yao you kexue, kexue yingdang xian shuo zhongguohua," *Tushu pinglun* 1, no. 3 (1932) and Lin Yutang, "Guanyu yiming tongyi de tiyi," in *Yuyanxue luncong* (Shanghai: Kaiming shudian, 1933). See also statements by "Dr. Fong F. Sec." (the translator and English expert Fong Foo Sec 鄺富灼, 1869–1931) in Fong Foo Sec, "The Co-operation of Chinese and Foreign Educationists in the Work of the Association," in *Records of the Sixth Triennial Meeting of the Educational Association of China, Held at Shanghai, May 19–22, 1909* (Shanghai: The Educational Association of China, 1909), 67–68: "No attempt, however, was ever made towards the unifying of the technical terms used [...] Since 1898 more translations have been made, more books have been issued, and more varied and confusing have been the terms used. The return of Chinese students from Japan has added much to the existing difficulty, for they at once began to translate from the Japanese text [...]," adding however that "[...] the day of unification of technical terms is in sight."

82 Fryer, "Scientific Terminology," 546.

83 FP, Mss C-B 968, carton 3, folder 34, "Miscellaneous Materials—Material of Chinese Scientific Terminology." Courtesy of The Bancroft Library, University of California, Berkeley. His wish would not be fulfilled, since in *Records of the Sixth Triennial Meeting*, Part II, Minutes, Reports, Etc., appendix "Catalogue of publications of the Education

Nevertheless, according to Fryer, the discrepancies listed above were not as evident as would be impossible to find a remedy. In the fourth and final part of the speech, he provided different solutions to the problems stressed, with the view of "preparing a comprehensive Chinese Scientific Dictionary [...]."[84] The work for the standardisation of scientific and technical nomenclature was going on in various parts of China, comprising what he already accomplished with the publication of the *TVM*.[85] The nine suggestions advanced by Fryer included the creation of a committee of experienced people, including but not limited to missionaries, and to promote the standardisation of scientific and technical language, particularly through the publication of glossaries in English and Chinese, based on the materials existing in the field. Glossaries to be considered as references comprised those already published or compiled as manuscripts by Protestant missionaries, native literature, terms compiled by Jesuits and by those who were familiar with specific nomenclatures for their work, including lists from Japan and other countries. These recommendations are particularly important: first of all, Fryer restated the role that lexicographical works compiled by Jesuits and native texts could have in the standardisation of late Qing scientific and technical nomenclature; secondly, he once more suggested that lists of terms to be adopted as reference or published should also come from outside China, as he personally did while editing the *TVM*.

The purpose of these suggestions was the drafting of rules, to which translators should strictly adhere, for the compilation of different glossaries, both from English to Chinese and from Chinese to English, as well as Chinese definitions alone. The *TVM* format, arranged alphabetically on the basis of the English or scientific Latin lemmata, corresponds to the first model. According to Fryer, the success of these glossaries should have been ensured by the approval of Chinese central authorities. The success of the *TVM* has been certified not only by them, but particularly by all the compilers of texts, glossaries, and lexicographical works that decided to use the *TVM* as a reference.[86]

Association of China," 11, only "Chemical Terms and Nomenclature. 化學名目," compiled "especially on the part of Drs. C. W. Mateer and W. M. Hayes," is mentioned.

84 Fryer, "Scientific Terminology," 547.

85 *Ibidem.*

86 Fryer, "Scientific Terminology," 547–48.

5 Conclusion

Fryer's theories on scientific and technical Chinese lexicon are closely connected to the practical problems encountered during translation work for the Jiangnan Arsenal, both by him and other translators and missionaries, in a concerted effort for the standardisation of the relevant nomenclature. An answer to these questions was provided by Fryer through the publication of the TVM. The TVM was not only composed of terms coined by Fryer and his Chinese colleagues, but also included lists provided by other translators and missionaries, of which some are clearly indicated in the manuscripts, while others can be located only indirectly. The choice of adopting a certain term, consulting a given list or refusing a translation principle did not only conform to purely linguistic and lexicological principles; the authority of a committee or an institution could not certainly alone decide the suitability of a term, but could help accelerate such a process. Fryer successfully combined these two factors: a deep understanding of the peculiarities of the Chinese language, with the historical significance of the institution supporting his activities, represented a key factor in the diffusion of the TVM.

Even though Fryer decided to abandon in principle his missionary intentions, his theories and work represent an example of the importance of the connections with the religious world of the late Qing dynasty. In a context in which missionaries' activities were still thriving, Fryer had to come to terms with them, and other colleagues, to ensure the success of his cause.

For a certain period of time, the lexicological solutions Fryer offered through the publication of the TVM had an influence on later glossaries, spreading through time and space. Nevertheless, the renderings provided by him and his Chinese colleagues were not all successful and were later superseded. Translations from Western books started to be gradually replaced by translations from Japanese sources, especially after 1895; consequently, the Chinese lexicon also started to be modernised accordingly.[87] However, the solutions listed in 1888 with the publication of the TVM must be considered

87 Refer to Benjamin A. Elman, "Toward a History of Modern Science in Republican China," in *Science and Technology in Modern China, 1880s–1940s*, ed. Jing Tsu and Benjamin A. Elman (Leiden, Boston: Brill, 2014), 24–29. It is important to highlight that in George A. Stuart, "Report on the Committee on Scientific Terminology," in *Records of the Sixth Triennial Meeting of the Educational Association of China, Held at Shanghai, May 19–22, 1909*. Part II, Minutes, Reports, Etc. (Shanghai: The Educational Association of China, 1909), 40, the creation of an "Advisory Committee," in which "at least two competent Chinese and one Japanese may be asked to serve," is suggested.

innovative and fundamental, not only from a lexicological and lexicographical point of view, but mainly from a methodological perspective. For Fryer, the *TVM*, as a collection of terms used and tried for his publications at the Jiangnan Arsenal, represented an answer to the need for homogeneity of scientific and technical nomenclature. Therefore, he can be considered a pioneer of standardisation work.

CHAPTER 4

Published Glossaries

1 Introduction

The analysis of Fryer's theories on Chinese scientific and technical lexicon shows how the lexicological ideas he put forward are tightly rooted in practice. When rendering concepts for the first time in Chinese, translators had two choices: to use newly-coined terms or to use lists provided by other scholars and missionaries. The first solution was in principle feasible, but the standardisation and uniformity, one of the prerequisites for a suitable nomenclature, would bear the brunt of it. The second solution could be apter particularly for Protestant missionaries. The technical innovation in printing and the relative easiness in communication, relative to the situation of the Jesuits a few centuries before, could greatly contribute to interactions and exchanges of views. Moreover, scholars in late Qing China had the undeniable advantage of being able to consult Japanese sources, as mentioned earlier in this work and analysed in detail later. Finally, there was one thing which could partially help the standardisation process, particularly pertaining to scientific and technical nomenclature: the terms translators were handling were not directly connected to God or sacred concepts, and this could help to avoid at least religious implications in the debate. Purely lexicological considerations could prevail, even if historical and personal circumstances, as well as power dynamics between different associations of scholars and missionaries, were involved, as seen in the case of Fryer's TVM.

2 *Vocabulary of Mineralogical Terms*

Fryer analysed the causes of discrepancies within the scientific and technical nomenclature of the Chinese language, with their implications for the spread of scientific notions; other than providing his suggestions to other translators, he personally did his part in order to solve this problem with the compilation of the TVM. The first glossary included in the text, in a chronological order based on the dates of the prefaces of the individual glossaries, is the *Vocabulary of Mineralogical Terms*. The full title is *Vocabulary of Mineralogical Terms Occurring in the Manual by J. D. Dana. A. M.*, with the Chinese title *Jinshi Zhong-Xi mingmu biao* 金石中西名目表. The work referred to in the title is

 | DOI:10.1163/9789004443211_006

Manual of Mineralogy, Including Observations on Mines, Rocks, Reduction of Ores, and the Application of the Science to the Arts, composed by James Dwight Dana. It was not possible, though, to ascertain the edition used by the Chinese mathematician and translator Hua Hengfang 華蘅芳 (also known as Hua Ruoting 華若汀, 1833–1902)[1] and Daniel Jerome Macgowan for their Chinese translation of the text, *Jinshi shibie* 金石識別 (Identification of minerals).[2]

A reference to the manual by Dana can be located in the FP. In the already quoted letter to Smith Elder & Co., dated March 18, 1868, about a book order for the Jiangnan Arsenal, Fryer only quotes a generic "Dana's Elements of Mineralogy" published by "Trubner."[3] Another reference to the text can be found within a list of "Works published at the translation department, Kiangnan Arsenal";[4] in this case as well, other details on the *Jinshi shibie* are not given.[5] The *Jinshi shibie* is unanimously recognised by scholars as the first translation in Chinese of a text specifically dedicated to mineralogy,[6] while the glossary compiled by Fryer, based on this work, is the first with English and Chinese parallel translation created in the field.[7] The *JS* is composed of a total

1 For more information of Hua Hengfang, refer to Li Nanqiu, "Shuxue wenxian fanyijia Hua Hengfang," *Zhongguo fanyi*, no. 6 (1982); Wang Yusheng, "Hua Hengfang: Zhongguo jindai kexue de xianxingzhe he chuanbozhe," *Ziran bianzhengfa tongxun—Journal of Dialectics of Nature*, no. 2 (1985); and Idem, "Hua Hengfang: Forerunner and Disseminator of Modern Science in China," in *Chinese Studies in the History and Philosophy of Science and Technology*, ed. Fan Dainian and Robert S. Cohen (Dordrecht, Boston, London: Kluwer Academic Publishers, 1996).

2 Daniel Jerome Macgowan and Hua Hengfang, trans., *Jinshi shibie* (Shanghai: Jiangnan zhizaoju, 1871). The text was translated in 1868 but was published only three years later, as can be read from the English and Chinese prefaces and as indicated also in Wright, *Translating Science*, 338. The preface signed by Hua Hengfang, though, is dated 1872, as stated in ivi, 217, footnote no. 89; in Wei, *Jiangnan zhizaoju ji*, II, 20B, the text is signalled as published in *Tongzhi shiyi nian* "同治十一年" (Eleventh year of the reign of Tongzhi), corresponding in reality to 1872.

3 Dagenais, *The John Fryer Papers*, I, 350.

4 Fryer, *An account*, 22–23. Information on the original text, *yuanzhu* "原著," provided in Wei, *Jiangnan zhizaoju ji*, II, 20B, is also limited to *Mei Daina* "美代那" (America, Dana); the *jiaodui* "校對" (reviser), is the Chinese scholar Jiang Heng 江衡 (1852–1926).

5 A general reference to the translation by Hua Hengfang can be found in Zhao Ziqin, "Zhongguo yiduo": "華若汀新譯地學書極詳載" (Hua Ruoting recently published books on geology that are extremely detailed). The text has been republished in Chen and Xia, *Tuxiang wan Qing*, 240.

6 On the topic, refer to Wang Genyuan and Cui Yunhao, "Guanyu *Jinshi shibie* de fanyi, chuban he diben," *Zhongguo keji shiliao—China Historical Materials of Science and Technology* 11, no. 1 (1990).

7 Ai Suzhen "Qingdai chuban de dizhixue yizhu ji tedian," *Zhongguo keji shiliao—China Historical Materials of Science and Technology* 19, no. 1 (1998): 12–13.

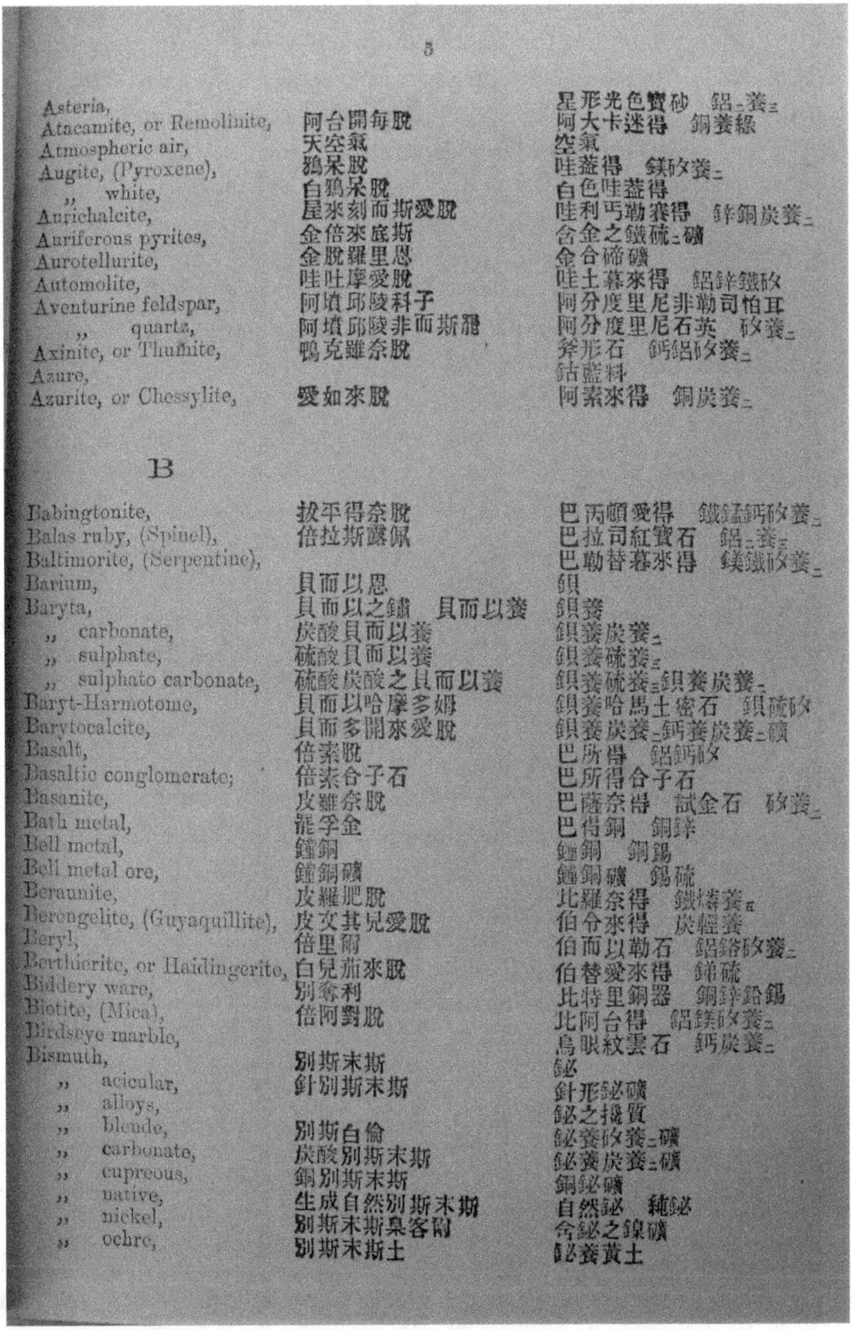

5

Asteria,		星形光色寶砂 鋁二養三
Atacamite, or Remolinite,	阿台開每脫	阿大卡迷得 銅養綠
Atmospheric air,	天空氣	空氣
Augite, (Pyroxene),	鴉呆脫	哇蓋得 鎂矽養二
,, white,	白鴉呆脫	白色哇蓋得
Aurichalcite,	屋來刻而斯愛脫	哇利丐勒賽得 鋅銅炭養二
Auriferous pyrites,	金倍來底斯	含金之鐵硫二礦
Aurotellurite,	金脫羅里恩	金合碲礦
Automolite,	哇吐摩愛脫	哇土幕來得 鋁鋅鐵矽
Aventurine feldspar,	阿墳邱陵科子	阿分度里尼非勒司怕耳
,, quartz,	阿墳邱陵非而斯羅	阿分度里尼石英 矽養二
Axinite, or Thumite,	鴨克雖奈脫	斧形石 鈣鋁矽養二
Azure,		鈷藍料
Azurite, or Chessylite,	愛如來脫	阿素來得 銅炭養二

B

Babingtonite,	拔平得奈脫	巴丙頓愛得 鐵錳鈣矽養二
Balas ruby, (Spinel),	倍拉斯露佩	巴拉司紅寶石 鋁二養三
Baltimorite, (Serpentine),		巴勒替幕來得 鎂鐵矽養二
Barium,	貝而以恩	鋇
Baryta,	貝而以之鏽 貝而以養	鋇養
,, carbonate,	炭酸貝而以養	鋇養炭養二
,, sulphate,	硫酸貝而以養	鋇養硫養三
,, sulphato carbonate,	硫酸炭酸之貝而以養	鋇養硫養三鋇養炭養二
Baryt-Harmotome,	貝而以哈摩多姆	鋇養哈馬土密石 鋇硫矽
Barytocalcite,	貝而多開來愛脫	鋇養炭養二鈣養炭養二礦
Basalt,	倍索脫	巴所得 鋁鈣矽
Basaltic conglomerate;	倍索合子石	巴所得合子石
Basanite,	皮雖奈脫	巴薩奈得 試金石 矽養二
Bath metal,	龍孚金	巴得銅 銅鋅
Bell metal,	鐘銅	鐘銅 銅錫
Bell metal ore,	鐘銅礦	鐘銅礦 錫硫
Beraunite,	皮羅肥脫	比羅奈得 鐵燐養五
Berengelite, (Guyaquillite),	皮文其兒愛脫	伯令來得 炭輕養
Beryl,	倍里爾	伯而以勒石 鋁鋊矽養二
Berthierite, or Haidingerite,	白兒茄來脫	伯替愛來得 銻硫
Biddery ware,	別奪利	比特里銅器 銅鋅鉛錫
Biotite, (Mica),	倍阿對脫	比阿台得 鋁鎂矽養二
Birdseye marble,		鳥眼紋雲石 鈣炭養二
Bismuth,	別斯末斯	鉍
,, acicular,	針別斯末斯	針形鉍礦
,, alloys,		鉍之攙質
,, blende,	別斯白倫	鉍養矽養二礦
,, carbonate,	炭酸別斯末斯	鉍養炭養二礦
,, cupreous,	銅別斯末斯	銅鉍礦
,, native,	生成自然別斯末斯	自然鉍 純鉍
,, nickel,	別斯末斯臬客爾	含鉍之鎳礦
,, ochre,	別斯末斯土	鉍養黃土

FIGURE 1 Page 5 of the *JS, Chinese and English Technical Vocabularies*, T10.F8, University of California, Berkeley

of thirty-six pages, with a page of errata.[8] The total number of English entries and of the corresponding translations provided by Fryer in the third column is 1,672, considering the single entries for which at least one translation is provided, while 1,592 translations are supplied in the second column.[9]

The prefaces and afterwords of books translated at the Jiangnan Arsenal are all fundamental sources of information, and not only about Fryer's theories on translation and his work. More importantly, they reveal the editions used and provide indications on the reasons for translation, mirroring the history of Western learning in late Qing China and of science and technology translation.[10] The principle naturally applies to the glossaries included in the *TVM*; hereafter is the preface to the first, the *JS*.

> Vocabulary of Mineralogical terms occurring in the manual by J. D. Dana A. M.
>
> Mr. Dana's "Manual of Mineralogy" was translated into Chinese by Dr. D. J. Macgowan in the year 1868. A vocabulary of the Chinese terms and their English equivalents does not appear to have been made. Students of Mineralogy, finding mostly Foreign names, are often at a loss to understand the book. Translators of works on Chemistry, Mineralogy and allied subject, having no vocabulary to refer to, are obliged to invent new names, and hence the confusion that has arisen is very vexatious. The present vocabulary has recently been drawn up with much pains by comparing this translation with the original. The terms used by Dr. Macgowan will be found in the middle column. Wherever he has omitted to translate any term, either one of his synonyms is used in its stead or the

8 Another description can be consulted in Li, *Zhongguo kexue*, 382–90.

9 The glossary is included among the texts of the library of the German linguist Paul Georg von Möllendorff (1847–1901), as explicated in the section "Dictionaries" of "East Asiatic Languages" in Paul Georg von Möllendorff, *Catalogue of P. G. von Möllendorff's Library* (Shanghai: American Presbyterian Mission Press, 1905), 31. In the entry, the text is quoted as "*Dana, J. D. A. M.*—Vocabulary of mineralogical terms occurring in the manual, 1883. 1 fasc.," thus directly ascribing the work to Dana. This and other glossaries analysed in this chapter are also quoted as "Fryer, J.—*Translator's Vademecum: Chemistry, Mineralogy, Materia Medica, Proper Names, Steam Engine*. Shanghai, Presbyterian Mission Press. 1883–89. 8°." in Idem, *Praktische Anleitung zur Erlernung der Hochchinesischen Sprache* (Shanghai: Kelly and Walsh, 1900), 188, in the fourth section, "Die Schriftsprache" (the written language) of its "Bibliographie." Therefore, the *TVM* is explicitly described as printed in various years, starting from 1883 and ending in 1889, the year of publication of the *QJ*. Moreover, a note under the entry describes it as "Meist Umschreibungen der fremden Ausdrücke" (mostly periphrases of foreign expressions).

10 As pointed out in Wen Shufang, Cui Ruifeng, and Qiao Yaming, "Jiangnan zhizaoju yishu xuba wenxian jiazhi kaoshu," *Tushuguan lilun yu shijian—Library theory and practice*, 3 (2018): 69.

> place is left blank. The terms now generally used in works on Chemistry, Mineralogy etc. are placed in the right hand column for comparison. This vocabulary will enable students to find out the English names of the rocks and minerals, the corresponding names in other Chinese works, and the chief elements of which they are composed. It may also be of assistance to Translators generally.
>
> April, 1883.

The Chinese version is as follows:

> 美國代那作《金石識別》書，同治八年瑪高温譯以漢文，所定金石之名，初時未曾列表，故考究礦學者，往往既得金石，祗有西名而無華名，即不能從已譯之書索其底蘊，且後人續譯化學礦學等書，因無金石名表，故不免另立新名. 由是金石家更以名目不同為憾. 茲將西名列於左行，瑪氏所定之名列於中行，其有遺漏者，則考其原有之別名代之. 其竟無別名可代者，闕之. 續譯化學礦學等書，所定金石之名與其最要之原質列於右行，異同是非可比較而得之. 金石家，從礦石而得西名，從西名而得華名，求之於已譯之金石礦學等書，亦足有裨實用也. 光緒九年三月.[11]

Dana's original text was translated in 1868, even if the English version of the preface incorrectly indicates the year 1869. The main flaw of the *Jinshi shibie* was represented by the absence in the appendix of a glossary; it is exactly from this shortage that the confusion in the terminology within the translations of texts about chemistry, mineralogy, and other relevant fields derived.[12]

Fryer's modus operandi, presented in the two prefaces above, is reflected in the structure of the glossary. A feature that differentiates the *JS* from other glossaries gathered in the *TVM* is the presence of three columns: the first includes the original terms, namely "Terms used in J. D. Dana's manual"; the second the "Terms used in Dr. Macgowan's translation," comprising synonyms and leaving a blank space when translations were not provided in the *Jinshi shibie*; the third column, finally, corresponding to the "Terms in general use and chief elements," lists Chinese translations, when present, of the English term, and the chemical elements of the minerals or rocks considered.

When comparing the terms included in the other glossaries gathered in the *TVM*, it can be easily inferred how the translations provided in the third column correspond exactly to Fryer's renderings:

11 Punctuation has been added by the author for the reader's convenience.

12 According to Ai, "Qingdai chuban," 12, the glossary will be later published as an appendix to the editions of the *Jinshi shibie* published in 1883, 1889, 1896, and 1899.

TABLE 1 *JS* and *HXCL*

Original term	*JS*	*HXCL*
Brucite	布路歲得	布路歲得
Chrysoberyl	金色伯而以勒石	可里蘇伯而以勒石，即金色伯而以勒石
Periclase	圍劈石	派利可賴斯石，即圍劈石

The main differences between renderings provided by Macgowan and Hua Hengfang and those by Fryer are essentially represented by the translation choices adopted. Even if a consistent portion of terms listed in the third column was not destined to be later absorbed into modern Chinese lexicon, especially in the case of phonemic loans, it is clear how renderings Fryer supplied are not phonemic loans as much as possible. When phonemic loans were already provided in the *Jinshi shibie*, Fryer tried to substitute unsuitable characters.[13]

Consider the example of "Agate, (Quartz)," on page 1 of the *JS*. In the first column the translation provided by Macgowan and Hua Hengfang is a phonemic loan, *yadaituo* 鴨呆脫. In the third column the term *hei bai ma'nao* 黑白瑪瑙 is listed; the latter is composed of the contemporary Chinese term used to label the mineral, *ma'nao*, preceded by the characters indicating respectively "black" and "white," *hei* and *bai*, two possible colours in which the mineral is present in nature. After this term, its chemical formula is also listed: *xiyang*$_{er}$ 矽養$_{二}$; this translation, based on the chemical structure of the mineral, SiO_2, is also listed on pages 2 and 55 of the *XYDCYP*, and on page 31 of the *HXCL* precisely as the translation of "silice," or silicon dioxide.

Examining another example, "Perofskite," a variant for "Perovskite," is translated in the *Jinshi shibie* again with the phonemic loan *piluofugaituo* 皮落夫蓋脫. The term in the third column of the *JS*, on page 25, is also a phonemic loan, but with different characters, *bilufusigaide* 比路夫斯蓋得. Finally, again in the

13 As described in *Descriptive Catalogue and Price List of the Books, Wall Charts, Maps, Etc. Published or Adopted by the Educational Association of China* (Shanghai: Presbyterian Mission Press, 1907), 10: "1 vol. 6mo. 15 cents. This vocabulary, consisting of about 1,500 terms in English and Chinese, was published in 1873. Mr. Dana's 'Manual of Mineralogy' had been translated by Dr. D. J. Macgowan and published at the Kiangnan Arsenal in 1870. An entirely new and systematic arrangement of the nomenclature was afterwards made by Dr. Fryer, harmonizing with the terms used in native and foreign works on that and allied subjects. The terms used in Dr. Macgowan's translation are placed side by side with the new ones, and in addition the chief elements in each mineral are given."

TABLE 2 *JS*, *HXCL*, and chemical formulas

Original term	*JS*	*HXCL*	Chemical formula
Agate, (Quartz)	鴨呆脫, 黑白瑪瑙, 矽養$_{二}$	黑白瑪瑙	SiO_2
Perofskite	皮落夫蓋脫, 比路夫斯蓋得, 鈣鐟養$_{二}$		$CaTiO_3$

third column, the principal elements that compose the mineral are presented as *gaitiyang$_{er}$* 鈣鐟養$_{二}$; the actual chemical formula of calcium titanate, the main constituent of the mineral, is $CaTiO_3$, not $CaTiO_2$ as could be inferred from Fryer's translation. Nevertheless, it should be noted that uncertainties in basic chemical knowledge at that time were not uncommon; original texts in which Chinese translations were based did not often represent the latest editions available.

It was not possible to locate the *JS* manuscript among the volumes of the EAL.[14] Nevertheless, another confirmation that those listed in the third column represent translations provided by Fryer can be obtained from the example provided above. The character *zan* 鐟, employed in one of the renderings for "Perofskite," has another reading, *ti*; the character was included among those listed by Fryer and Xu Shou in the *Huaxue jianyuan* as the translation of "titanium." In the *Dai kanwa jiten* 大漢和辞典 (Great Sino-Japanese dictionary), this character, listed as number 40919, also has a *kun* reading in Japanese corresponding to the term titanium, namely *chitaniumu* チタニウム.[15] More important is the fact that in the same entry of the dictionary, the Chinese character dictionary *Zhonghua da zidian* 中華大字典 is quoted as a reference. In the explanation of the corresponding entry of the *Zhonghua da zidian*, the character is indicated as an element, *yuanzhi* "原質," easily oxidisable, concept rendered as *yanghua* "養化." Both terms, *yuanzhi* and *yanghua*, particularly the second, are patently based on those used by Fryer and Xu Shou in the *Huaxue jianyuan*; the oxidation is caused indeed by the reaction of a metal and oxygen and the last is indicated in the *Huaxue jianyuan* as *yangqi* 養氣 (nurturing

14 The volume that is the most similar to the *JS* in content is number seven, namely the VGT; a quick analysis, though, excludes any correlation between the two.

15 Morohashi Tetsuji, *Dai kanwa jiten* (Tokyo: Taishūkan shoten, 1971).

gas), not corresponding to the contemporary Chinese term *yanghua* 氧化.[16] The fact that the first edition of the *Zhonghua da zidian* was published in 1915 represents an important hint that its authors used Fryer's translations for the chemical terminology adopted in the dictionary.

3 *Vocabulary of Names of Chemical Substances*

In the first glossary included in the TVM, Fryer adopted a structure that is not applied in the others: the presence of three columns, even if serving as a good instrument for comparing the terminologies of various sources, was not put into practice in the other TVM glossaries, e.g. the second, namely the *Vocabulary of Names of Chemical Substances*, with the Chinese title *Huaxue cailiao Zhong-Xi mingmu biao* 化學材料中西名目表. The full English title indicated on the first page of the glossary is *Vocabulary of names of substances occurring in various works on chemistry; Chiefly in "Bloxam's Chemistry," first edition.* As can be inferred from this description, and as is also explained in the preface, in the HXCL terms used in different handbooks and treatises on chemistry, particularly ones drawn from "Bloxam's Chemistry," are included. The latter is a reference to *Chemistry Inorganic and Organic*[17] by the English chemist Charles Loudon Bloxam (1831–1887).[18] The section of inorganic chemistry of this volume had been translated by Fryer and Xu Shou with the title *Huaxue jianyuan bubian* 化學鑑原補編 (Supplement to the Mirror of chemistry), published in 1882,[19] and the organic part had been rendered in the *Huaxue jianyuan xubian* 化學鑑原續編 (Sequel to the Mirror of chemistry), printed in 1875.[20]

---

16 The term *yuanzhi* was previously employed by the Presbyterian missionary William Alexander Parsons Martin (Ding Weiliang 丁韙良, 1827–1916), as indicated in Wright, *Translating Science*, 408, footnote no. 12. On Martin, consult Ralph Covell, *W. A. P. Martin: Pioneer of Progress in China* (Washington: Christian University Press, 1978).

17 Charles Loudon Bloxam, *Chemistry Inorganic and Organic: With Experiments and a Comparison of Equivalent and Molecular Formulæ* (London: John Churchill and Sons, 1867).

18 For more information, refer to D. I. Davies, D. C. Lyon, and R. J. Spring, "Charles Loudon Bloxam—A Victorian University and Military Academy Chemistry Teacher," *Ambix* 33, 1 (1986).

19 Scholars have proposed different hypotheses; as highlighted in Wright, *Translating Science*, 50, and Li Yadong, "Xu Shou suo yi huaxue zhuzuo de yuanben," *Huaxue tongbao*, no. 3 (1985): 54, it might also have been printed in 1879.

20 For this date, refer to Wang Yangzong, "Jiangnan zhizaoju," 7. In Wei, *Jiangnan zhizaoju ji*, II, 21A, the date for the *Huaxue jianyuan bubian*, shortened as *Huaxue bubian* 化學

Hereafter is the English preface to the *HXCL*:

This vocabulary was commenced in 1869 during the translation of the First edition of "Bloxam's Chemistry," and was primarily intended only as an appendix to that work; but a few terms have subsequently been added making it more generally useful.

A system had to be invented to express the names of substances unknown to the Chinese and not fixed by previous translators.

For most of the elementary substances new terms of course had to be coined, one Chinese character being found sufficient for each. Some of these new terms are descriptive while others are merely phonetic and represent the prominent sound in the Foreign name.

For compound inorganic substances generally, the elements with the number of their equivalents are combined in the same way as in the Symbols used in Foreign countries.

For organic substances either the meaning of the term or a description of the substance has been translated or where this was impracticable the term itself has been transliterated. In some cases the terms thus formed are rather cumbersome and will no doubt eventually be much shortened.[21] The same general plan has been carried out in rendering terms belonging more strictly to Mineralogy and Materia Medica, of which separate and more complete lists are also published.

It is unfortunate that the work of translation had to be commenced during the period of the transition from the old to the new notation when only text books in the old notation were obtainable. When the new system had become fully established in Western countries several books had been published in Chinese according to the old notation. It was feared that a change of system in succeeding works would only lead to confusion, and hence the old notation has been retained throughout the whole series of works relating to Chemical subjects. A little trouble however is all that is needed to enable the student to change from the old to the new and *vice-versa*.

Kiangnan Arsenal.

January, 1885.

補編, is reported as *Guangxu ba nian* "光緒八年" (Eighth year of the reign of Guangxu), corresponding indeed to 1882, while the publication of the *Huaxue jianyuan xubian* is signalled in *Guangxu yuannian* "光緒元年" (First year of the reign of Guangxu), that is, 1875.

21 For a specific study on the topic, refer to Wang Yangzong, "A New Inquiry."

Below is the Chinese preface, following the English one:

> 化學材料中西名目表. 小序. 是表於同治九年，在江南製造總局，繙譯化學鑑原，續編，補編，時所作，原意祇將此表，附於本書之後，但因陸續加入別種化學書內之名目，冀其用處，更覺寬廣. 各種化學材料，有中國尚未知者，有前繙譯家尚未定名者，無奈必設公法，特命新名. 所有原質，多無華名，自必設立新者，而以一字為主，或按其形性大意而命之，或照西字要聲而譯之. 所有雜質之名，率照西國之法，將其原質之名與數併而成之，中國有者，另為釋註.所有生物質之名，或將其原意譯其要略，或按其西音譯以華字，因此不免字多，名似過長，內有屬於礦學與藥品之名目，亦歸同法譯之，惟另有此兩種學之細目，刊印成表，故此表不多及之. 本局初譯化學類之書時，西國所用各質之分劑數，適在廢舊法而興新法之間，當時祇能得用舊法之書為底本，後雖新法盛行，而本局已刻化學諸書，均依舊法，如今改用新法，則前後不應，恐誤學者，故仍前分劑，以歸一律，然以舊法變為新法，或以新法變為舊法，觀化學書內所設公法，即易明悉. 光緒十年十二月十一日識.[22]

From the preface, important information on the *TVM* can be derived. The compilation of the *HXCL* started in 1869. Nevertheless, the choice to add other terms led to the decision to expand the glossary, consequently postponing its printing. This is exactly why Fryer and his colleagues decided to include in the *TVM* not only terms rendered by them, but also, for the sake of uniformity, translations originating from other sources. From Fryer's words it is also possible to conclude that he already had the original plan for the compilation of the *TVM* in mind at least since 1869, one year after the beginning of his collaboration with the Jiangnan Arsenal.

From a lexicological perspective, in the preface Fryer notes that "new terms" had been coined for the great majority of "elementary substances." In this case, Fryer is speaking of characters, *zi*, not words, *ci*; the reference is to the nomenclature of chemical elements, for which Fryer and Xu Shou had provided an adequate solution fourteen years before the printing of the *HXCL*. In his essays and speeches, Fryer hardly mentioned the topic of disyllabicity, even less so the difference between *zi* and *ci*. This is not surprising, since Chinese grammarians also analysed the difference at a later stage. The confusion that Fryer, who was not a linguist in the strictest sense of the term, has on the topic may

22 Here and in the prefaces to the *XYDCYP* and the *QJ* punctuation is included, contrary to the case of the *JS*. Nevertheless, in the original text only commas are used, substituted with full stops whenever a new paragraph was created in the original text.

also derive from the suggestions provided by his Chinese colleagues, of whom the great majority were scientists.[23]

In the two prefaces to the *HXCL*, a hint to two other glossaries included in the *TVM* can be located, but a discrepancy in the publication dates arises. Fryer stated that the same translation modality used in the *HXCL* had been applied in the *JS* and the *XYDCYP*, "of which separate and more complete lists are also published." If for the former, published in 1883, the information is coherent from a chronological point of view, for the second, published in 1887, a contradiction with the date of the prefaces of the *HXCL*, compiled in 1885, stands out.[24] The only feasible hypothesis is that the *XYDCYP* had already been published before 1885 in a reduced or incomplete format, possibly by Fryer personally, and not through the official channels of the Jiangnan Arsenal, and that later Fryer continued to add terms and carry out corrections, finally printing the glossary in 1887 as a monograph and one year later in the *TVM*.

The compilation of the *HXCL* was realised in a period of transition from the old to the new notation, during which only texts in the former were available.[25] However, this peculiarity did not affect the value of the glossary, as described years after its publication:

> 1 vol. 6 mo. 18 cents. The names of about 1,800 Chemical Substances that are most commonly known and met with are here arranged in alphabetical order, in English, with Chinese equivalents. These are the terms used in the various translations by Dr. J. Fryer of works on Chemistry and allied subjects. There are in the old notation, but a small table of such alterations as were made when the new notation was established, and which anyone can easily construct for himself, will render this vocabulary

---

23 As pointed out in personal communication by Professor Giorgio Francesco Arcodia.

24 According to Wang Yangzong, "A New Inquiry," 279, the text was published in 1884; the same information comes from Joseph Clarke Thomson, "Medical Publications in Chinese," *The China Medical Missionary Journal* 1, no. 3 (1887) (Shanghai: Kelly & Walsh), 121. In the section of the latter where "Medical Publications in Chinese, and Works Bearing on Medical Science, by John Fryer, Esquire" are listed, the "化學材料中西名目表 *Vocabulary of Names of Chemical Substances*. 1 vol., pp. 37" is also listed among works already printed, indicated as published in 1884. For the *China Medical Missionary Journal* and the China Medical Missionary Association, see Zhang Daqing, "Zaoqi yixue mingci tongyi gongzuo: Boyihui de nuli yu yingxiang," *Zhonghua yi shi zazhi* 24, no. 1 (1994).

25 On the Berzelian notation, refer to Wright, *Translating Science*, 257 and 276, and to Gabriele Tola, "Between Phonemic Loans and Descriptive Labelling: Principles, Examples and Lexical Innovation in the *Huaxue cailiao Zhong-Xi mingmu biao*," *Rivista degli Studi Orientali* 90, no. 1–4, New series (2018): 265–66.

> equally serviceable for either, until such time as the new edition of the vocabulary, in the new notation, is ready for use.[26]

The layout of pages and terms in the *HXCL* is different from that of the *JS*: the text has two columns, in which terms are listed in alphabetical order, with some mistakes and exceptions. The English lemma always precedes the Chinese rendering which can be more than one. The *HXCL* amounts to thirty-seven pages, for a total of 3,362 entries.[27] The glossary is mostly devoted to the chemical field, as indicated by "chiefly" in its title, though terms from mineralogy, botany, pharmacology, and natural science in the broader sense are also included.[28]

The manuscript of the *HXCL*, the VCT, can be identified with volume six stored in the EAL. Its title, according to the library catalogue description, is "Vocabulary of chemical terms." In reality, the *HXCL* manuscript is gathered in the first part of the volume; in the same volume, two other incomplete glossaries are also included, clearly separated by many blank pages, pertaining to botany and zoology.[29] The *HXCL* manuscript is composed of a total of 102 pages. Two blank pages follow, after which the section of botanical terms starts, for a total of ninety-six pages. The part dedicated to zoology is made up of seventeen pages.

The terms of the *HXCL* manuscript are generally listed in alphabetical order according to the original lemma, annotated from left to right; Chinese translations are usually written on the same line of the original term but, on the contrary, from top to bottom. By analysing the *HXCL* manuscript and comparing it with the printed version, it is possible to identify certain constants, of which some are common to other glossaries, while others are peculiarities of the VCT

26 *Descriptive Catalogue*, 6.

27 In this case as well the reference is to the original terms, the vast majority in English and scientific Latin. For most of them, actually, more than one Chinese translation is provided.

28 For a specific analysis of chemical terminology in Chinese, refer to Wright, *Translating Science*, and Alleton and Alleton, *Terminologie de la chimie*. For a history of chemistry in China in the time frame considered in this work, consult Reardon-Anderson, *The Study of Change*; Cao Yuanyu, *Zhongguo huaxue shi hua* (Nanjing: Jiangsu kexue jishu chubanshe, 1979); Yuan Hanqing, *Zhongguo huaxue shi lunwenji* (Beijing: Sanlian shudian, 1956); and Zhang Zigao, *Zhongguo huaxue shigao (gudai zhi bu)* (Beijing: Kexue chubanshe, 1964).

29 In this section, composed of even fewer pages than the part dedicated to botany, terms are included such as "Acipenser," translated first as *axipensa* 阿西噴薩 and to which another translation, *xunlong yu* 鱘龍魚 was later added; "Hymenoptera," rendered with the hybrid *haimenwapotila ke* 海门哇坡替拉科, followed by a second translation, *mochi* 膜翅; "Mollusca," first translated with the phonemic loan *moluosikai* 摩羅司楷, combined with the translation *jie* 介, to which the definition *luobang bu* 螺蚌部 was added, later further modified as *jie bu* 介部, and many others all belonging to zoology.

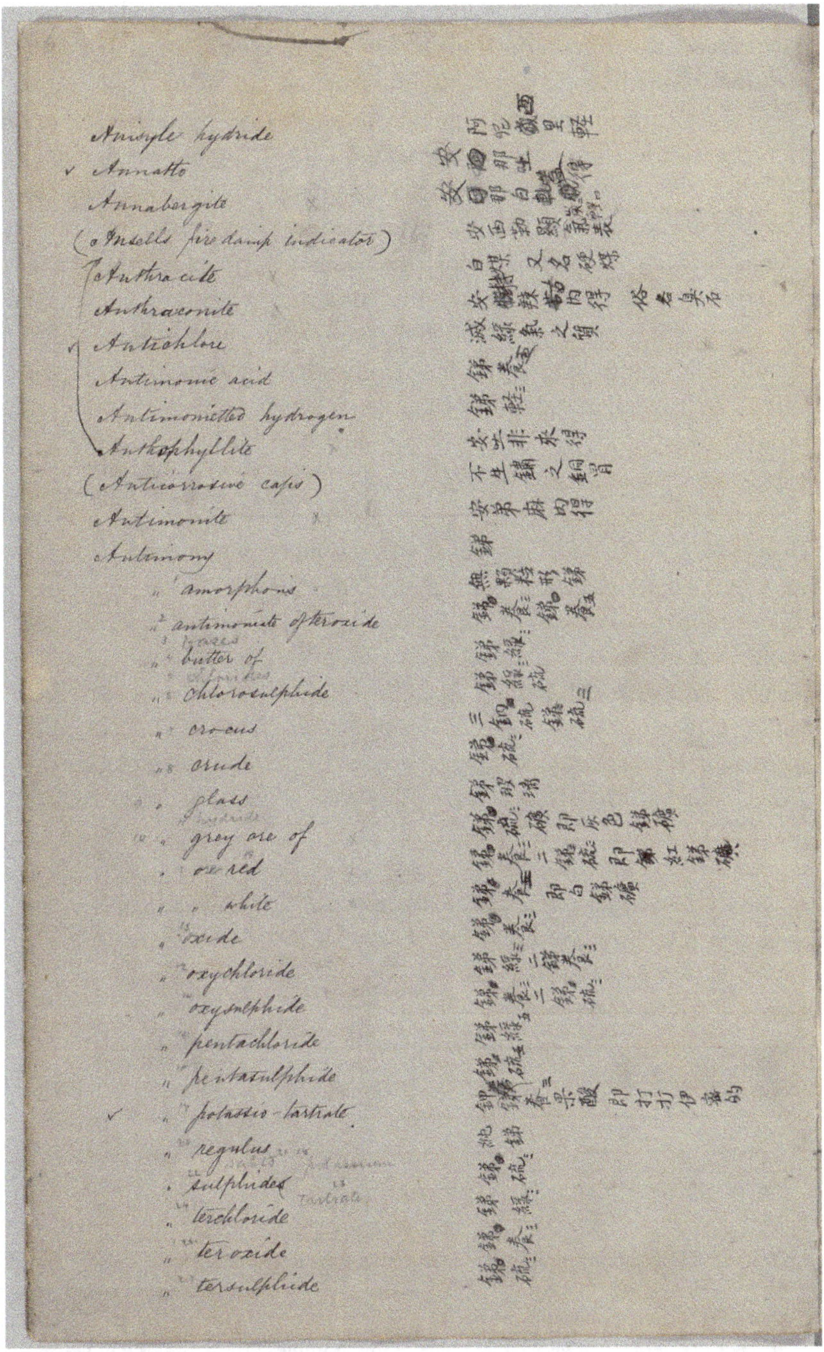
Anisyle hydride
Annatto
Annabergite
(Ansells fire damp indicator)
Anthracite
Anthraconite
Antichlore
Antimonic acid
Antimonietted hydrogen
Anthophyllite
(Anticorrosive caps)
Antimonite
Antimony
" amorphous
" antimoniate of teroxide
" butter of
" chlorosulphide
" crocus
" crude
" glass
" grey ore of
" ore red
" " white
" oxide
" oxychloride
" oxysulphide
" pentachloride
" pentasulphide
" potassio-tartrate
" regulus
" sulphides
" terchloride
" teroxide
" tersulphide

FIGURE 2 Page 6 of the manuscript of the *HXCL*
COURTESY OF THE C. V. STARR EAST ASIAN LIBRARY, UNIVERSITY OF CALIFORNIA, BERKELEY

and the *HXCL*. In both cases, they are fundamental for comprehending the evolution of Fryer's theories on the translation of Chinese scientific and technical nomenclature, and how they were put into practice.

The first and most important tendency in the passage from the VCT to the *HXCL* is the addition in phonemic loans, as much as feasible, of elements that could help the reader locate the referent and semantic field of the lemma itself.

Consider "Reaumur's porcelain" on page 29 of the VCT: the term is first translated as *Liuma ci* 六麻瓷, to which a second rendering was later added, *bu touguang boli* 不透光玻璃.[30] The latter is a descriptive labelling that explains a peculiarity of this porcelain, devised by the French physicist René-Antoine Ferchault de Réaumur (1683–1757). The revision made by Fryer in the passage to the *HXCL* consisted in the addition of the character *shi* 氏, in order to underline

TABLE 3 *HXCL* manuscript and its printed version

Original term	VCT, first version	VCT, second version	*HXCL*
Reaumur's porcelain	六麻瓷, 即不透光玻璃		六麻氏瓷, 即不透光玻璃
Campholic acid	嵌末夫里克	加暮夫里克	加暮非哇里克酸
Carbuncle	卡朋可辣	卡朋可勒	卡朋可勒石
Ancaster stone	安卡司得		安卡司得石
Euclase	由可賴斯礦		由可賴斯石, 即易斷石
Apatite	阿八台得	阿巴台得	阿巴台得
Calcite	鈣勒賽得	丐勒賽得	丐勒賽得
Magnetite	鎂軋尼台得		馬軋尼台得, 即八面形鐵礦
Free stone	易鑿鋸之石		軟砂石
Actinic rays	光内化物氣		光内化物線
Actinolite	阿克替奴來得	阿各替奴來得	阿格替奴來, 即光線石

30 Numerous translations in the *TVM* are introduced by terms such as *ji* 即 (also, that is to say, otherwise), *you* 又, *you ming* 又名, *yi ming* 亦名, or *huo* 或 (also, also called, or). At times Fryer simply indicates the vernacular term for a given lemma by adding *suming* 俗名. These terms will be included only in the examples provided in the tables of this work, not of those of its body text.

that the previous two characters, *Liuma*, are referring to an anthroponym, Réaumur. The final result is thus *Liuma shi ci* 六麻氏瓷, a hybrid more easily understandable for the glossary user, keeping the second rendering unaltered.

Other characters constantly added at a later phase to phonemic loans include *suan* 酸, indicating acids, and *shi* 石, referring to minerals. For example, in the case of the lemma "Ancaster stone," listed on page 3 of the *HXCL*, the term had been originally rendered with the phonemic loan *ankaside* 安卡司得. In the VCT Fryer used a symbol, similar to a small circle, to indicate the need to add a character before printing the *HXCL*; in the latter, the translation corresponds to the hybrid *ankaside shi* 安卡司得石.

Substitution of characters in the passage from the manuscript to the printed version also took place in hybrids: "Euclase" had first been translated in the VCT with the hybrid *youkelaisi kuang* 由可賴斯礦, in which the character *kuang* 礦 had been employed, not *shi* 石. In the published version another hybrid, *youkelaisi shi* 由可賴斯石, can be found, together with the descriptive labelling *yiduan shi* 易斷石, recalling the tendency of this mineral to chip, as also indicated in the Greek etymology of the term, κλάσις.

The second constant in the passage from the VCT to the *HXCL* is represented by modifications in the characters adopted for phonemic loans. An example is the entry "Apatite," on page 4 of the *HXCL*. This class of minerals was originally rendered by Fryer with the phonemic loan *abataide* 阿八台得, successively modified in the VCT with the version destined to be printed in the *HXCL*, namely *abataide* 阿巴台得. Fryer avoided the use of the character *ba* 八, bearer of its own specific meaning, substituting it with the homophone *ba* 巴, still employed in Chinese mainly in phonemic loans.[31]

Even more emblematic is the example of "Calcite" on page 7 of the *HXCL*. The mineral was originally rendered as *gailesaide* 鈣勒賽得, but Fryer cancelled the radical *jin* from the first character in the manuscript, providing the one that will be in the final version of the *HXCL*, namely *gailesaide* 丐勒賽得. It is important to remember that even though *gai* 鈣 and *gai* 丐 are homophones, the first one bears its own specific meaning, the chemical element "calcium."[32]

Besides the aforesaid examples, it should be added that in the VCT Fryer frequently gathers groups of entries, sharing a common chemical feature, by

31 In the final version, Fryer will add the rendering *qiren shi* 欺人石; the term can be considered a calque from the Greek root απατεῖν, "deceive, to be misleading," as apatite is easily confused with other minerals.

32 It is necessary to add that in the VCT, next to this phonemic loan, a second translation, *gai kuang* 鈣礦, can be read. These three characters are encircled with a pencil; this second translation will not be included in the *HXCL*.

using different graphical signs; these signs represented an indication to substitute one or more characters in all the entries referred. Fryer actually highlighted the importance that the connection existing between terms belonging, in a broader sense, to the same semantic field, should be clearly pointed out, whether phonemic loans or other translation choices. Within this constant, it is also possible to pinpoint Fryer's will to reduce, as much as possible, the number of characters used in phonemic loans. An example can be appreciated in the entry "Schorl," on page 31 of the HXCL, firstly translated in the VCT as *sishe'erle shi* 司舍耳勒石, later modified as *she'erle shi* 舍耳勒石.

The third tendency is the substitution of syntagma or descriptive labelling with other translation choices. Consider the entry "Free stone," on page 15 of the HXCL, where its rendering is a descriptive labelling, *ruan sha shi* 軟砂石. The first draft of the translation was the syntagm *yi zao ju zhi shi* 易鑿鋸之石, a term including two nouns with verbal function and a structural particle, which makes the translation hardly absorbable in Chinese lexical texture. Such a term cannot comply with two of the four characteristics required for a "modern" language, to wit "productivity" and particularly "transmission capability."[33]

The fourth tendency is the correction of those translations, or at least part of them, which proved to be insufficiently scientific, avoiding ambiguous terms as much as possible. Consider "Actinic rays"; in the VCT it had been translated as *guang nei huawu qi* 光内化物氣, conveying the concept of "ray" through the character *qi*, which includes among its meanings "gas," therefore a patently inadequate choice. The rendering was modified on page 1 of the HXCL with the definition *guang nei huawu xian* 光内化物線; here the concept of "rays" is rendered with the term under which they are still known in contemporary scientific and technical Chinese. There are also exceptions, represented by mistakes found in the printed version but absent in the manuscript. To give an example, the term "Ammonium sulphide," on page 3 of the HXCL, has been translated as *danqing*$_{si}$*liu* 淡輕$_{四}$硫, based on the chemical structure of the compound, $(NH_4)_2S$. In the manuscript, the solution adopted was closer to the actual chemical formula of the salt, rendered as *er danqing*$_{si}$*liu* 二淡輕$_{四}$硫.

The fifth feature of the passage from the VCT to the HXCL is the inclusion of additional translation choices. For example, the first version of the entry "Actinolite" was represented by the phonemic loan *aketinulaide* 阿克替奴來得;

33 The other two being "diffusion ability" and "egalitarianism," as indicated in Shen Guowei, "Chinese Language and the Modern: Contemplating Chinese in the East Asian Linguistic Environment," in *A Selection of Essays on Oriental Studies of ICIS* 1, ed. Kansai daigaku (Osaka: Institute for Cultural Interaction Studies, Kansai University, 2011), 101–102.

the first correction concerned the second character, modifying the translation into *agetinulaide* 阿各替奴來得. The final version on page 1 of the *HXCL* is also a phonemic loan, with the second character modified again, *agetinulaide* 阿格替奴來得. Fryer, though, will also add a second rendering to it, a descriptive labelling, namely *guangxian shi* 光線石.[34] He explicitly shows the reader that the referent is a mineral, disambiguating a phonemic loan that could not provide the user with any information, and was hardly absorbable in the Chinese lexicon, particularly because of the numerous characters composing it.[35]

The last feature is a peculiarity of this glossary—the removal of numerous entries in the passage from the VCT to the *HXCL*. In the former, the original terms in the entries studied above have been written down in most cases with a pencil, within brackets, and in the majority of cases followed by a sign "(b)," in contrast to the body text of the manuscript. It is plausible to suppose that Fryer was planning to include these entries in a later revised edition of the glossary, since none were later listed in the *HXCL*.[36]

The glossary, as will be shown later in this work, represented an important step in the formation of modern Chinese chemical nomenclature. The universally renowned scholar Liang Qichao clearly attributed great value to it: "中西文並列最要" (Original text with Chinese parallel translation; indispensable).[37]

---

34 The term can also be analysed as a calque, since it reflects the Greek etymology of "Actinolite," namely ἀκτίς, with the meaning "ray."

35 The choice Fryer adopted of keeping in any case the phonemic loan and adding a second rendering as an alternative choice, probably reflects his uncertainty on the accuracy of the latter: the term referring to the mineral in contemporary Chinese is indeed *yangqi shi* 陽起石, not *guangxian shi*.

36 To quote a few examples, it is possible to read terms such as "Groves' battery," *Guluwu fadianqi* 古路物發電器; "Acetification," *cheng cusuan* 成醋酸; "Marsh's test," *Maershi shi shenyang$_{san}$ zhi fa* 瑪而石試鉮養$_{三}$之法; "Refinery," *tichun tie zhi lu* 提純鐵之爐; and many others.

37 Liang, "Xixue shumu biao," in the entry where the *HXCL* is listed. The *HXCL* is also listed in von Möllendorff, *Catalogue*, 31, in the section "Dictionaries" of "East Asiatic Languages." In the entry, the text is quoted as "*Fryer, John*, LL. D.—Vade Mecum vocabulary of names of chemical substances. Shanghai, 1885. 2 parts in 1 vol., 8vo, half calf"; the reference to the two parts, though, is not completely clear. It is important to note that again in von Möllendorff, *Catalogue*, 12, among the "Original Chinese works," in the "Miscellaneous" category, the entry "化學材料名目 Hua-hsüeh-ts'ai-liao-ming-mu. 1 vol. Vocabulary of names of chemical substances" can be read; this text thus appears to be listed as a different autonomous work.

4 *Vocabulary of Names of Materia Medica*

A structure similar to that of the HXCL can be seen in the third glossary included in the TVM, the *Vocabulary of names of Materia Medica occurring in the translation of "Royle's Manual of Materia Medica, and Therapeutics." With lists of names of persons and places occuring in the same work, and in various treatise on Chemistry and allied subjects*; in Chinese, *Xiyao dacheng yaopin Zhong-Xi mingmu biao* 西藥大成藥品中西名目表. Analogously to the JS and the HXCL, the preface supplies important clues to the planning of the glossary:

> This vocabulary contains the names of all the Materia Medica mentioned in Dr. Royle's well known treatise together with all the names of chemicals plants and animals. Both Latin and English names are given in Alphabetical order so as to make the vocabulary more generally useful.
>
> The Latin name[38] of each genus of plant and animal has generally been phoneticised while the descriptive term denoting the species or variety has been translated literally where possible, and placed before the generic term. Thus Chinchona rotundifolia[39] appears as the round leaved Chin-chi-na. Where a descriptive term is the name of a person or place or is otherwise untranslateable [sic] the whole Latin name has been phoneticised. In every case where the Chinese term could be found with any degree of certainty it has been added.
>
> The English botanical and zoological names are translated on the general principles above mentioned except that they are not phoneticised, as a rule, when the Chinese equivalent is known.
>
> The derivative terms retain the sounds of the root word or a prominent portion of it.[40] Thus Cinchonia, Cinchonine, Cinchonicine, Cinchonidia, &c., retain the common root Chin-chi-na; while Quinine, Quinicine, Quinidia, Quinone, and other derivatives retain Chi-na as part of the

---

38 Similarly, in Perny, *Appendice du dictionnaire français-latin-chinois*, 85–121, within the appendix "Table des matières de l'histoire naturelle. Partie latine," lemmata are listed according to the Latin nomenclature.

39 In the body text of the glossary, it will actually be listed with the correct spelling, "Cinchona rotundifolia."

40 Two examples from the glossary will help clarify: the term "Cinchonine," an alkaloid found in the Cinchona officinalis, is rendered on page 15 as *jinji'nayiya* 金雞那以亞, keeping the rout *jinji'na* 金雞那; "Quinicine," an alkaloid extractable from the bark of the Cinchona, is in turn translated on page 51 as *ji'nayixini* 雞那以西尼, keeping for phonetic reasons only a portion of the root, *ji'na* 雞那.

root. The organic alkaloids, acids &c., are always expressed by the same terminal characters analogous to our ic, ia, ine &c.

The inorganic terms follow the general rules laid down in the vocabulary of chemical terms already published.[41]

Two lists of proper names are appended. They contain the whole of the names of persons and places referred to in Dr. Royle's work as well as in various other books on Medicine, Chemistry and kindred subjects. No definite system for phoneticising has been employed; and no new proper name has been invented where a satisfactory one could be found in any previous translation.

This vocabulary, commenced twelve years ago, with the translation of Dr. Royles' work is only an experiment and no one can perhaps be more sensible of its imperfections than the compilers. Some features in it could easily be improved separately but then they would not harmonize with the rest. It remains for some one to invent a system that shall be more perfect throughout.

Kiangnan Arsenal
May, 1887.

Shown below is the Chinese preface, following the English one:

西藥大成藥品中西名目表附人名地名兩表. 此表載英國醫士來拉著西藥大成一書內各種藥品名目，並化學料與植物動物名，其中臘丁與英文俱依字母排列，便於用此書者查考，令其用處更廣. 凡植物動物分類所有之臘丁名目，平常譯其音，尚有分種之名，則譯其意，而列於類名之前，如圓葉金雞哪，其金雞哪為類名，圓葉為種名是也，如其種名因原為人名或地名或因他故無法譯其意，則仍譯其音，凡能察得中華已有常用之名目，亦並記之. 凡植物動物之英文名目，亦照前欵之意譯之，如確知中華名目者，則不譯其音.凡藥料變成之名目，必存其原音之根，或原音根之要分，如金雞哪以亞，金雞哪以尼，金雞哪以西尼，金雞哪以第亞等，俱存金雞哪為音之根，又如雞哪以尼，雞哪以西尼，雞哪以第亞，雞哪哇尼等，俱存雞哪為音根之要分，凡生物鹼類酸類等，其各名之末字，常歸一例記之，如以克以尼以亞等是也，與西名同法. 凡死物質之名，俱依前印化學材料中西名目表所載之公法而定之.另附人名地名二表，此不但有來拉所作西藥大成一書之人名地名，兼有醫學化學等書內常遇之人名地名，此各名不用一定之華字，代一定之西音，又如在已有之中國書內，得合用人名地名，則必從之，不敢另設新法記之. 初譯此書兼造名目，自起手迄今，已逾

41 This is additional confirmation that the glossary was published before the *TVM*.

> 十二載，衹為試作之意，故不免有弊，且其弊有試作者所預知，而比他人知之更詳者，然如改其一弊，又恐有他弊由此而生，所以改弊之全法，以俟後之君子.光緒十三年夏四月江南製造總局排印.

The *XYDCYP*, published in 1887, comprises terms related to pharmacology included in the work by "Dr. Royle," in addition to those pertaining to chemistry, botany, and zoology.[42] The text referred to by Fryer, composed by the British botanist John Forbes Royle (1798–1858)[43] and the physician Frederick William Headland (1830–1875), is *A Manual of Materia Medica and Therapeutics*.[44] Even though Fryer acknowledges that some features of the glossary could have been easily improved, the glossary would still be for sale many years later, demonstrating its value.[45]

As in the case of the *HXCL*, the layout of the pages of the *XYDCYP* comprises two columns, each listing the entries in alphabetical order, with the original terms preceding the Chinese translation. The number of entries is 6,073, divided into sixty-eight pages.[46]

The *XYDCYP* manuscript can be located in number five of the volumes housed in the EAL. Entries in this volume are not directly written on the pages,

---

42 In Thomson, "Medical Publications in Chinese," the "西藥大成藥品中西名目表 *Vocabulary of Names of Materia Medica*, etc. 1 vol, pp. 68" is also included; nevertheless, the publishing date diverges from the one unanimously recognised by scholars, backdating it to 1886. With respect to von Möllendorff, *Catalogue*, the glossary is mentioned on page 12, among "Original Chinese works" in the category of "Miscellaneous," as "西藥名目 Hsi-yao-ming-mu. 1 vol. Vocabulary of names of meteria [sic] medica. 2 copies." In this case as well it is considered an autonomous text, published as a single volume.

43 See also Danda Venkata Subba Reddy, "John Forbes Royle, Botanist, Medical Historian, Teacher, and Benefactor of British Empire," *Bulletin of the Institute of Medicine Hyderabad* 3, no. 2 (1973).

44 John Forbes Royle and Frederick William Headland, *A Manual of Materia Medica and Therapeutics; Including the Preparations of the British Pharmacopœia, and Many Other Approved Medicines*, 5th ed. (London: John Churchill & Sons, 1867).

45 *Descriptive Catalogue*, 9. As described in the text: "1 vol. 6mo. 15 cents. This is a vocabulary in English and Chinese, combined with Latin and Chinese, embracing all the terms used provisionally in the translation of Royle and Headland's Materia Medica and Therapeutics, numbering about 6,500. To native medical students and medical missionaries this vocabulary ought to prove of much service. The appendix contains a list in English and Chinese of over 2,000 names of persons and places occurring in works on Medicine, Chemistry, and allied sciences."

46 In this case as well, the number of entries refers to the original terms in English or scientific Latin; for many of them, more than one Chinese translation is provided. In this glossary as well, though, incongruities and errors are present, however rare, in the alphabetical order.

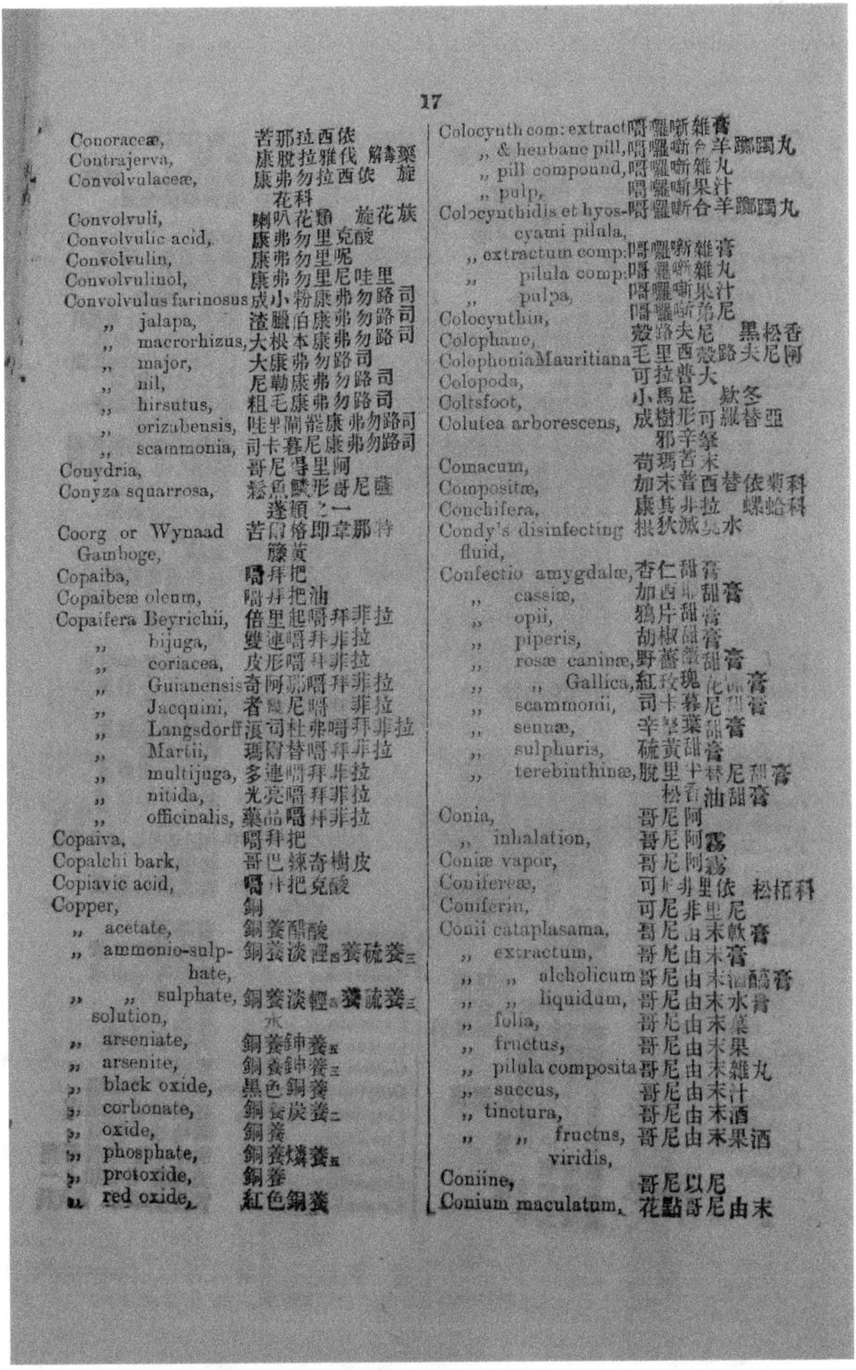

17

Conoraceæ, 苦那拉西依
Contrajerva, 康脫拉雅伐 解毒藥
Convolvulaceæ, 康弗勿拉西依 旋花科
Convolvuli, 喇叭花類 旋花族
Convolvulic acid, 康弗勿里克酸
Convolvulin, 康弗勿里呢
Convolvulinol, 康弗勿里尼哇里
Convolvulus farinosus 成小粉康弗勿路司
,, jalapa, 渣臘伯康弗勿路司
,, macrorhizus, 大根本康弗勿路司
,, major, 大康弗勿路司
,, nil, 尼勒康弗勿路司
,, hirsutus, 粗毛康弗勿路司
,, orizabensis, 哇里闌龍康弗勿路司
,, scammonia, 司卡暮尼康弗勿路司
Conydria, 哥尼導里阿
Conyza squarrosa, 鬆魚鱗形哥尼薩蓬類之一
Coorg or Wynaad Gamboge, 苦閣格即韋那特籐黃
Copaiba, 嗬拜把
Copaibeæ oleum, 嗬拜把油
Copaifera Beyrichii, 倍里起嗬拜非拉
,, bijuga, 雙連嗬拜非拉
,, coriacea, 皮形嗬拜非拉
,, Guianensis 奇阿那嗬拜非拉
,, Jacquini, 者[illegible]尼嗬[illegible]非拉
,, Langsdorffi 浪司杜弗嗬拜非拉
,, Martii, 瑪尉替嗬拜非拉
,, multijuga, 多連嗬拜非拉
,, nitida, 光亮嗬拜非拉
,, officinalis, 藥品嗬拜非拉
Copaiva, 嗬拜把
Copalchi bark, 哥巴辣奇樹皮
Copiavic acid, 嗬拜把克酸
Copper, 銅
,, acetate, 銅養醋酸
,, ammonio-sulphate, 銅養淡輕四養硫養三
,, ,, sulphate, solution, 銅養淡輕四養硫養三水
,, arseniate, 銅養鉮養五
,, arsenite, 銅養鉮養三
,, black oxide, 黑色銅養
,, corbonate, 銅養炭養二
,, oxide, 銅養
,, phosphate, 銅養燐養五
,, protoxide, 銅養
,, red oxide, 紅色銅養

Colocynth com: extract 嗬囉嘶雜膏
,, & henbane pill, 嗬囉嘶合羊躑躅丸
,, pill compound, 嗬囉嘶雜丸
,, pulp, 嗬囉嘶果汁
Colocynthidis et hyoscyami pilula, 嗬囉嘶合羊躑躅丸
,, extractum comp: 嗬囉嘶雜膏
,, pilula comp: 嗬囉嘶雜丸
,, pulpa, 嗬囉嘶果汁
Colocynthin, 嗬囉嘶𦬁尼
Colophane, 殼路夫尼 黑松香
ColophoniaMauritiana 毛里西殼路夫尼阿
Colopoda, 可拉普大
Coltsfoot, 小馬足 款冬
Colutea arborescens, 成樹形可羅替亞邪辛拏
Comacum, 苛瑪苦末
Compositæ, 加末普西替依菊科
Conchifera, 康其非拉 螺蛤科
Condy's disinfecting fluid, 根狄滅臭水
Confectio amygdalæ, 杏仁甜膏
,, cassiæ, 加西耶甜膏
,, opii, 鴉片甜膏
,, piperis, 胡椒甜膏
,, rosæ caninæ, 野薔薇甜膏
,, ,, Gallica, 紅玫瑰花甜膏
,, scammonii, 司卡暮尼甜膏
,, sennæ, 辛拏葉甜膏
,, sulphuris, 硫黃甜膏
,, terebinthinæ, 脫里卞替尼甜膏 松香油甜膏
Conia, 哥尼阿
,, inhalation, 哥尼阿霧
Coniæ vapor, 哥尼阿霧
Coniferеæ, 可尼非里依 松栢科
Coniferin, 可尼非里尼
Conii cataplasama, 哥尼由末敷膏
,, extractum, 哥尼由末膏
,, ,, alcholicum 哥尼由末酒醕膏
,, ,, liquidum, 哥尼由末水膏
,, folia, 哥尼由末葉
,, fructus, 哥尼由末果
,, pilula composita 哥尼由末雜丸
,, succus, 哥尼由末汁
,, tinctura, 哥尼由末酒
,, ,, fructus, viridis, 哥尼由末果酒
Coniine, 哥尼以尼
Conium maculatum, 花點哥尼由末

FIGURE 3 Page 17 of the *XYDCYP*, *Translator's vade-mecum*, T10.F8 c.2, University of California, Berkeley

instead annotated on white strips of paper glued onto them, without a legend. The original terms, for the most part written in purple and, more rarely, black ink, precede the corresponding Chinese translation, in turn written in black ink. The total number of pages is 210. Contrary to the VCT, this volume can be considered a real proof;[47] apart from very little difference in content, this hypothesis can be confirmed as it is possible to note some graphical signs indicating the beginning of a new page in correspondence with those destined to be the last entries of single pages in the *XYDCYP*.[48] More evidence supporting the aforesaid hypothesis is the presence of signs in the manuscript vertically cutting groups of entries. In the overwhelming majority of cases, these entries are connected by a term repeated exactly the same way in the following entry. Such a term will be substituted in the *XYDCYP* with a specific graphical sign in order to avoid graphical repetitions. Finally, the employment of different kinds of ink for amendments and additions affirms that the manuscript has been drafted in different working stages.[49]

When analysing the volume, the scarcity of discrepancies between the manuscript and the printed version is clear. The main difference is represented by the addition directly on the pages in the manuscript of some entries, in most cases using purple ink for the original term and black ink for the corresponding Chinese rendering. Nevertheless, these additions were not included in the *XYDCYP*. The remaining few differences between the manuscript and the published version concern some orthographic revisions, for instance the use of totally or partially simplified characters; this is a feature it has in common with the other EAL manuscripts as well.[50] When comparing the *XYDCYP*

---

47 At the beginning of page I 1, a date can be read: "23/2/86." It is not possible to say with certainty what it is referring to, the volume being void of other relevant indications. Nevertheless, as the glossary was printed in 1887, it is possible to infer it is the date of specific corrections, of compilation, or even of publication.

48 These indications are not consistently precise or correspondent to the printed version. For instance, after the entry "Ferula assofoetida," in the manuscript a red sign written in pencil reading "stop" can be noted. Even so, the entry will be the second to last of page 24 of the *XYDCYP*. Other information relating to printing can be found in numerous entries for which Fryer provided more than one translation; in those cases, the different translations are separated by a small circle.

49 Another confirmation comes from the fact that "Amygdaleæ," translated as *amigedaliyi* 阿米格大里依 and *xing ke* 杏科, and "Amygdalæ, pulvis compositus," rendered as *xingren zasan* 杏仁雜散, are preceded by the numbers "2" and "1" in order to point out that the disposition should have been inverted, following alphabetical order, as will be the case of the published version.

50 Even rarer are cases of corrections of whole characters. An example can be found on page C 22; the entry "Convolvulin" is rendered in the manuscript as *kangfuwulini* 康弗勿里尼, while in the printed version it will be translated as *kangfuwulini* 康弗勿里呢, with the

and the *HXCL*, an overlap in the entries is evident, perhaps unsurprising given the partial commonality of the scientific branches covered by the two glossaries. Nevertheless, Fryer did not limit himself to copying translations from one to the other—he also conducted actual revisions in the almost two years between the publication of the two glossaries. To cite one example, the entry "Crotonic acid" is rendered in the manuscript volume as *keluodunyike suan* 可羅敦以克酸; the same translation is to be found on page 18 of the *XYDCYP*, but on page 12 of the *HXCL* it is presented with a different character, resulting in *geluodunyike suan* 各羅敦以克酸.

Given the nature of the manuscript, almost identical to its printed version, it is not possible to observe the differences of the published version, as shown in the preceding section. However, from the few annotations in the volume housed in the EAL it can be noted how Fryer continued, in this glossary as well, to put his theories on Chinese scientific and technical lexicon into practice. For instance, on page P 2, in the entry "Paramenispermia," rendered with the phonemic loan *balaminisibeiermoyiya* 巴辣米尼司卑耳摩依亞, the four characters *beiermoyi* 卑耳摩依 have been circled by Fryer; above them, he added the note *baimi* 白迷. This is probably a reminder of the possibility to substitute the four characters with the other two; the original phonemic loan given is composed of an excessive number of characters and could never have been included in the Chinese lexicon. It is clear how Fryer, according to his numerous statements on the topic, tried to reduce the number of characters in phonemic loans he coined, eliminating those he believed unnecessary for an exact phonetic representation of the original term.[51]

---

change of *ni* 尼 into *ni* 呢 through the addition of the radical *kou* 口. Another discrepancy between the manuscript and the *XYDCYP* consists in the entry "Acetum ipecacuanhae." In the former, it is rendered as *yibige cu* 叱嗶喀醋, and in the latter as *yibige cu* 叱嗶格醋. This instance can plausibly be considered a mistake in the manuscript, since in all the subsequent entries of the *XYDCYP* that comprise the term "ipecacuanha" and its derivatives, the character *ge* 喀 is present. Finally, it is important to note that the entries on page C 12 "Cherry laurel," *luoerwusi yingtao* 羅耳烏司櫻桃, page C 18, "Citrullus colocynthis," *geluoxin xituoluoluosi* 哥囉嘶西脫羅羅司 and "Citrus limetta," *limuta xituoluosi* 里暮塔西脫羅司, page C 27, "Cuminum cyminum," *jiminumo gumini* 幾米奴末古米尼 and *maqin* 馬芹, page L 5, "Linaceæ," *li'naxiyi* 里那西依 and *ma ke* 麻科, and page P 11, "Polygonaceæ," *boligunaxiyi* 波里谷那西依 and *liao ke* 蓼科, have all been cancelled in the manuscript, but will be listed in the *XYDCYP*.

51 On the topic, refer to the statement in the preface to the STFNC, section 5.8. It should be added, though, that in the *XYDCYP* the final version of the term will be the original one; the hypothesis is that the revision has not been accepted, or it may have been carried out after the printing for a possible revised edition.

As indicated in the previous section, one that can be considered a partial version of the *XYDCYP* can also be located in volume six of the EAL, right after the VCT. This section, though, does not represent a complete version of the glossary, but at the most a preparatory draft, plausibly employed as a basis for volume five, analysed above. More specifically, it can be interpreted as a draft for the botanical terms of the *XYDCYP*. In this section included in volume six of the EAL, entries are listed in alphabetical order according to the original term, either in English or scientific Latin for the overwhelming majority, as was the case with the VCT. Original terms are written from left to right, as well as their Chinese translations. Compared with the section on which the *HXCL* is based, entries are much less organised and decrease in number as pages advance, while corrections with distinct kinds of ink increase. As with the section of the volume dedicated to chemistry, the hypothesis is that revisions were done at different stages.

When comparing this part of the manuscript with the *XYDCYP*, some previously discussed constants become evident:

TABLE 4 *XYDCYP* manuscript and volume six of the EAL

Original term	*XYDCYP* manuscript, first version	*XYDCYP* manuscript, second version	*XYDCYP*
Coriandreæ	可里安特里依, 胡荽科	可里安特里依科, 胡荽科	可里安特里依科 胡荽科
Chenopodæ	雞奴布提, 即鵞掌菜	幾奴布提依, 即藜科	幾奴布提依, 藜科
Convolvuli	喇叭花類, 鐘形科	喇叭花類, 旋花族	喇叭花類 旋花族
Cryptogameæ	暗生植物, 又名無花之部		暗生植物, 無花部
Acacia albida	阿楷西耶阿勒比達	略白色阿楷西耶	略白色阿揩西耶
Ichthycolla	依克弟可拉		魚肚膠
Acipenser	阿西噴薩	阿西噴薩, 即鱘龍魚	阿西噴薩, 鱘龍魚

The first tendency in the passage from the manuscript to the printed version is the addition, in phonemic loans, of some elements that could help the user locate the referent and its semantic field. In the same way as in the first section of the volume, the choice is performed through the addition of characters. As an example, the entry on page 23 of the manuscript, "Coriandreæ," a tribe, was first translated with the phonemic loan *keli'anteliyi* 可里安特里依, to which Fryer added *husui ke* 胡荽科, erroneously adopting the character for families *ke*

科.[52] Later, Fryer believed it necessary to add the character *ke* in the phonemic loan as well, thus creating a hybrid, *keli'anteliyi ke* 可里安特里依科, in order to make explicit the correspondence between the two renderings provided.

The second tendency lies in the revisions of characters used in phonemic loans. For instance, the entry on page 20, "Chenopodæ," indicating the contemporary family of Chenopodiaceae, had been translated in the first version with the phonemic loan *jinubuti* 雞奴布提, to which Fryer later added a second rendering. In the final version, in addition to modifying the second translation, Fryer also altered the phonemic loan, changing the character *ji* 雞 into *ji* 幾, and adding another character, an extremely rare decision. The final printed translation will thus result in *jinubutiyi* 幾奴布提依.

The third constant is the replacement of syntagma or descriptive labelling with other translation choices: see the example "Convolvuli," on page 22, referring to all plants ascribable to the genus Convolvulus. The first version Fryer provided was a descriptive labelling, *laba hua* 喇叭花, denoting the shape of the flower, similar to the musical instrument suona, to which he added the character *lei* 類 to indicate that the term was generally referring to all plants with the aforementioned peculiarity. The final result was *laba hua lei* 喇叭花類, to which Fryer joined a second descriptive labelling, *zhongxing ke* 鐘形科. In this case, the prominent feature of the plants was compared to a bell, with the addition of the character *ke*, denoting families. In the second and printed version, Fryer will keep the first descriptive labelling unaltered, replacing the second with another, *xuanhua zu* 旋花族. The latter is the contemporary term referring to the Convolvuleae tribe, belonging to the family Convolvulaceae.

The fourth tendency is the correction of translations, or parts of them, which proved to be scarcely scientific, thus eliminating all components that made their inclusion in Chinese scientific and technical lexicon more difficult. With this tendency, Fryer frequently replaced phonemic loans, or parts of them, with descriptive labelling, more concise or other translation choices. In the case of the entry "Acacia albida," a plant nowadays known as Faidherbia albida, the first translation provided was a phonemic loan, *akaixiye alebida* 阿楷西耶阿勒比達. As Fryer affirmed in the preface to the *XYDCYP*, the Latin name of the genera of plants should have been rendered with a phonemic loan, while the "descriptive term" denoting the species or variety should have been transliterated. The second translation Fryer gave reflects this principle; the concept

52 Contrary to the aforementioned manuscript included in volume five, numeration is not indicated here; the author opted to add it for the reader's convenience.

of "albida" will be rendered by him as *lüebai se* 略白色 (slightly white), the final translation resulting in a hybrid, namely *lüebai se akaixiye* 略白色阿楷西耶.[53]

Finally, the fifth constant in the passage from the manuscript to the printed version is the addition of more translations. As an example, the entry "Acipenser," on page 1 of the zoology section, is translated with the phonemic loan *axipensa* 阿西噴薩. Fryer decided to add, in a subsequent phase of correction, the second translation *xunlong yu* 鱘龍魚. Even though the latter generally indicates fish belonging to the contemporary family of Acipenseridae, it surely bears a more precise semantic connotation compared with the phonemic loan supplied in the first place.

Pertaining to the links with other contemporary lexicographical materials, their mutual influence, and the transcultural flows of terminologies and notions, Doolittle's *VHCL* can be identified as one of Fryer's sources. As stated in the preface to volume 1,[54] completed in Fuzhou and dated January 1, 1872, the third part of the *VHCL* "[…] consists largely of Tables and Lists of Terms and Phrases contributed to this work by various Gentlemen connected with the Consular and with the Customs services, by Missionaries belonging to American, English and German Societies, and by other residents in China. […] The Topics of these Tables and Lists are very numerous and relate to subjects of general interest and importance."[55]

Therefore, within the *VHCL*, texts drawn from other authors are also included. Starting from page 295 of the dictionary, terms listed in the "Classification of Medicines. 藥分類 yao fên lei. By J. G. Kerr, M. D." can be read. As explained in the *VHCL*, "These Terms are found in the Manual of Materia Medica (西藥畧釋) by Dr. Kerr, and are furnished by him for insertion in this work. […] For a full explanation of such terms, the student is referred to Dr. Kerr's Chemistry 化學初階." Doolittle thus affirms that terms included in his work derive from the

53 At times Fryer simply modifies those that can be considered oversights in the translation of terms in scientific Latin. For instance, the entry "Acacia Adamsonii," translated on page 1 of the *XYDCYP* as *Adansun akaixiye* 阿但孫阿揩西耶, in the VCT section pertaining to botany, had been firstly rendered as *akaixiye Adansun* 阿楷西耶阿但孫, later modified as *Adansun akaixiye* 阿但孫阿楷西耶; in the printed version, only the character *kai* 楷 has been substituted with *kai* 揩.

54 Doolittle, *Ying-Hua cuilin yunfu*, 1, preface, 11.

55 In the preface to volume 2, on page 11, signed "Shanghai, South Gate, Dec. 25, 1872," the text is slightly different, providing more generic information: "[…] consists of Contributions from Gentlemen living in China, and Lists and Tables, either selected and arranged from the works mentioned above and in the Preface to first volume, or prepared by myself."

Xiyao lüeshi 西藥略釋 (Brief explanations of Western medicine). The text[56] had been translated by the American medical missionary John Glasgow Kerr (Jia Yuehan 嘉約翰, 1824–1901) and the Chinese scholar Kong Jiliang 孔繼良 (n.d.),[57] published for the first time in Guangzhou in 1871 by Boji yiyuan 博濟醫院,[58] successively republished in 1876 and, as a revised edition, in 1886.[59] In the fourth volume of the revised and enlarged third edition of the text, a small glossary of medical terms with parallel Chinese-English translation is included.[60]

The other text Doolittle mentioned as a reference, for those desiring to obtain more exhaustive information than what is included in a concise glossary, is the *Huaxue chujie* 化學初階 (First steps in chemistry).[61] The text, published by the Boji yiyuan in Guangzhou in 1871, is the translation by Kerr and He Liaoran 何瞭然 (n.d.)[62] of David Ames Wells' work, *Principles and Applications of Chemistry*.[63]

A comparison of the same entries in the *XYDCYP* and in the *VHCL* will allow the reader to appreciate the almost absolute correspondence between the two:

56 As stated in Pan Jixing, *Zhong wai kexue zhi jiaoliu* (Hong Kong: Xianggang zhongwen daxue, 1993), 537.

57 Information on Kong Jiliang's life is almost untraceable.

58 Consult William Warder Cadbury and Mary Hoxie Jones, *At the Point of a Lancet: One Hundred Years of the Canton Hospital, 1835–1935* (Shanghai: Kelly and Walsh, 1935).

59 Refer also to the advertisement by Wenyutang 文裕堂 in the Universal Circulating Herald, or *Xunhuan ribao* 循環日報, as reported in Uchida, *A Study of Cultural Interaction*, 154–56.

60 After a few years, John Glasgow Kerr, *A Vocabulary of Diseases, Based on Thomson's Vocabulary and Whitney's Anatomical Terms, Prepared by J. G. Kerr, M. D., for the Committee on Nomenclature of the Medical Missionary Association of China* (Shanghai: American Presbyterian Mission Press, 1894), would be published. A later important text in the field will be Philip Brunelleschi Cousland, *An English-Chinese Lexicon of Medical Terms Compiled for the Terminology Committee* (Shanghai: Medical Missionary Society of China, 1908); on Cousland, refer to Zhang Daqing, "Gao Silan: yixue mingci fanyi biaozhunhua de tuidongzhe" *Zhongguo keji shiliao—China Historical Materials of Science and Technology*, no. 4 (2001).

61 John Glasgow Kerr and He Liaoran, trans., *Huaxue chujie* (Boji yiju: Guangzhou, 1870). See also Zhang Zigao, "He Liaoran de *Huaxue chujie* zai huaxue yuansu yiming shang de lishi yiyi," *Qinghua daxue xuebao* 9, no. 6 (1962) and Zhang Zigao and Yang Gen, "Cong *Huaxue chujie* he *Huaxue jianyuan* kan wo guo zaoqi fanyi de huaxue shuji he huaxue mingci," *Ziran kexue shi yanjiu—Studies in the History of Natural Sciences* 1, no. 4 (1982).

62 Information on He Liaoran is extremely scarce; refer to Pan, *Zhong wai kexue*, 520.

63 Wells, *Principles and Applications of Chemistry*.

TABLE 5 *XYDCYP* and *VHCL*

Original term	*XYDCYP*	*VHCL*
Alteratives	改血藥	改病藥
Anæsthesiants[a]	蒙藥	朦藥
Antacids	解酸藥	解酸藥[b]
Anthelmintics	殺蟲藥	殺蟲藥[c]
Aperients		輕瀉藥
Antispasmodics	安肚腹腦筋藥, 治轉筋藥	治轉筋
Astringents	收斂藥	斂藥
Cathartics	瀉藥	瀉劑

a "Anaesthetics" in the *VHCL*.
b In the appendix to the fourth juan of John Glasgow Kerr and Kong Jiliang, trans., *Xiyao lüeshi*, 3rd ed. (Guangzhou: Boji yiju, 1886), the term will be changed to *jie suan zhi ji* 解酸之劑. On the text, refer to John Fryer, "Lun xinyi *Xiyao lüeshi*," *Gezhi huibian* 8 (1876): 8–9.
c In *ibidem*, the term will be changed to *sha chong* 殺蟲.

In this case as well, Fryer did not limit himself to copying names from the list Kerr provided, but often integrated and modified them whenever he believed necessary. For example, in the entry "Anæsthesiants" the term—Fryer also revised the original English word—had been rendered by Kerr as *mengyao* 朦藥. On page 4 of the *XYDCYP*, Fryer will modify it into *mengyao* 蒙藥, meaning *meng* 朦 "hazy, dim" and *meng* 蒙 "cover, overspread."[64] The latter, however not perfect, was a more accurate and scientifically appropriate translation.[65]

In other cases, Fryer believed it necessary to revise scarcely accurate translations from a semantic point of view. The entry "Alteratives," for instance, rendered by Kerr with the descriptive labelling *gaibing yao* 改病藥, has been translated on page 3 of the *XYDCYP* as *gaixue yao* 改血藥, since these substances were believed purifiers of the blood, in addition to being able to alter the process of a disease. In some cases, Fryer's revisions can be seen in the

64 For a discussion on the rivalry between Chinese chemical elements nomenclature adopted by Fryer and Xu Shou, and Kerr and He Liaoran, refer to Wright, *Translating Science*, 337–50.

65 In the appendix to the fourth juan of Kerr and Kong, *Xiyao lüeshi*, the character will be changed to *meng* 蒙.

substitution of the characters used as right-hand components of words:[66] the entry "Cathartics," rendered by Kerr as *xie ji* 瀉劑, has been modified on page 13 of the *XYDCYP* as *xieyao* 瀉藥.[67] Some doubt, though, arises, pertaining to the connections between the aforementioned texts. Given the intense correspondence between Fryer and Kerr, and the dispute on the authorship of the nomenclature of chemical elements, for which Fryer had provided Kerr himself a list of terms he personally coined, it has not been possible to ascertain whether Fryer also supplied his translations to Kerr in this case, or whether he indirectly tapped from Kerr's text, through the mediation of Doolittle's *VHCL*.[68]

5 *Vocabulary of Terms Relating to the Steam Engine*

The last glossary in the *TVM* from a chronological point of view, considering the *RMDM* as an appendix, is the *Vocabulary of Terms Relating to the Steam Engine*, with the Chinese title *Qiji Zhong-Xi mingmu biao* 汽機中西名目表. In reality, the glossary was not in the *TVM* edition published in 1888, as the *QJ* was printed in 1889. Nevertheless, it was included in both *Chinese and English Technical Vocabularies* and the *Translator's vade-mecum*.

Hereafter is the preface to the *QJ*:

> The basis of this Vocabulary is the translation of Main's Manual of the Steam Engine which was made in 1868. During the translating of various subsequent treatises such as "Bourne's Catechism," "Bourne's

66 Even if its morphology is present, Chinese is essentially an isolating language, therefore, strictly speaking, it is not possible to speak of affixes. As indicated in Jerome Lee Packard, ed., *New Approaches to Chinese Word-Formation: Morphology, Phonology and the Lexicon in Modern and Ancient Chinese* (Berlin, New York: Mouton de Gruyter, 1998), 17, it is possible, though, to find in Chinese a large class of morphemes with affixal properties encoding lexical, not grammatical information, which can appear as left-hand or right-hand components of a word. The characters analysed here and further on can be considered a typical example. See also Wei-wen Roger Liao, "Morphology," in *The Handbook of Chinese Linguistics*, ed. C.-T. James Huang, Y.-H. Audrey Li, and Andrew Simpson (Malden: Wiley Blackwell, 2014); Huang Shuanfan, "Chinese Morphology: Anatomy of a Headless Language," in Zhongyang yanjiuyuan di er jie guoji hanxue huiyi lunwenji: yuyan yu wenzi zu, ed. Zhongyang yanjiuyuan, two volumes (Taipei: Academia Sinica, 1989); and Zdenka Hermanova-Novotná, *Affix-Like Word-Formation Patterns in Modern Chinese* (Dissertationes Orientales, vol. 21. Prague: Oriental Institute in Academia, 1969).

67 In the appendix to the fourth juan of Kerr and Kong, *Xiyao lüeshi*, the term will be changed to *xieyao* 瀉藥.

68 On the list of terms provided by Fryer, see Wright, *Translating Science*, 341 and passim.

> Introduction," "Burgh's practical Rules," &c, a much more extensive nomenclature was accumulated.
>
> Recently, a Chinese rendering of "Sennett's Manual of Marine Engineering" having been called for, the present more comprehensive Vocabulary was compiled, to be in readiness. It embraces all the terms that had been fixed up to the beginning of 1889. An appendix will contain such as may afterwards be found necessary.
>
> It will be noticed that the terms made use of are descriptive, and that in order to be systematic, they do not always follow the exact meaning of the English. Some of them were invented before the great improvements in the Steam Engine were effected, and are, perhaps, not so applicable now as then. Having come into current use, however, they have necessarily to be retained to prevent the confusion that entirely new terms would cause. No attempt has been made in this Vocabulary to phoneticise any English words, except proper names.
>
> In some cases it is difficult to determine whether a term belongs to Marine Engineering or to Naval Architecture. There are necessarily words that are common to the two subjects, and will therefore be found also in the Vocabulary of terms in Naval Architecture which is to follow.
>
> Kiangnan Arsenal, October 1889.

Shown below is the Chinese preface, following the English one:

> 汽機中西名目表. 小序. 是表以汽機發軔所定名目為主，因發軔譯於同治十年，為汽機之第一書後，更續譯汽機，必以汽機新制等書名目亦逐漸增多，今擬譯兵船汽機一書，恐前後名目或有互異，故先將光緒十五年以前，所有成書內已定汽機名目，輯成中西名目表，嗣後有新出名目，擬另加坿表，至於是表名目，皆指形象物，亦有言其功用，而與英文本義，不甚脗合者，間有數名目，為後前所定，祇合於當時之用，揆之於今，稍有不稱，然歷經習熟勢不能二更新，致前後不符，故皆仍其舊，便於通行，若人名地名，祇能譯以英子之音，其餘則均解意義，使閱者易於了然，惟有機件名目，或屬汽機，或屬船體，為兩項所公用者，茲亦列入，俟後再輯入船體名目表，彼此互見，庶幾便於查檢云爾. 光緒十五年十月. 江南機器製造總局序.

Different works mentioned in the preface, also presented in a schematic way in the tables of the appendices, provide a glimpse of the most significant technical and scientific Western texts translated in late Qing China. The compilation of the *QJ* commenced with the translation of the "Manual of the Steam Engine," a text composed by "Main." In Fryer, *An account*, within the "List

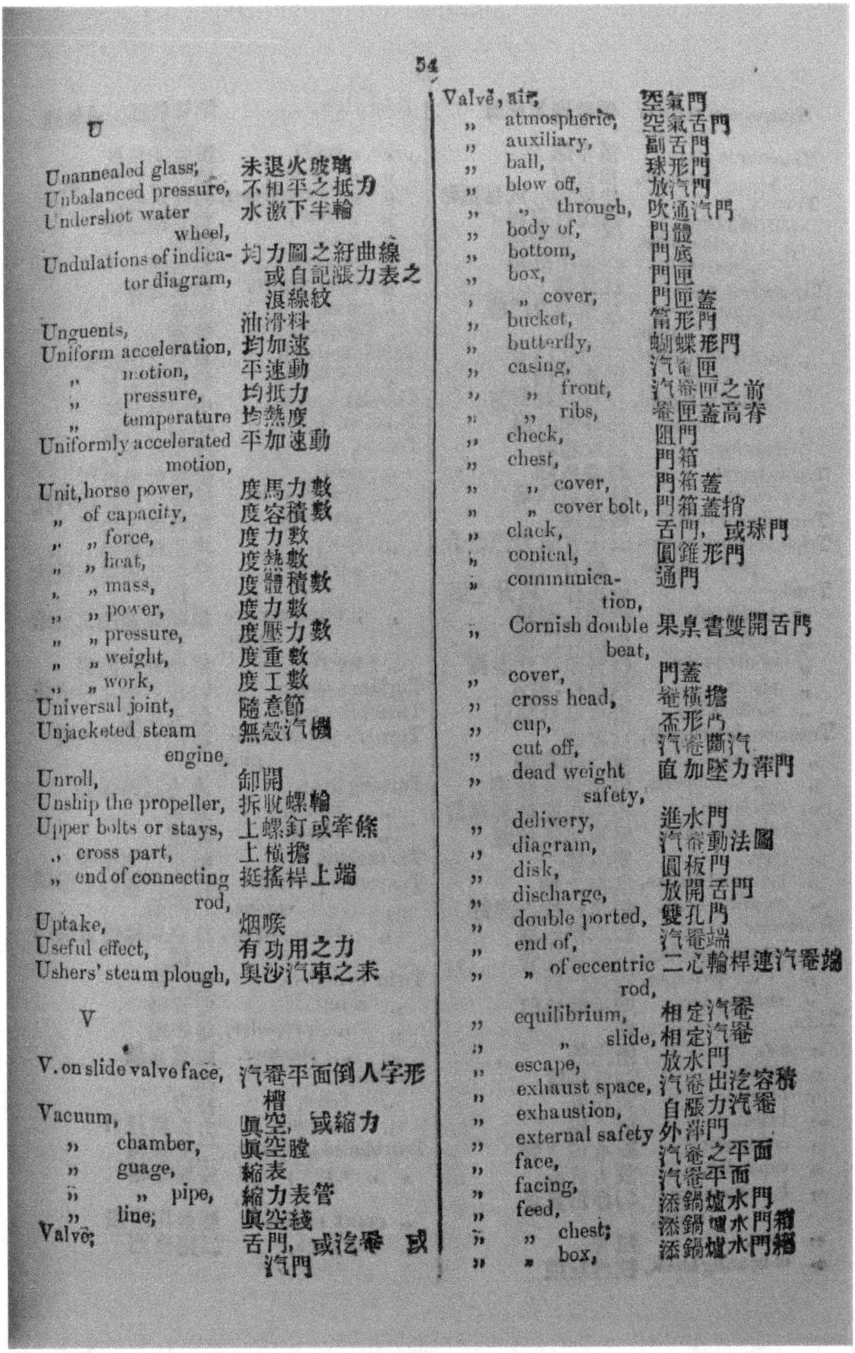

54

U

Unannealed glass;	未退火玻璃
Unbalanced pressure,	不相平之抵力
Undershot water wheel,	水激下半輪
Undulations of indicator diagram,	均力圖之紆曲線或自記漲力表之浪線紋
Unguents,	油滑料
Uniform acceleration,	均加速
,, motion,	平速動
,, pressure,	均抵力
,, temperature	均熱度
Uniformly accelerated motion,	平加速動
Unit, horse power,	度馬力數
,, of capacity,	度容積數
,, ,, force,	度力數
,, ,, heat,	度熱數
,, ,, mass,	度體積數
,, ,, power,	度力數
,, ,, pressure,	度壓力數
,, ,, weight,	度重數
,, ,, work,	度工數
Universal joint,	隨意節
Unjacketed steam engine,	無殼汽機
Unroll,	卸開
Unship the propeller,	拆脫螺輪
Upper bolts or stays,	上螺釘或牽條
,, cross part,	上橫擔
,, end of connecting rod,	挺搖桿上端
Uptake,	烟喉
Useful effect,	有功用之力
Ushers' steam plough,	奥沙汽車之耒

V

V. on slide valve face,	汽鞲平面倒人字形槽
Vacuum,	眞空，或縮力
,, chamber,	眞空膛
,, guage,	縮表
,, ,, pipe,	縮力表管
,, line;	眞空綫
Valve;	舌門，或汽鞲 或汽門
Valve, air,	空氣門
,, atmospheric,	空氣舌門
,, auxiliary,	副舌門
,, ball,	球形門
,, blow off,	放汽門
,, ,, through,	吹通汽門
,, body of,	門體
,, bottom,	門底
,, box,	門匣
,, ,, cover,	門匣蓋
,, bucket,	筩形門
,, butterfly,	蝴蝶形門
,, casing,	汽鞲匣
,, ,, front,	汽鞲匣之前
,, ,, ribs,	鞲匣蓋高脊
,, check,	阻門
,, chest,	門箱
,, ,, cover,	門箱蓋
,, ,, cover bolt,	門箱蓋梢
,, clack,	舌門，或球門
,, conical,	圓錐形門
,, communication,	通門
,, Cornish double beat,	果臬書雙開舌門
,, cover,	門蓋
,, cross head,	鞲橫擔
,, cup,	盃形門
,, cut off,	汽鞲斷汽
,, dead weight safety,	直加墜力萍門
,, delivery,	進水門
,, diagram,	汽鞲動法圖
,, disk,	圓板門
,, discharge,	放開舌門
,, double ported,	雙孔門
,, end of,	汽鞲端
,, ,, of eccentric rod,	二心輪桿連汽鞲端
,, equilibrium,	相定汽鞲
,, ,, slide,	相定汽鞲
,, escape,	放水門
,, exhaust space,	汽鞲出汽容積
,, exhaustion,	自脹力汽鞲
,, external safety	外萍門
,, face,	汽鞲之平面
,, facing,	汽鞲平面
,, feed,	添鍋爐水門
,, ,, chest;	添鍋爐水門箱
,, ,, box,	添鍋爐水門箱

FIGURE 4 Page 54 of the *QJ*, *Chinese and English Technical Vocabularies*, T10.F8, University of California, Berkeley

no. 1" comprising the "Works published at the Transltaion [sic] Department, Kiangnan Arsenal," entry number 23 reports the Chinese title *Qiji faren* 汽機發軔 (A primer on the steam engine);[69] the "Foreign work used as basis" is represented by the "Manual of the Steam Engine."[70] The latter text can be identified with *The Marine Steam-Engine*, written by the mathematician Thomas John Main (1818–1885) and Thomas Brown (1812–1887)[71] and translated by Alexander Wylie and Xu Shou in 1871.[72]

The second text mentioned in the preface is "Bourne's Catechism," corresponding to entry number 24 in *An account*, namely "Catechism of the Steam Engine."[73] The title of the original text is *A Catechism of the Steam Engine*,[74] composed by John Bourne (n.d.); its Chinese version is indicated as *Qiji biyi* 汽機必以 (Necessary instruments for the steam engine), translated by Fryer and Xu Jianyin.[75]

A reference to three other texts follows in the QJ preface; the first one is "Bourne's Introduction." In this case as well information of the original text comes from Fryer, *An account*, in which the text is listed as entry number 25, "Introduction to the Steam Engine."[76] The title of its Chinese translation is *Qiji biyi fujuan* 汽機必以附捲 (An appendix to the Necessary instruments for the steam engine). In reality, the entry, as indicated in the two columns with information pertaining, among others, to the author, translators, and publishing date, represents the translation of a single original volume, composed by Bourne, which is nonetheless listed as two individual entries in *An account*.[77]

---

69 Alexander Wylie and Xu Shou, trans., *Qiji faren* (Shanghai: Jiangnan zhizaoju), 1871.

70 Fryer, *An account*, 22–23.

71 The author has consulted Thomas John Main and Thomas Brown, *The Marine Steam-Engine: Designed Chiefly for the Use of the Officers of Her Majesty's Navy*, 4th ed. (London: Longman Brown, Green and Longmans, 1860).

72 As stated in Sun Lei and Lü Lingfeng, "Jiangnan zhizaoju zheng qiji yizhu diben kao," *Wakumon*, no. 20 (2011): 38–39, since the *Qiji faren* had already been translated into Chinese by 1868, translators should have used one of the previous editions of *The Marine Steam-Engine* as a basis; the text was published for the first time in 1849 and later reprinted in 1852, 1855, 1860, 1865, and 1879. On the basis of a comparison of the contents of the original text and of its translation, Sun and Lü suppose it may be the third edition, published in 1855 in London by Brown, Green, and Longmans.

73 Fryer, *An account*, 22–23.

74 Dagenais, *The John Fryer Papers*, II, 374, and Sun and Lü, "Jiangnan zhizaoju," 41–43, affirm it is the eleventh edition, printed in London in 1865.

75 Fryer John and Xu Jianyin, trans., *Qiji biyi* (Shanghai: Jiangnan zhizaoju, 1872).

76 Fryer, *An account*, 22–23.

77 As pointed out in Sun and Lü, "Jiangnan zhizaoju," 41–43, the edition of the text by Bourne printed in 1865 is composed of two parts: the first is the "Introduction to catechism of the steam engine," published as a monograph in John Bourne, *Recent Improvements in*

The following work Fryer annotated as a reference for the compilation of the *QJ* is "Burgh's practical Rules." In this case as well, in *An account* information on the original volume and its translation can be located.[78] The Chinese rendering is listed as *Qiji xinzhi* 汽機新製 (New patterns in the steam engine),[79] printed in 1873 in two volumes, while the starting text is described as "Practical rules for Steam Engine"; the latter can be identified with *Pocket-book of Practical Rules for the Proportions of Modern Engines and Boilers for Land and Marine Purposes*, by the British engineer Nicholas Procter Burgh (1834–1900).[80]

The last work mentioned in the preface to the *QJ* is "Sennett's Manual of Marine Engineering." As Fryer explained, the rendering of the text had only been solicited at the time of the writing of the preface, and perhaps had not started yet; the translation will be printed in 1894, under the title *Bingchuan qiji* 兵船汽機 (The warship steam engine).[81] The work referred to in the preface is *The Marine Steam Engine*, composed by the naval engineer officer Richard Sennett (1847–1891).[82] It is exactly for the translation of the latter work that Fryer started the compilation of the *QJ*; in the *QJ* all terms established as of 1889 are included. All the texts mentioned, other than contributing to delineating the focus of the manuals chosen at the Jiangnan Arsenal, are an important element for sketching the transcultural flows of knowledge in which translations, terminology, and notions all played a fundamental role.

From Fryer's words, it is possible to affirm with certainty that the glossary was updated at approximately the date of its preface, signed in October 1889. Nevertheless, the *QJ* was completed at an earlier date, some months having passed before its printing. Later on in the preface, though, the allusion to a possible appendix comprising terms that Fryer would find necessary to add

---

the Steam Engine (Philadelphia, J. B. Lippincott & Co.; London: Longmans & Co., 1868); the second corresponds to the "Catechism of the steam engine." It is therefore possible to conclude that the first part of *A Catechism of the Steam Engine* had been rendered in the *Qiji biyi fujuan*, the second in the *Qiji biyi*.

78 Fryer, *An account*, 22–23.

79 John Fryer and Xu Jianyin, trans., *Qiji xinzhi* (Shanghai: Jiangnan zhizaoju, 1872).

80 In Sun and Lü, "Jiangnan zhizaoju," 43–45, there is a comparison of the editions published in 1864 and 1868; the authors identified the former, published in London by E. and F. N. Spon, as the one employed by Fryer and Xu Jianyin for their translation.

81 John Fryer and Hua Beiyu, trans., *Bingchuan qiji* (Shanghai: Jiangnan zhizaoju, 1894).

82 The author has consulted Richard Sennett, *The Marine Steam Engine: A Treatise for the Use of Engineering Students and Officers of the Royal Navy*, 2nd ed. (London: Longmans, Green, and Co., 1885). Sennett's text is not included in Fryer, *An account*, since the first edition of *The Marine Steam Engine* was printed in 1882. Nevertheless, among books included in Dagenais, *The John Fryer Papers*, II, 650, as the sources used by Fryer for his translations, it is the second edition of Sennett's text that is listed. Another confirmation of this hypothesis can be drawn from Sun and Lü, "Jiangnan zhizaoju," 45–46.

to the *QJ*, is unclear. Even if he planned it, this appendix has not been located, unless one supposes the CT represents it.

As Fryer himself was plainly aware of, numerous terms in the field of the steam engine had been coined in a much earlier phase compared with the recent development in this field. Since such terms were already an integral part of the Chinese scientific and technical lexicon, they have been kept by Fryer to avoid confusion, even if they were probably unsuitable for the glossary. It should be noted that the purpose of the glossaries included in the *TVM* was the uniformity of such a lexicon, even more than its accuracy; to sacrifice a term already extensively employed, due to its inaccuracy, to the detriment of standardisation, was not conceivable for Fryer. The aim of his work was indeed that of conveying and transmitting notions, helping others to also accomplish this task. In this respect, therefore, the stress was on terminological uniformity, rather than on purely linguistic considerations.

In the preface, Fryer then highlights one of the *QJ*'s major features: with the exception of anthroponyms and toponyms, in the glossary no attempt has been made to "phoneticise" the original terms, as was attempted with the other three glossaries. It is thus possible to appreciate the ongoing evolution of Fryer's ideas on translation, from the first glossary published in 1883 to the one presented in this section, printed in 1889. While the presence of phonemic loans in the *JS*, the *HXCL*, the *XYDCYP*, and even more so in the *RMDM* is rather substantial, in the case of the *QJ* this solution represents an extremely low percentage in the total of Chinese translations provided. The tendency to substitute phonemic loans as much as possible, seen in the passage from the manuscripts to the printed glossaries—a constant in Fryer's work—is most significant in the *QJ*.

In the conclusive part of the *QJ*'s preface, Fryer finally affirms that he resolved to compile a "Vocabulary of Terms in Naval Architecture," whose publication should have been subsequent to the *QJ*. The text will likely never be printed, but I identified its manuscript, the CT, among the volumes stored in the EAL.

In the body text of the *QJ*, the glossary is composed of fifty-eight pages, for a total of 5,061 entries.[83] The *QJ* manuscript is included in volume nine of the EAL, and the total number of pages is 158, followed by twelve blank pages and another glossary, the CT.

83 The number refers to the original terms in English—for many of them, more than one Chinese translation is provided. In this glossary as well, though, incongruities and errors, however rare, are present in alphabetical order.

Analogously to the *XYDCYP* manuscript, entries in the *QJ* manuscript are written on strips of paper glued onto right-hand pages, usually in alphabetical order, with the original term directly followed by its Chinese rendering. The layout of the manuscript is similar to the printed version: the text is composed of two columns, to be read from top to bottom and from left to right. Nevertheless, the legend corresponding to the colours blue, sky blue, wine, red, and green, is not included. Through a comparison between the entries of the manuscript, texts stated in the preface to the *QJ*, sectoral works, and other scientific and technical dictionaries, I was able to reconstruct the sources almost entirely. The strips in blue are drawn from *From Keel to Truck*, composed by the German captain Heinrich Paasch (1835–1904);[84] entries in sky blue come from the *VHCL* by Doolittle; terms in green are obtained from *Pocket-book of Practical Rules for the Proportions of Modern Engines and Boilers for Land and Marine Purposes*, by Burgh;[85] and finally, those in red are taken from *A Catechism of the Steam Engine*, by Bourne.[86] In order to better highlight the interactions with other contemporary lexicographical materials, and therefore, indirectly, of their respective compilers, hereafter the second of the sources Fryer employed mentioned above, Doolittle's *VHCL*, is briefly analysed, due to the similarity of its contents with the glossary presented here. As already mentioned, in the preface to the third part of his dictionary, Doolittle states that texts by other authors have been used in his work. From pages 175 to 178 of the *VHCL*, "Terms used in Mechanics, with special reference to the Steam Engine. By Mr. A. Wylie" are listed. Conducting a comparison between some of the *QJ* entries and the relevant terms in the *VHCL*, it is possible to perceive their almost total equivalence:

---

84 Heinrich Paasch, *From Keel to Truck—A Marine Dictionary in English, French and German Amply Illustrated by Explanatory Diagrams of the Most Important Details—For the Use of Ship-Owners, Builders, Brokers, Insurance-Societies, Average-Staters, Barristers, Solicitors, Translators, Surveyors, Masters, Mates, Engineers, Naval-schools, Nautical Journalists, etc.* (Antwerp: Ratinckx Frères, 1885). The book was very influential, as shown by countless later editions published even nowadays. See also Marc Van Campenhoudt, "Un apport du monde maritime à la terminologie notionnelle multilingue: étude du dictionnaire du capitaine Heinrich Paasch *De la quille à la pomme de mât* (1885–1901)," two volumes (PhD diss., Paris XIII, 1994).

85 The edition the author of this work had access to, namely Nicholas Procter Burgh, *Practical Rules for the Proportions of Modern Engines and Boilers for Land and Marine Purposes* (Philadelphia: Henry Carey Baird, 1865), was published one year later than the one identified in Sun and Lü, "Jiangnan zhizaoju."

86 John Bourne, *A Catechism of the Steam Engine in its Various Applications to Mines, Mills, Steam Navigation, Railways, and Agriculture* (London: Longman, Green, Longman, Roberts, & Green, 1865).

TABLE 6 *QJ* and *VHCL*

Term	*QJ*	*VHCL*
Accelerating force	漸加速力	漸加力
Acceleration	漸增速	漸加速
Acting, perpendicularly	直加力	直加
Air	風，或氣	風, 氣
Air gun	風鎗	風鎗
Air pump	恒升車，或抽氣筩	恒升車
Air pump cross head	恒升車橫擔	恒升橫擔
Air pump side rod	恒升車邊搖桿	恒升搖桿
Condenser	凝水櫃	凝水櫃
Double eccentric[a]	雙兩心輪	雙兩心輪
Friction wheel	磨阻力輪, 或減阻力之輪	減阻力之輪
Inertia	永靜性, 或不肯動之性	不肯動之性
Extrados	甃行之上面	甃行上面
Head of water	水得速當過路	水得速當過路
Non-condensing steam engine[b]	不凝水汽機	不凝水汔機

a Listed as "Double excentric" in the *VHCL*.
b Listed as "Non-condensing steam engine" in the *VHCL*.

The extreme similarity, at times equivalence, between the translations, and in most cases of the original terms as well, confirms the *VHCL* was a source for the *QJ*. Nevertheless, as inferred from table 6, Fryer also made revisions and corrections to Doolittle's text first by adding numerous terms drawn from other sources and integrating them with those Wylie provided through the mediation of the *VHCL*. This is clear when comparing the limited amount of terms in the latter with the much more conspicuous quantity in the *QJ*. Even more important is that Fryer modified Chinese translations whenever he considered it necessary. For example, the entry "Accelerating force" is translated as *jian jiasu li* 漸加速力 on page 1 of the *QJ*, while among terms supplied by Wylie in the *VHCL*, it is listed as *jian jiali* 漸加力. Another instance is the entry "Air pump side rod." Fryer believed it indispensable to modify Wylie's rendering, *heng sheng yaogan* 恒升搖桿, to *heng sheng che bian yaogan* 恒升車邊搖桿, thus reinstating the use of the character *che* 車, also keeping uniformity with the translation of "Air pump," and adding the character *bian* 邊 for a more specific rendering of "side." In other cases, Fryer provided alternative solutions to Wylie's renderings, as with "Air pump."

Analysing the *QJ* manuscript once more, there are cases in which entries are written directly on the pages, as with "Admiralty rules," translated as *Yingguo zhanchuanbu zhangcheng* 英國戰船部章程, or "Air pump delivery valve guard," *heng sheng che jinshui men dang* 恒升車進水門擋; both were later included in the printed version of the glossary. In other circumstances, entries are added on the left-hand pages of the glossary, which are normally blank. In table 7, the entries added on the left-hand page facing the first of the manuscript, with their relevant final renderings, are listed as an example.

TABLE 7 *QJ* manuscript and its printed version

Original term	Manuscript	*QJ*
Adze[a]	橫斧頭	
Air lock	空氣閘	
一 Absolute pressure	全壓力	全壓力
二 Accelerated motion	漸加動	漸加動速
三 Accumulation of pressure	聚壓力	聚壓力
四 Action of propeller	螺輪之力	螺輪施力
五 Actual pressure	實壓力	實壓力
六 Adequate strength	足用之力	足用之力
七 Adjusting	配准	配准
八 Admission port of slide valve	汽罨進汽孔	汽門進汽孔
九 After-crank shaft	後曲拐	後曲拐
After water ballast tank	後水壓儎箱	後水壓儎箱
十 Ahead motion	前行動	前行動
十一 Air casing of funnel	烟通空氣壳	路通空氣壳
Air cook	空氣塞門	空氣塞門
十二 Air extractor	抽空氣門	抽空氣機
十三 Air pipe	空氣管	空氣管
Air pressure	空氣壓力	空氣壓力
Air bucket		
Air barrel		

a The first two entries in the table are not preceded by numbers and are separated from the others; they will not be included in the *QJ*. All subsequent entries until "Air Pressure" are written with a purple pencil for the original term and black ink for the Chinese translations, all preceded by numbers in ascending order expressed in characters and written with a thick stroke. The two groups of entries "Air casing of funnel" with "Air cook" and "Air pipe" with "Air pressure" are included in the same number two by two. The numbers mentioned were then reported, written with the same thick stroke, on the facing page, beside or above the strips of paper, in order to indicate the position where the terms should have been inserted at the time of printing. Finally, the last two entries are written with a purple pencil and have been cancelled; they will not be included in the *QJ*.

The almost total coincidence of the *QJ* manuscript and its printed version can be deduced from the table above and those in the appendices. The principal differences lie in the eliminations of entries originally annotated in the manuscript, with or without translation; discrepancies in the orthography of original terms; errors in the sequence of Chinese characters; different Chinese renderings for the same original entry, later merged in a single translation; and additions after the first draft.

Nevertheless, in the passage from the manuscript to the *QJ* a few constants also exist, even if to a lesser extent than with other glossaries.[87] The first one is the reduction in the number of characters in the Chinese translations. This choice is performed, for example, in the elimination of the structural particle *zhi* 之 whenever deemed unnecessary. For example, "Injection valve spindle" is rendered in the manuscript as *pen lengshui men zhi gan* 噴冷水門之桿; on page 28 of the *QJ*, the character *zhi* will be eliminated, as Fryer believed it unnecessary, making *pen lengshui men'gan* 噴冷水門桿 the final translation.[88]

In rarer circumstances, though, Fryer conducted the opposite procedure, if considered necessary, by adding elements useful for a more precise rendition of the original term. For example, the entry "Insertion cloth joint" was originally translated in the manuscript as *sai bu cheng jie* 塞布成節, later modified as *bu qian cheng zhi jie* 布嵌成之節. On page 28 of the *QJ* the Chinese solution will finally be *bu qian cheng zhi jie* 布嵌成之節, as Fryer considered the addition of the structural particle *zhi* instrumental for better pointing out the value of *jie* as the head of the syntagm.

The second tendency is the revision of the translations, or parts of them, that proved to be scarcely scientific. For example, the entry "Inside cut off slide valve" is rendered as *nei duan qi zhi qi yan* 內斷汽之汽罨 in the manuscript. The concept of "valve" was conveyed through the character *yan*, indicating nets for catching birds and fish—scarcely adequate from a semantic point of view. The final translation on page 28 of the *QJ* will in turn be *nei duan qi zhi qi men* 內斷汽之汽門, with *yan* substituted by *men*; in contemporary Chinese, the latter still bears among its meanings that of "valve."

Finally, the third constant is the addition of supplementary solutions. However, in the *QJ*, contrary to the three glossaries presented above, phonemic loans being few, the renderings provided will be alternative to calques or descriptive labelling already given. For example, "Indicator, steam" is firstly

87 Graphical discrepancies in the characters, such as the tendency to employ simplified ones, for example *ji* 机 and *lu* 炉 rather than *ji* 機 and *lu* 爐, will not be considered.

88 On the topic, refer to Arcodia, *Lexical Derivation*, 131 and passim.

translated in the manuscript as *zhi li qi* 指力器. The first revision modified the Chinese solution into *zhi qili qi* 指汽力器, but on page 28 of the *QJ* another can be read, *ziji qi zhangli biao* 自記汽漲力表, a descriptive labelling as well.

6 *Vocabulary of Proper Names*

Finally, the last glossary gathered in the *TVM* is the *RMDM*. As indicated by the full title of the *XYDCYP*, the *RMDM* had been originally planned as an appendix to the *XYDCYP* as a list of anthroponyms and toponyms listed in *A Manual of Materia Medica and Therapeutics*, and other texts pertaining to chemistry and relevant branches.

An important element for better sketching the transcultural flows of knowledge considered in this work, as well as the policy of selected works translated at the Jiangnan Arsenal, is that, according to preliminary research, it is plausible that the original terms in the *RMDM* are drawn from *Encyclopædia Britannica*. A mention of the availability of the eighth edition of the text among the members of the Jiangnan Arsenal can be found: "The original idea was [...] to prepare an encyclopædia that should bear some resemblance to the 'Encyclopædia Britannica.' It was soon found, however, that many of the treatises in the eighth edition of that valuable work were too elementary and too far behind the times."[89] The eighth and ninth edition of the *Encyclopædia Britannica* represented indeed one of the sources for Fryer's translations.[90] This hypothesis is substantiated by a letter signed June 3, 1869, sent by Fryer to "Henry S. King & co": within a list of books, instruments, and other materials ordered for the Arsenal, the entry "The Encyclopedia Britannica. Complete to the present date. Latest edition, with all the Supplements. Halfbound in Calf" is included.[91]

The full title of the glossary, as can be read on its first page, is *Vocabulary of proper names occurring in Royle and Headland's* Materia Medica and Therapeutics, *and in various works on medicine and chemistry*. The glossary is composed of thirteen pages and 1,872 entries for the first part, namely "Names

89 Fryer, *An account*, 11; nevertheless, further analysis is necessary to identify the edition Fryer employed.

90 As indicated in Dagenais, *The John Fryer Papers*, II, 645–52.

91 Dagenais, *The John Fryer Papers*, I, 395.

of persons." The second one, "Names of places,"[92] includes nine pages and 1,195 entries, for a total of twenty-two pages and 3,067 entries in the *RMDM*.[93]

The glossary's manuscript is stored in the EAL, corresponding to volume number four. The original terms with their translations are annotated on strips of paper glued onto the pages. The strips of paper present different colours: mustard, dark sky blue, light sky blue, purple, olive green, and wine. Strips without any colour are also present; in any case, the relevant legend is not reported. Moreover, a group of strips do not present colour; their paper is more yellowed than the others, presumably since they were glued in a distinct phase of working.[94]

Revisions in the manuscript are few. Apart from some oversights on the sequence of characters or in the alphabetical order of the entries, some names cancelled with a line are present at times. This line is usually traced with the same ink of the entries, at times with a sky blue pencil, in turn sometimes retraced with black ink.[95] It can be concluded that this manuscript, analogously to the *XYDCYP* manuscript, represents the proof copy before the printing of the *RMDM*; graphical signs indicating the change of page can actually be read in this volume as well. For example, on page 11,[96] level with the entry "Homburg," written without the legend and translated as *Hanbaige* 汗白格, later corrected

92 Lemmata are for the most part represented by foreign toponyms; for a glossary devoted mainly to Chinese ones, refer to George Macdonald Home Playfair, *The Cities and Towns of China: A Geographical Dictionary* (Hong Kong: Noronha & Co., 1879).

93 For this glossary as well, the reference is to the original terms; for most, only a Chinese translation is provided. Exceptions are represented by "Crichton," on page 3 of the "Names of persons," and "Sonsonate coast," on page 8 of the "Names of places," for whom a rendering is not supplied.

94 Some of the strips of paper have the same colour as the legend, or at times another one, also after the Chinese translation, as if the original term and its rendering originated from various sources.

95 The strips of paper cancelled only with the blue pencil were later kept in the *RMDM*, contrary to those deleted with black ink or with the blue pencil retraced with ink; most of the time the latter were one or more variants or reproductions of an entry. It therefore can be concluded that, in the majority of circumstances, the entries eliminated from the manuscript were merely repetitions of the immediately preceding one, which was written and translated in the same way but with a different colour for the legend, and thus a different source.

96 The numeration of the glossary pages is not linear. After the first unnumbered page, the count continues until page 16, where a new letter, D, beings; numeration therefore starts again from 1 until page 6 and from there on, with the letter F, recommencing from 1, reaching 43. In the successive page, with the letter T, count restarts from 1, arriving at 9, with the last words corresponding to Z. From the successive page, in the first of the "Names of Places," numeration proceeds from 1 to 17, in turn followed by a blank page; count then restarts from 19 to arrive at 56. The total number of pages is therefore 129.

into *Hongbaige* 烘白格, a sign written with red marker can be found.[97] The entry is precisely the last of page 6 of the *RMDM*, thus confirming that in this manuscript as well Fryer provided information for people in charge of publication. Another instance of corrections and annotations can be read on page 32: the entry "Lussac," marked with purple and rendered as *Luosage* 羅薩格, is put side by side with two characters, *renming* "人名" (anthroponym). The term will be listed on page 8 of the "Names of persons" in the *RMDM*. At any rate, corrections in the passage from the manuscript to the printed version are extremely rare.

To convey an idea of the abundance of sources Fryer gathered, in the table below translations listed in the manuscript of the entries "Italian" and "Italy," rendered on page 18 of the *RMDM* as *Yidaliguo zhe* 意大里國者 and *Yidaliguo* 意大里國, are shown. The table provides further confirmation of the fact that Fryer was not limited to coining terms or adopting translations that were at his disposal, without conducting a comparison. On the contrary, he carried out extremely detailed research, deciding in most cases only when certain of the suitability of the choice employed, often with reconsiderations immediately preceding publication. It should indeed be reminded that the final purpose

TABLE 8 Entries "Italian" and "Italy" in the *RMDM* manuscript

Term and colour[a]	Manuscript
Italian	意大里者
Italian (purple)	意大里國者
Italy (wine)	意大里
Italy	意大里国
Italy (dark sky blue)	意大里国
Italy (wine)	以大利
Italy (mustard)	以大利亞
	以大利

a The first, third, fourth, sixth, and seventh entries were already deleted in the manuscript, contrary to the second and the fifth. This indicates that, at times, reconsiderations of the translations occurred in the phase immediately preceding the printing.

97 This sign was originally annotated with another red marker, level with the entry "Homberg." The term, marked with red, had been rendered as *Hangboke* 杭拨克; the translation was finally deleted, substituted by the printed version *Hanbaige* 罕白格.

of his work was the uniformity of lexicon for the sake of scientific divulgation; only by accomplishing the former, the latter could also proceed swiftly.

7 Conclusion

Even though the *TVM* was printed after the glossaries were included within the text, its structure is coherent, both from a methodological and linguistic perspective, and covers many branches of science. The publication of the text was a milestone in the translation of scientific and technical terms in late Qing China, its influence also extending further on in time. The first collection of scientific and technical English-Chinese bilingual glossaries was finally published and could contribute to the standardisation of such nomenclature.

Analogously to what happens at times with Chinese contemporary terms expressing new concepts, for which at different stages of development there can be several variants, in late Qing China numerous scientific and technical terms were translated in various ways. Even though manuscripts circulated before the publication of the first edition of the *TVM*, the role of a printed edition of the text could surely be more influential among translators, and in the task of conveying notions. The linguistic adaptability of some of the terms adopted in the *TVM*, presented in such a form, could better overcome personal contrasts between late Qing scholars and missionaries, contributing to scientific progress through the Chinese language. Despite its success, a second volume of the *TVM* was never published, due to bigger historical circumstances and personal considerations by Fryer. Even so, as was the case with the handwritten versions of the glossaries analysed above, the other unpublished manuscripts also played a fundamental role in the future of the Chinese scientific and technical lexicon, both from a lexicological perspective, in the circulation of ideas conveyed, and in the creation of a community of scholars and missionaries concerned. By analysing these manuscripts, it is possible to get a better outlook on the translation process in late Qing China, the way ideas and concepts were transmitted from original texts into the Chinese language among Chinese scholars, how missionaries and scholars exchanged terms and ideas in direct or indirect ways, and how lexical interactions would become even more complicated with the consultation of Japanese sources.

CHAPTER 5

Unpublished Glossaries

1 Introduction

Every translation process requires different steps. With scientific and technical texts, the accumulation of an extended and standardised nomenclature, rooted in the linguistic knowledge of the translator, is all the more necessary. It also needs the agreement of different committees, particularly in the intense translation of texts from another language, as was the case in late Qing China. Considering personal and institutional animosities and the diversity of objectives of foreign scholars and missionaries in China at that time, it is not surprising that numerous volumes were destined to go unpublished.

Even though the manuscripts of other glossaries, for the most part presumably destined for the second volume of the *TVM*, were not printed, a careful investigation of their contents and linguistic peculiarities contributes firstly to the description of lexicographical interactions at work in late Qing China, which involved earlier sources, such as Jesuit ones, and later on Japanese texts. Although the role of these manuscripts in spreading ideas and notions was less significant compared to printed documents, their inclusion in this work is essential for sketching a well-rounded outlook on the transcultural flows of knowledge that took place in late Qing China and that involved Western and East Asian sources of different epochs and origins. Secondly, the manuscripts cast a brighter light on the linguistic theories Fryer endorsed and the working methodology adopted by him and his Chinese colleagues. Finally, by stressing some lexicographical and compositional tendencies evident within the manuscripts, the analysis contributes to a better description of their interactions with other sources compiled by other missionaries and scholars.

2 "Vocabulary of Terms in Naval Architecture"

The fifth glossary considered in the previous chapter is the *QJ*. Even though it was published in 1889, one year after the first edition of the *TVM*, it can be considered a fully-fledged integral portion of it. Its compilation is strictly connected to another manuscript glossary, presented here, that contributes to the description of the cultural and linguistic interactions analysed in this work. In the preface to the *QJ*, Fryer affirmed: "In some cases it is difficult to determine

 | DOI:10.1163/9789004443211_007

whether a term belongs to Marine Engineering or to Naval Architecture. There are necessarily words that are common to the two subjects, and will therefore be found also in the Vocabulary of terms in Naval Architecture which is to follow." The reference is to a future publication that has never been identified by scholars.[1]

Within the first part of volume nine of the manuscripts housed in the EAL, the handwritten version of the *QJ* is stored.[2] This section is formed of seventy-nine pages—158 if the left-hand pages, where missing items were later inserted, are considered. The section is then followed by twelve blank pages. Next comes the "Vocabulary of terms in Naval Architecture which is to follow" that Fryer referred to in the *QJ* preface. The Chinese title can be inferred from the Chinese preface, *Chuanti mingmu biao* 船體名目表.

The CT is much better preserved than the first part of the volume, and is composed of eighty-nine pages, only considering right-hand pages; it would be 178 if left-hand pages, on which only scattered definitions are included, are counted.[3] The first line of the CT reads "English Index"; the whole glossary, analogously to the manuscript of the *QJ*, is almost entirely written on right-hand pages. However, in the CT original lemmata are written directly on the volume, without strips or colour key. The original lemmata, the great majority of them in English, and their Chinese translations are divided into two columns, arranged in alphabetical order, to be read from top to bottom and left to right. The English lemmata precede the Chinese translations; both are, in most cases, written in black ink.[4]

The overlaps between the *QJ* and the CT are sporadic. Even in those few cases where the English lemma is the same in both, the Chinese translation is almost always different. I have identified the source texts from which original lemmata within the CT are drawn, for the vast majority represented by English terms, as *From Keel to Truck*, composed by Heinrich Paasch. Among the appendix of the book, there is an "English index," corresponding almost entirely to

1 A detailed presentation of the glossary can be read in Tola, "John Fryer." The only mention before this article can be found in Dagenais, *The John Fryer Papers*, II, 497, in a reproduction of a page of the preparatory *QJ* manuscript. It is true that Fryer also referred to the vocabulary in Fryer, "The Present Outlook," 157–58; another reference can be found in "Educational Notes," 33.

2 In the CT pages are unnumbered; page numbers have been added for the reader's convenience.

3 After the last page of the CT, there are 153 blank pages, where only a few entries and abbreviations at the top of the pages are inserted.

4 It should also be noted that the calligraphy of entries listed in the CT, particularly of English lemmata, is slightly different from that of other EAL manuscripts.

the original lemmata listed in the CT. The only general exceptions are some of the entries that were later added by Fryer in the manuscript.

From the point of view of translation choices, Fryer constantly tried to avoid the use of phonemic loans as much as possible. In the *QJ* and the CT this principle achieved its best enactment, as phonemic loans in the two glossaries are extremely infrequent, also in line with the statement in the preface to the *QJ*. From the perspective of compositional features of the CT, some are difficult to interpret. The first one is the presence of a thick red dash before numerous original lemmata—others are instead preceded by numbers written with a pencil. By analogy with the other manuscripts stored in the EAL, two hypotheses can be proposed. The most plausible is that numbers represent information in the order of the lemmata in the glossary or indications for the printing layout. Less plausible is that numbers refer to sources from which the lemmata were drawn, or overlaps with analogous lemmata in the *QJ*. Moreover, in the CT, when entries starting with the same initial letter end, those with a different initial letter continue on the same page; the separation is represented by a larger letter written in more refined calligraphy above the first entry. This is the difference with the majority of other manuscripts Fryer compiled housed in the EAL, in which a new page is used when a new initial letter begins. At the same time, this peculiarity of the CT derives from the "English index" of *From Keel to Truck*, in which the layout is precisely the same.

The CT is neatly written and there are not many insertions or corrections. This may imply that the glossary was meant for immediate publication, since missing Chinese translations are few. In some entries, the Chinese translation is simply not given. In other lemmata, on the other hand, the original term has been written with ink traced over a preceding draft in pencil, and there are also cases where insertions between two existing terms are made in purple pencil or, more commonly, in black ink. This also suggests that the CT, analogously to other manuscripts already presented, underwent later phases of correction and revision.

Fryer was not limited to just employing the "English index" of *From Keel to Truck*. As an example, on page 80 of the CT, he indicates a whole group of entries to be added after "- Timber," *leigu* 肋骨, *muliao* 木料 and before "- Timber and space," *leigu ji kongxi zhi kuan* 肋骨及空隙之寬, which were actually listed one after the other in the "English index."[5] This group includes: "- Timber 2nd futtock," *leigu zhi di er jie* 肋骨之第二節, "- Timber 4th futtock," *leigu zhi di si jie* 肋骨之第四節, and "- Timber 6th futtock," *leigu zhi di liu jie* 肋骨之第六節. The last three lemmata are listed on page 18 in the body text of *From Keel to Truck*,

5 A dash before the English definition represents a thick red dash added on the manuscript.

as "Second futtock," "Fourth futtock," and "Sixth futtock." Fryer looked at the original work in detail to draw up a glossary as correct and complete as possible, integrating entries from the index and others from the body text, analogously to what he accomplished for other glossaries published in the *TVM*. Furthermore, the fact that some Chinese translations are not provided implies that Fryer was unsure about which renderings to list; he needed more time to ponder them, in order to avoid giving inadequate solutions.

The incompleteness of the CT and the resulting unsuccessful publication is not surprising but conclusions as to why cannot be drawn. The possible reasons are indeed multiple, including the appointment of Agassiz Professor waiting for Fryer in Berkeley and his consequent departure from China. Changes in the publication plans of the Educational Association of China and financial cuts in the ship industry itself, though, must be considered the most important factors.[6]

The last compositional peculiarity of the CT considered here is the presence in the whole volume, in which both the manuscript of the *QJ* and the CT are included, of two abbreviations at the top of each page, composed of three, or more rarely, four letters. These abbreviations, though, do not correspond with the entries or the alphabetical order of either glossary. It may be inferred that the volume had been initially meant for another purpose and was later destined for the CT and for the *QJ* manuscript. This hypothesis is confirmed by some lemmata, not strictly pertaining to shipbuilding or naval architecture, added before the volume was adopted for the *QJ* and the CT. For example, in the last section of the volume, it is possible to read three lemmata; the most interesting is "Sulphuric ammonide," under the column "Sul," rendered as *danqing$_{san}$ liuyang$_{san}$* 淡輕$_{三}$硫養$_{三}$. On page 3 of the *HXCL*, the term is instead translated as *er danqing$_{san}$ liuyang$_{san}$* 二淡輕$_{三}$硫養$_{三}$, a solution correspondent to the chemical formula of the compound, $(NH_3)_2So_3$, as indicated in some contemporary chemistry manuals.[7]

3 "List of Botanical Terms"

The second unpublished glossary stored at the EAL is the LBT—it is one of the most useful for identifying Fryer's intellectual itinerary in the translation process of scientific and technical nomenclature, which led to the compilation

6 Refer to Elman, *A Cultural History*, 184.

7 For example, in Charles Loudon Bloxam, *Chemistry, Inorganic and Organic: With Experiments* (Philadelphia: Henry C. Lea, 1873), 302 and 665. On the other hand, in Bloxam, *Chemistry Inorganic and Organic*, 269 and 639, the formula is annotated as $NH_3.So_3$, thus corresponding to the solution listed in the CT.

of the *TVM*. At the same time, it is a key instrument for the reconstruction of late Qing China transcultural flows of knowledge and the lexicological interactions with other contemporary sources presented in this work. The glossary, included in volume eight of the manuscripts housed in the EAL, is indeed the only one in which the sources Fryer employed are explicitly annotated. Nevertheless, while some can be easily identified, for others cross-reference research is necessary.

It is important to first highlight that the LBT does not present any close connection with the *XYDCYP* or the relevant subjects of other glossaries included in the *TVM*. Apart from discrepancies in contents, the dating of the volume, 1892, comes four years after the first edition of the *TVM*. Therefore, the first question on the LBT pertains to its purpose, in which two hypotheses can be advanced. The first and more plausible one is that Fryer, five years after the publication of the *XYDCYP*, the closest glossary to the LBT from the point of view of contents, became aware of the incompleteness of the *XYDCYP* itself. He therefore considered another publication necessary, according also to what he personally affirmed in numerous circumstances on the standardisation of Chinese scientific and technical nomenclature. The second and less probable hypothesis is that Fryer planned to use the LBT purely as a personal instrument of consultation; supposing this, though, the specific reference to the sources used would be unlikely indispensable. In any case, the volume was not destined to be printed.

The volume in which the LBT is included is composed of 172 pages. On the first page, the sources from which Fryer drew his translations are introduced as follows: "In this list of Botanical terms, the following colours represent the books from which the terms are taken." The sources are listed on strips of paper placed on the side with a colour key:

> White.—Stonehouse from Williams
> Red.— " "Shai & Sz chuen list."[8]
> Purple.—Japanese list.
> Purple and red.—Bretschneider's list.
> Ink.—"Bridgmans Chrestomathy."
> Ultramarine blue.—"Bowra (Doolittle)."
> Yellow ochre.—Porter Smiths' "Materia Medica."
> Green and red. "Monthly Flowers" (Doolittle).
> Sepia. "Custom's catalogue."
> Indigo green. A. Henry's list.

8 The symbol """ represents the corresponding term in the line above, namely "Stonehouse"; "Shai" is a shortened form indicating "Shanghai."

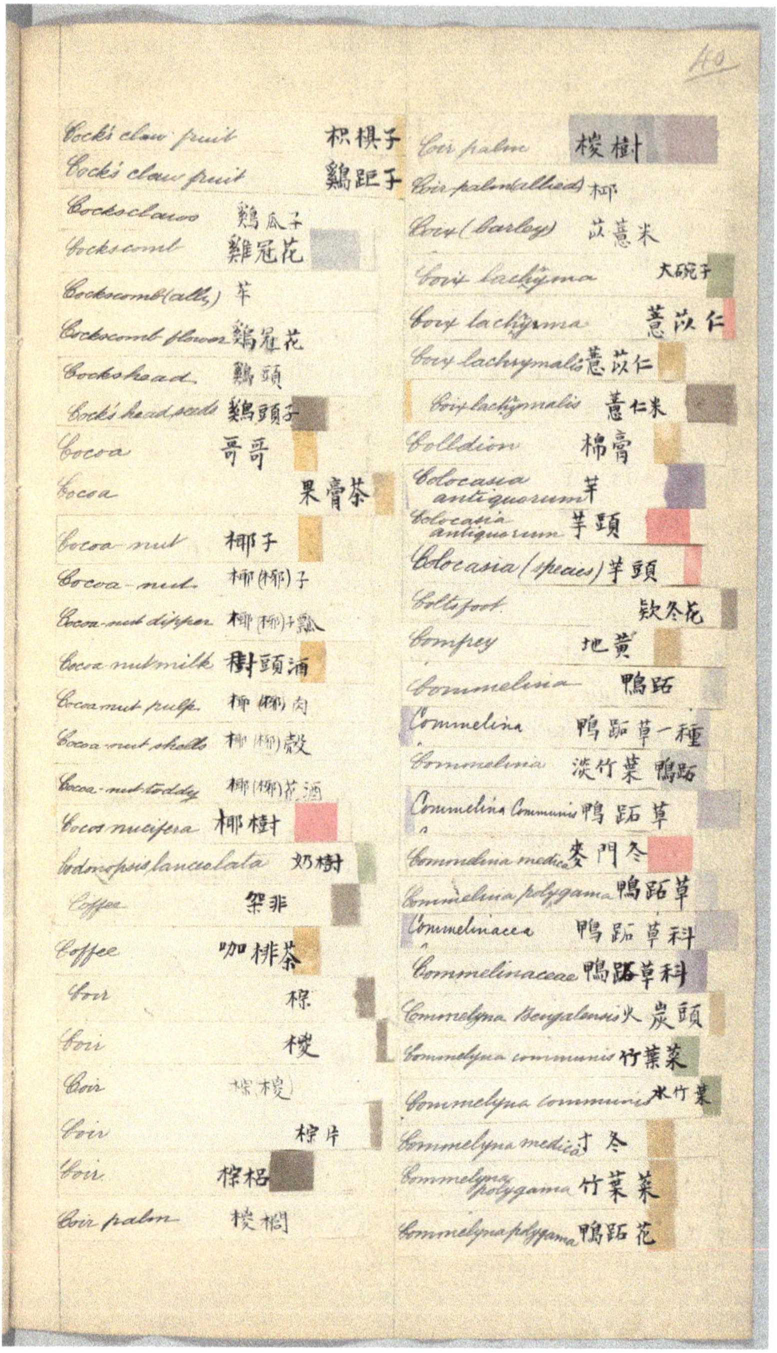

FIGURE 5 Page 40 of the "List of botanical terms"
COURTESY OF THE C. V. STARR EAST ASIAN LIBRARY, UNIVERSITY OF CALIFORNIA, BERKELEY

The first page of the LBT ends with the date and place of compilation of the volume: "Shanghai, China April 25th., 1892."

The first source is denoted by Fryer with white. The relevant entries are actually included in the volume without any colour and correspond to the first part of volume three of manuscripts stored in the EAL, to wit the BTWDCL. The latter can be identified with *A Syllabic Dictionary of the Chinese Language* by the American missionary Samuel Wells Williams (Wei Sanwei 衛三畏, 1812–1884).[9] As for "Stonehouse," the reference is to Joseph Stonehouse (1854–1901), a missionary of the London Missionary Society.[10] In all likelihood, Fryer hinted to a list of terms in Williams' work provided by Stonehouse.

When comparing some of the entries concerning botany in the *SDCL* with those indicated in the LBT, as drawn from "Stonehouse from Williams," it is possible to remove all doubt on the identification of the text:

TABLE 9 LBT and *SDCL*

Chinese term	LBT	*SDCL*
樓	Birch (a sort of), 21[a]	"[...] It is a sort of birch," 258
樺皮局	Birch bark shop, 21	"A birch bark shop," 241
苙	Carex (spec.), 30	"A species of sedge (*Carex*) [...]," 820
芙明子	Cassia tora seeds, 31	"Seeds of the *Cassia tora* [...],"447
棕樱筍	Cluster of palm buds, 39	"Clusters of the flower buds of palms [...],"1022

a Henceforth, numbers in all the tables listed in this work will indicate the page in which the annotated term appears in the relevant text, unless other annotations such as "vol." and "no," are reported. Different cases will be pointed out with specific footnotes.

9 Samuel Wells Williams, *A Syllabic Dictionary of the Chinese Language; Arranged According to the Wu-Fang Yuen Yin, With the Pronunciation of the Characters as Heard in Peking, Canton, Amoy and Shanghai* (Shanghai: American Presbyterian Mission Press, 1874).

10 Details on Stonehouse are few. In Robert Coventry Forsyth, *The China Martyrs of 1900: A Complete Roll of the Christian Heroes Martyred in China in 1900 with Narratives of Survivors* (New York, Chicago, Toronto: Fleming H. Revell Company, 1904), 478, some indications on his life can be found: "Mr. Stonehouse was a missionary of the London Missionary Society, stationed at Peking at the time of his death. He was born in Middlesborough in 1854, and was trained at Rotherham College. He was ordained at Saddleworth, Yorkshire, on July 27, 1882, and sailed three months later for his station at Shanghai, China. Two years later he married Miss Gertrude E. Randle of Huntingdon. In 1886 he was removed to Peking, and engaged in work in the East city there for the next six years, when he came to England on furlough, returning to Peking in 1894, where he remained till his death." Stonehouse was killed in March 1901 during the Boxer rebellion.

The last term from the table above provides irrefutable proof that the first source of the LBT is represented by the *SDCL*. In the entry, listed on page 39 of the manuscript, the two graphical variants of a single character, *zong* 棕 and *zong* 椶, come one after the other, precisely as they were listed one under the other in Williams' *A Syllabic Dictionary*. In the latter, nevertheless, in the entry translated as "Clusters of the flower buds of palms […]," they were in reality meant as an alternative to each other.

The entries in the LBT marked with red are in turn drawn from the second and third part of the aforesaid volume three of the EAL manuscripts, the BTWDCL. In the final section of the BTWDCL, two lists are included: the first one is "Vegetable kingdom ect [sic] of Shanghai," drawn from the *General Description of Shanghae and its Environs* by the English missionary Walter Henry Medhurst (Mai Dusi 麥都思, 1796–1857), published in 1850.[11] The second is the "Sz-chiuen list of Plants," in reality divided into two lists. They correspond to the section "Sz Ch'uan plants," from page 301 to page 308 of the text *Up the Yang-tse*, written by the English sinologist Edward Harper Parker (Zhuang Yanling 莊延齡, 1849–1926) and published by China Mail Office in 1891, even if it was originally reported in the *China Review*.[12]

In this case as well, through a comparison of LBT entries indicated with red and those of the two aforesaid lists included in the BTWDCL, the coincidence between them is patent; the only slight discrepancies lie in the order or in some slight corrections of the original Chinese terms.

TABLE 10 LBT and "Sz-chiuen list of Plants"

Chinese term	LBT	"Sz-chiuen list of Plants"
水莧菜	Ammannia rotundifolia, 7	"Ammannia rotundifolia, Roxb.," 304
雀雀草 (or 菜) (or 包) (or 台) or 牛屎花 or 牛石花 or 鴨屎條	Crepis japonica, Benth., 44	"Crepis japonica, Benth.," 304
麥豆草	Eclipta alba, 53	"Eclipta alba, Linn.," 304

11 Walter Henry Medhurst, *General Description of Shanghae and its Environs, Extracted from Native Authorities* (Shanghai: Mission Press, 1850).

12 Edward Harper Parker, *Up the Yang-tse* (Hong Kong: China Mail Office, 1891).

TABLE 11 LBT and *GDSE*

Chinese term	LBT	*GDSE*
北瓜	Cucumber (northern), 45	"[…] northern cucumber […],"69
地丁草	Earth nail plant, 53	"[…] earth-nail plant […]," 70
石薢	Dendrobium ceraia, 49[a]	"The dry land herbs and flowers, common in this district, are such as […] 石斛 shĭ hŏ, […]," 70

a In this entry, annotated in the LBT and also listed in Williams' *SDCL*, the following description can be found: "one of the dry land herbs a flos wh are com here." On page 233 of Williams' dictionary it can actually be read: "石薢 a tonic medicine (*Dendrobium ceraia*), growing on the rocks in southern China […]."

Leaving the third source aside for the moment, the fourth one, indicated by Fryer as "Bretschneider's list," corresponds to *Early European Researches into the Flora of China*[13] by the sinologist Emil Bretschneider (Bei Le 貝勒, 1833–1901),[14] published in 1881.[15] In turn, from page 135 to page 184 of the text, a list of terms comprised in *Flora Cochinchinensis*,[16] composed by the Portuguese Jesuit missionary João de Loureiro (1710–1791) and printed in 1790, is given.[17] *Flora Cochinchinensis* is introduced in *Early European Researches into the Flora of China* as follows:

> Chinese Plants Described in Loureiro's Flora Cochinchinensis—In what follows I shall give a list of all the plants Loureiro mentions for China and include also those species which he had gathered only in Cochinchina

13 Emil Bretschneider, *Early European Researches into the Flora of China* (Shanghai: American Presbyterian Mission Press, 1881).

14 For a description of his role as a botanist, refer also to Fa-ti Fan, *British Naturalists in Qing China: Science, Empire, and Cultural Encounter* (Cambridge, MA: Harvard University Press, 2004), 107–12 and passim.

15 The text appeared for the first time in the *Journal of the North China Branch of the Royal Asiatic Society*, New Series, no. 15, 1–194.

16 João De Loureiro, *Flora Cochinchinensis: sistens plantas in regno Cochinchina nascentes; quibus accedunt aliæ observatæ in Sinensi imperio, Africa Orientali, Indiæque locis variis; omnes dispositæ secundum systema sexuale Linnæanum* (Lisbon: Real Academia de Ciências, 1790).

17 More information on De Loureiro can be found in Laurent Gaide, "La médecine européenne en Annam, autrefois et de nos jours," *Bulletin des Amis du Vieux Hué*, no. 4 (1921): 195–96.

> but which subsequently have been observed also in China [...] I arrange the list [...] giving for each plant at first the Loureirian name, the habitat and the Chinese name as noticed by Loureiro, and quoting the page in the Flora cochin. [...] Many of Loureiro's plants have by the progress of science received other names and mistakes made by that author have been occasionally corrected. In my researches with respect to Loureirian plants I depend upon *De Candolle's Prodromus, Kunth's Enumeratio plantarum, Bentham and Hookers' Genera Plantarum, Bentham's Flora honkongensis*, the botanical papers published by *Maximowicz*, Dr. *Hance* and others. I shall also give the Chinese names of plants noticed by Loureiro, as far as I have been able to ascertain them.[18]

This passage confirms the temporal, conceptual, and lexicological line connecting Fryer's *TVM*, texts by Protestant authors, and texts by the Jesuits. The *TVM* indeed adopted the terminology of other contemporary lexicographical materials, in turn connecting on different levels with Jesuit and native sources. This incontrovertible lexicological connection, and consequently that of the ideas behind such texts, is further complicated when considering that, from Bretschneider's last words, we can infer that he not only used the text composed by de Loureiro as a reference, but other sources too.[19] The first is *Prodromus systematis naturalis regni vegetabilis* by the Swiss botanist Augustin Pyrame de Candolle (1778–1841), his son Alphonse Louis Pierre Pyrame de Candolle (1806–1893), and the German botanist Heinrich Wilhelm Buek (1796–1879).[20] The second is the *Enumeratio plantarum omnium hucusque cognitarum* by the German botanist Karl Sigismund Kunth (1788–1850),[21] followed by *Genera plantarum* by the English botanists George Bentham (1800–1884) and Joseph Dalton Hooker (1817–1911).[22] The fourth is another volume by Bentham, *Flora*

18 Bretschneider, *Early European Researches*, 135.

19 Consult for example those quoted in Idem, *Botanicon Sinicum: Notes on Chinese Botany from Native and Western Sources* (London, Trübner & Co., 1882), 32–50.

20 Augustin Pyrame de Candolle, Alphonse Louis Pierre Pyrame de Candolle, and Heinrich Wilhelm Buek, *Prodromus systematis naturalis regni vegetabilis, sive enumeratio contracta ordinum, generum, specierumque plantarum hucusque cognitarum, juxta methodi naturalis normas digesta* (Paris, and others: Treuttel et Würtz, and others, 1824–1873).

21 Karl Sigismund Kunth, *Enumeratio plantarum omnium hucusque cognitarum, secundum familias naturales disposita, adjectis characteribus, differentiis et synonymis* (Stuttgart and Tübingen: J.G. Cotta, 1833–1850).

22 Joseph Dalton Hooker and George Bentham, *Genera plantarum ad exemplaria imprimis in herbariis kewensibus servata definita* (London: A. Black, 1862–1883).

Honkongensis: A Description of Flowering Plants and Ferns.[23] Finally comes a reference to unspecified "papers" published by the Russian botanist Karl Ivanovič Maksimovič (1827–1891) and the British botanist Henry Fletcher Hance (1827–1886). Such indications are extremely important because they show the textual interactions in action in the EAL manuscripts and in the *TVM*. Not only were contemporary texts by other Protestant missionaries employed to draw these manuscripts, but those composed by earlier Jesuit missionaries, such as João de Loureiro, and other European sources indirectly as well. This is a typical example of the connections linking late Qing texts composed by Protestants and earlier works produced by Jesuits. By analysing this link, it is possible to better describe the interactions of the textual production produced by foreign translators and missionaries in China, their efforts for the scientific and technical nomenclature and, in a broader sense, their contribution to the *Xixue Dongjian*.

Going back to *Early European Researches into the Flora of China*, in table 12 some entries listed in manuscript eight, identified by the colours "purple and red," and those included in text by Bretschneider, are compared. Moreover, the phonetic representation of the pronunciation of characters provided for the three entries above within *Flora Cochinchinensis* and *Early European Researches into the Flora of China* is almost coincident, further confirming Bretschneider' direct adoption of De Loureiro's earlier text; the only differences are represented by the addition of some diacritics in *Flora Cochinchinensis*.[24]

TABLE 12 LBT and *Early European Researches into the Flora of China*

Chinese term	**LBT**	***Early European Researches into the Flora of China***
韮菜	Allium angulosum, 5	"*Allium angulosum* [...] Sin: *kieu tsai* (韮菜)," 178
落葵	Basella nigra, 18	"Basella nigra [...] Sin: *lo quei* (落葵)," 169
紅藍花	Carthamus tinctorius, 30	Carthamus tinctorius [...] Sin: *hum lan hoa* (紅藍花)," 160

23 George Bentham, *Flora Honkongensis: A Description of Flowering Plants and Ferns* (London: Lovell Reeve, 1861).

24 Respectively on pages 203, 183, and 481.

TABLE 13 LBT and *A Chinese Chrestomathy in the Canton Dialect*

Chinese term	LBT	*A Chinese Chrestomathy in the Canton Dialect*
紫蘭	Bletia hyacinthina, 21	"Bletia hyacinthina," 452
紫荊	Cercis siliquastrum, 34	"Cercis siliquastrum," 452
文樹蘭	Crinum asiaticum, 44	"Crinum Asiaticum," 453
酸枝	Ebony like wood, 53	"Ebony-like wood," 441

The fifth source of the LBT, indicated by Fryer as "Bridgmans Chrestomathy," can be identified with *A Chinese Chrestomathy in the Canton Dialect*, composed by Elijah Coleman Bridgman.[25] Fryer had already quoted the text as one of the sources for his work on the standardisation of Chinese scientific and technical lexicon, with a view to printing relevant glossaries;[26] this is further confirmation that the manuscripts analysed were meant for publication. In chapter 16 of *A Chinese Chrestomathy in the Canton Dialect*, dedicated to botany, from page 436 to page 459 different sections are included, all devoted to descriptions, from different points of view, of plants, trees, and flowers. Shown in table 13 is a comparison between entries in the LBT drawn from "Bridgmans Chrestomathy" and those listed directly in Bridgman's work, ascertaining Fryer's use of *A Chinese Chrestomathy in the Canton Dialect*.

The subsequent entry within the list of sources cited in the LBT is labelled in ultramarine blue and corresponds to "Bowra (Doolittle)." This is a reference to Doolittle's *VHCL*[27] that, analogously to Bridgman's *A Chinese Chrestomathy in the Canton Dialect*, had already been quoted by Fryer as one of the texts consulted for his translation work.[28] In the "Contents of part III" of the *VHCL*, on page v, the "Index Plantarum: Sinice et Latine. By E. C. Bowra" is in turn listed, reported from page 419 to page 437 of the *VHCL*. The "Index Plantarum" was composed by the British sinologist and botanist Edward Charles Macintosh Bowra (Bao La 包腊, 1841–1874) expressly for the *VHCL*.[29] The table below is a comparison between some entries of the LBT taken from "Bowra (Doolittle)" and those within the "Index Plantarum" included in the *VHCL*.

25 Elijah Coleman Bridgman, *A Chinese Chrestomathy in the Canton Dialect* (Macao: Samuel Wells Williams, 1841).
26 Fryer, "Scientific Terminology," 536.
27 Doolittle, *Ying-Hua cuilin yunfu*.
28 Fryer, "Scientific Terminology," 541.
29 For more information on Bowra, refer to Leslie Mitchell, *Maurice Bowra: A Life* (Oxford: Oxford University Press, 2009), 4.

TABLE 14 LBT and *VHCL*

Chinese term	LBT	*VHCL*
常山	Celastrus orixa, 32	"Celastrus orixa […]," 419
赤楊	Alnus japonicus, 6	"Alnus Japonica […]," 421
老鼠	Argemone Mexicana, 11	"Argemone Mexicana […]," 427

Furthermore, on page 419 of the *VHCL*, in the introduction to the "Index Plantarum" by Bowra, it is possible to read: "This list […] is merely a compilation from many scattered sources […]." This is further reference to other sources used in Doolittle's dictionary, in turn used by Fryer as a reference for the compilation of the LBT, as with *Early European Researches into the Flora of China* by Bretschneider, analysed above.

The seventh source listed in the first page of the LBT, "Porter Smiths' "Materia Medica," represented by an ochre colour key, corresponds to *Contributions Towards the Materia Medica and Natural History of China*, composed by the English doctor and missionary Frederick Porter Smith (Shi Weishan 師惟善, 1833–1888).[30] In table 15, some terms listed in the LBT as belonging to this source, compared with the relevant translations drawn from Smith's work, are shown.

TABLE 15 LBT and *Contributions Towards the Materia Medica and Natural History of China*

Chinese term	LBT	*Contributions Towards the Materia Medica and Natural History of China*
麥門冬	Aneilema commelyna medica, 8	"Aneilema commelyna medica […] (*Meh-men-tung*)," 17
假沒藥	Bdellium, 18	"Bdellium […] (*Kia-muh-yoh*)," 34
長菓子樹	"Cassia fistula," 30	"Cassia fistula […] (*Chang-kwo-tsz'-shü*)," 53

30 Frederick Porter Smith, *Contributions Towards the Materia Medica and Natural History of China: For the Use of Medical Missionaries & Native Medical Students* (Shanghai: American Presbyterian Mission Press; London: Trübner, 1871). The preface to the text, from pages v to vii, comprises useful indications on the state of the art of Chinese scientific and technical nomenclature in the second half of the nineteenth century.

Nevertheless, *Contributions Towards the Materia Medica and Natural History of China*, as other texts analysed before, presented references to other texts, at times corresponding to other books presented in this work. To give an example, the translation supplied by Smith for the entry "Aneilema commelyna medica" reads: "The dried tubers of this plant, according to Loureiro, are included under this name [...]. Morrison gives 火炭頭 (*Ho-tan-t'u*), as the name of a wild plant, called Commelyna Bengalensis."[31] Smith was referring to other works, in particular to passages in the aforementioned *Flora Cochinchinensis*[32] by de Loureiro and in the milestone lexicographical text by the Protestant missionary Robert Morrison (Ma Lixun 馬禮遜, 1782–1834), *A Dictionary of the Chinese Language*.[33] In turn, the term "Commelyna Bengalensis" can be located on page 40 of the LBT, listed as "Commelyna Bengalensis 火炭頭" in ochre, indicating *Contributions Towards the Materia Medica and Natural History of China*.

The same applies to "Cassia fistula," the third entry in table 15. On page 30 of the LBT, there are two Chinese renderings for the term—"Cassia fistula 槐花青" and "Cassia fistula 長菓子樹"—both marked with ochre. More specifically, the second translation is the one Smith provided in *Contributions Towards the Materia Medica and Natural History of China*, while the first is included in the description of the lemma, where it is possible to read: "Dr. Williams gives 槐花青 (*Hwai-hwa-ts'ing*) as the name of the fruit which is exported."[34] The last is a reference to another author, in this case Samuel Wells Williams and, more specifically, to *A Chinese Commercial Guide*.[35] The latter is the fourth, enlarged, and revised edition of the namesake text originally composed by the English interpreter John Robert Morrison (Ma Ruhan 馬儒翰, 1814–1843), son of the famous missionary Robert Morrison.[36] In the first edition of the text, the Chinese translation for "cassia fistula" cannot be found and the same applies

31 Smith, *Contributions Towards the Materia Medica*, 17.

32 De Loureiro, *Flora Cochinchinensis*, 38–41.

33 Robert Morrison, *A Dictionary of the Chinese Language, in Three Parts* (Macao: East India Company's Press, 1815–1823), I, part 1, 533.

34 Ivi, 53.

35 Samuel Wells Williams, *A Chinese Commercial Guide, Consisting of a Collection of Details and Regulations Respecting Foreign Trade with China, Sailing Directions, Tables, &c.*, 4th ed., revised and enlarged (Canton: Office of the Chinese Repository, 1856).

36 John Robert Morrison, *A Chinese Commercial Guide, Consisting of a Collection of Details Respecting Foreign Trade in China* (Canton: Albion Press, 1834).

to its second[37] and third edition.[38] On the other hand, in its fourth edition, enlarged and revised, and indicated as composed by Williams, the plant is rendered exactly as "槐花青, *kwai hwá tsing*."[39] It is not superfluous to note here that more and more Chinese translations for terms included in *A Chinese Commercial Guide* were gradually supplied by Williams in various editions of the text originally composed by Morrison. Although Williams already appears in the second edition as the publisher, his name only substitutes that of John Robert Morrison as the author in the fourth one.

Therefore, it is possible to affirm that Fryer did not just point out the translations included in Smith's work. On the contrary, he selected and integrated them with those supplied by Smith himself in the body text of *Contributions Towards the Materia Medica and Natural History of China*, within the descriptions for the specific lemmata. The same working modality was also adopted by Fryer for the other LBT sources and in the glossaries gathered in the *TVM*. The examples illustrated above show once more the complex interactions of textual production within the *TVM* and its manuscript. In broader terms, the analysis of the work carried out by missionaries and translators of late Qing China, within the theoretical frame of the "peripheral approach," implies a reference to earlier and contemporary sources, which in turn can cast a brighter light on the specific text researched. We should not forget that Fryer's purpose in editing the *TVM* and its manuscripts was to gather a nomenclature as complete as possible. The lexicographical work carried out consisted in integrating direct sources consulted personally, as well as other indirect sources, in turn included more or less explicitly in the former. This is further confirmation of the importance of the *TVM*, as well as its manuscripts, as instruments for a better and clearer analysis of lexicographical production, textual interaction, and intercultural flows of knowledge in late Qing China.

Going back to the LBT sources, the next entry in the list is "'Monthly Flowers' (Doolittle)," represented with green and red. As pointed out by Fryer himself, the reference is to Doolittle's *VHCL* in this case as well. The text, indicated on page vii of the *VHCL*, corresponds to item 81 of the third part of Doolittle's vocabulary, "Flowers and Fruits According to Their Time of Blossoming. Compiled

37 John Robert Morrison, *A Chinese Commercial Guide, Consisting of a Collection of Details and Regulations Respecting Foreign with China*, 2nd ed. (Macao: Samuel Wells Williams, 1844).

38 John Robert Morrison, *A Chinese Commercial Guide, Consisting of a Collection of Details and Regulations Respecting Foreign with China*, 3rd ed. (Canton: Office of the Chinese Repository, 1848).

39 Williams, *A Chinese Commercial Guide*, 169.

TABLE 16 LBT and "Monthly Flowers"

Chinese term	LBT	"Monthly Flowers"
石栗花	Aleurites, 4	"Aleurites [...] shih li 'hua," 658
鶴頂蘭, 鶴頂花	Bletia tankervilleae, 21	"Bletia tankervilliae [...] 'ho ting lan; [...] 'ho ting 'hua," 658
海棠	Cydonia japonica, 46	"Cydonia japonica [...] 'hai 'tang," 659

and Arranged by the Editor. Under 12 monthly Parts," listed from page 657 to page 660. The introduction to this list reads: "The English names have been chiefly taken from the works of Morrison, Medhurst, and Williams. It is more applicable to Southern and Central than to Northern China [...]."[40] Therefore, once again, in the manuscripts stored at the EAL a reference to one text, in this case the *VHCL*, reveals other sources: here the names indicated by Doolittle are those of famous missionaries and translators already mentioned in this work. A comparison between the lemmata within the list "Monthly Flowers" will make the coincidence of the entries of the latter and those included in the LBT patent.

By carefully investigating the translations provided in this section of the *VHCL*, we can locate the works Doolittle was referring to, mentioning "Morrison, Medhurst, and Williams" and specifically whose terminology he adopted. For example, the fruit of the Aleurites, a genus of plants, was translated as *shili* 石栗 by Morrison,[41] and the same can be said for Medhurst[42] and Williams.[43] Analogously, the Bletia tankervilleae, a species of orchid, was rendered as *hedinglan* 鶴頂蘭 in the first,[44] as *hedinghua* 鶴頂花 in the second, while in the third the term is not present.[45] Finally, the terminology adopted for the third item in the table above, "Cydonia japonica," or Japanese quince, corresponds to *haitanghua* 海棠花 in Williams' but cannot be found in the other two authors' work.[46]

The eighth source for the LBT, indicated by Fryer as "Custom's catalogue" and marked with sepia, is the *Catalogue of the Chinese Imperial Maritime Customs*

40 Doolittle, *Ying-Hua cuilin yunfu*, II, 657.
41 Morrison, *A Dictionary of the Chinese Language*, part III, 20.
42 Walter Henry Medhurst, *English and Chinese Dictionary: In Two Volumes* (Shanghai: Mission Press, 1847–1848), I, 42.
43 Williams, *A Syllabic Dictionary of the Chinese Language*, 539.
44 Morrison, *A Dictionary of the Chinese Language*, part III, 257.
45 Medhurst, *English and Chinese Dictionary*, I, 137.
46 Williams, *A Syllabic Dictionary of the Chinese Language*, 861.

TABLE 17 LBT and *Catalogue of the Chinese Imperial Maritime Customs Collection*

Chinese term	LBT	*Catalogue of the Chinese Imperial Maritime Customs Collection*
昆布	Algaeplant, a sort of, 4	"A sort of algae plant [...]," 56
水片	Camphor baroos, 27	"Camphor baroos," 75
龍胆草	Centaury, 32	"Gentian Root, Centaury *Erythræa* Sp. (Gentianaceæ) [...]," 58

Collection.[47] In the latter, the "Catalogue of the Chinese Section" can be read, divided into different parts. In particular, within "Class 272.—*Medicines, Officinal and Unofficinal*," from pages 50 to 83 of the text, the corresponding Chinese characters of numerous listed entries are also reported. In table 17, a comparison between some of the entries included in the *Catalogue of the Chinese Imperial Maritime Customs* and in the relevant list of the LBT is shown.

It should be noted that in the *Catalogue of the Chinese Imperial Maritime Customs Collection* a "Catalogue of Publications by Protestant Missionaries in China" is appended. It is composed of 1,040 different titles, divided according to the languages, or "dialects," in which they were compiled, and on the basis of the contents. Each item includes the Chinese and English title, author, format, pages, and place and year of publication. In the "Catalogue," texts in Japanese, Mongol, and Manchu are also included; after these, a list of English texts follows, for a total of 109. This list was prepared by Alexander Wylie and was mostly based on his personal collection.[48] It represents a useful instrument for the compilation of indexes of texts written by missionaries in China during the Ming and Qing dynasties.

The final LBT source, marked by Fryer with indigo and green and indicated as "Dr. Henry's list," corresponds to the text composed by the English botanist Augustine Henry (Han Erli 韓爾禮, 1857–1930), *Notes on Economic Botany of China.*[49] Table 18 includes the translations for the same Chinese term included in the latter, and those supplied in the LBT by Fryer:

47 *Catalogue of the Chinese Imperial Maritime Customs Collection, at the United States International Exhibition, Philadelphia, 1876: Published by Order of the Inspector General of Chinese Maritime Customs* (Shanghai: Statistical Department of the Inspectorate General of Customs, 1876).

48 As indicated in Hanan, *Chinese Fiction*, 79, note no. 2.

49 Augustine Henry, *Notes on Economic Botany of China* (Shanghai: Presbyterian Mission Press, 1893).

TABLE 18 LBT and *Notes on Economic Botany of China*

Chinese term	LBT	*Notes on Economic Botany of China*
臭椿	Ailanthus glandulosa, 4	"*Ailantus glandulosa*, *Desf.*, is the *Ch'ou-ch'un* (臭椿) […]," 38
山樝	Crataegus pinnatifida, 44	"*Shan-ch'a* (山楂) […] is really a kind of hawthorn, *Crataegus pinnatifida, Bge*," 7
升麻	Astilbe chinensis, 14	"*Shêng-ma* (升麻) in Hu-peh is a name applied loosely to various species of *Astilbe* and *Cimicifuga* […]," 8

Analysing on the whole the translations adopted in the LBT, it is clear that translations provided in the *XYDCYP* and in the manuscript presented in this chapter are extremely different, the two texts having been composed in different periods and having different sources. The discrepancies are particularly evident when observing the glossary included in the *TVM*; for terms in which he had already found an adequate Chinese rendition, Fryer frequently added alternative renderings, such as phonemic loans and hybrids. On the contrary, in the LBT sources, the use of hybrids, and even more so of phonemic loans, is extremely rare, but some examples can be located: the term "Datura," referring to the namesake genus of plant and indicated as drawn from "Bowra (Doolittle),"[50] is translated as *mantuoluo* 曼陀羅; in contemporary Chinese the term corresponds indeed to *mantuoluo shu* 曼陀羅屬, from "mandala."[51] Another example of phonemic loans in the LBT is the rendering of "Cotton," indicated by the legend as being drawn from Smith's *Contributions Towards the Materia Medica* and rendered as *jialuopojie* 迦羅婆劫. It is important to note that for this term *Contributions Towards the Materia Medica* is quoting a fundamental text of Chinese tradition: "The *Pen Ts'au* gives […] 迦羅婆劫 (*Kia-lo-p'o-kieh*),"[52] also implying that this phonemic loan aptly reproduces the original "Indian name *Karpasi* usually given in books."[53] Smith is therefore

50 Doolittle, *Ying-Hua cuilin yunfu*, II, 428.

51 Consult Nakayama Shigeru and Nathan Sivin, ed., *Chinese Science: Explorations of an Ancient Tradition* (Cambridge, MA: The MIT Press, 1973), 274.

52 Smith, *Contributions Towards the Materia Medica*, 76.

53 Refer to Abdur Rahman, ed., *History of Science, Philosophy, and Culture in Indian Civilization—Volume III Part 2: India's Interaction with China, Central and West Asia* (New Delhi, New York: Oxford University Press, 2002), 167, and Tan Chung and Geng Yinzeng,

providing in his text the translation for the term as indicated in the monumental botanical text *Bencao gangmu* 本草綱目 (Compendium of Materia Medica), completed in 1578 by the well-known herbalist Li Shizhen 李時珍 (1518–1593) but published for the first time in 1596. Therefore, another fundamental link, represented by traditional Chinese nomenclature, can be made to the connection, within the *TVM* and its manuscripts, between textual production by Protestant and Jesuit missionaries, as also shown in the next section.

Pertaining instead to the modality and purpose of the LBT compilation, the plan of accumulating to the best of his ability an extensive nomenclature in the botanical field, with parallel translation, was certainly not only an exclusive idea of Fryer's, or of missionaries and translators active in late Qing China. Almost the same pattern Fryer followed in the LBT can be found in the text *Hyakkō fu* 百綱譜 (The genealogy of a hundred genera). Composed in 1826 by the renowned Japanese botanist Udagawa Yōan 宇田川榕菴 (1798–1846), the *Hyakkō fu* is a manuscript comprising the Japanese translations of terms for plants included in *Anleitung zur Kenntniss der Gewächse*,[54] a treatise by the German botanist Kurt Polycarp Joachim Sprengel (1766–1833), for an approximate total of 1,600 terms.[55]

To conclude, it is possible to affirm that the LBT represents an important key for interpreting the intellectual itinerary covered in the adaptation process of Chinese scientific and technical nomenclature at the Jiangnan Arsenal. The volume enables the reader to infer a portion of the texts available for Fryer's work. These texts, consequently, can be considered the most representative at the end of nineteenth-century China in the scientific branches examined, within, but not exclusively, the milieu of foreign missionaries and translators.

Furthermore, through the analysis conducted, some questions can be answered that provide a better outlook on the work by Fryer and other foreign scholars and on the circumstances of lexicographical textual production in late Qing China. Firstly, the LBT was most likely compiled for publication,

---

History of Science, Philosophy, and Culture in Indian Civilization—Volume III Part 6: India and China; Twenty Centuries of Civilization Interaction and Vibrations (New Delhi: Project of History of Indian Science, Philosophy and Culture, Centre for Studies in Civilizations, 2005), 223.

54 Kurt Polycarp Joachim Sprengel, *Anleitung zur Kenntniss der Gewächse* (Halle: C. A. Kümmel, 1802–1804).

55 Endō Shōji, Katō Nobushige, Kōda Masataka, and Matsuda Kiyoshi, ed., *Udagawa Yōan shokubutsugaku shiryō no kenkyū: Kyōu shooku shozō* (Osaka: Takeda kagaku shinkō zaidan, 2014), 371. The manuscript is also carefully reproduced, and its terms annotated, in ivi, 401–519.

possibly to be included in the second volume of the *TVM* as mentioned earlier. Closely related to this topic are the reasons for the non-publication of the LBT; the motivations are the same for the non-publication of the second volume of the *TVM*. On the other hand, another question raised regarding the working methodology adopted is why, in almost all the other manuscripts stored in the EAL, Fryer did not indicate the Chinese translations' sources. In this case, apart from broader historical circumstances, there are two possibilities: the first and most plausible is that some volumes remained uncompleted due to the conclusion of his career as a translator for the Jiangnan Arsenal; the second, less probable, is that the printed version of the LBT should include such a legend, or that its aim was different from that of other manuscripts.

4 Case Study—Sources of the "List of Botanical Terms": An Analysis

As demonstrated in the previous section, the analysis of the *TVM* provides important insight into the lexicological connections between texts composed during late Qing China: manuscripts that referenced certain sources, which in turn adopted Jesuit materials, original European works, and native literature. Therefore, starting from the analysis of a single piece of lexicographical material, we can study other documents and the notions transmitted therein, contributing to the disentanglement of their complicated interactions.

This is epitomised by the "Japanese list," the third source employed by Fryer in the LBT; the identification of the texts included in it presents more difficulties. In the same breath, it is fundamental for a comprehensive analysis of the textual and lexicographical production behind the *TVM* and, more generally, of late Qing China. More specifically, the "Japanese list" offers a glimpse of Japanese texts available at the Jiangnan Arsenal for the compilation of Chinese scientific and technical glossaries, as well as their interactions with other Western texts. The analysis conducted here, integrating and completing that carried out in the previous section, can also be considered a further clarification of the textual, and consequently lexical, exchanges that took place between China and Japan, even if on a comparatively small scale, before the First Sino-Japanese War. By looking at analogous materials compiled by other Protestant missionaries and translators, it becomes possible to better contribute to the description of a period in which such lexical exchanges were still not flourishing but certainly existed, at least through the translation of texts. Finally, this section adds another link in the temporal chain connecting lexicographical production by Jesuit missionaries and Protestant missionaries and translators, extending far beyond the borders of China, as personally expressed by Fryer.[56]

The research conducted, which analysed entries in the LBT indicated as drawn from the "Japanese list," points to renowned texts of Japanese botanical tradition: the work composed by Udagawa Yōan, *Shokugaku keigen* 植学啓原 (Principles of botany);[57] the *Taisei honzō meiso* 泰西本草名疏 (Notes on Western names of herbs), written by the biologist Itō Keisuke 伊藤圭介 (1803–1901);[58] and finally the *Sōmoku zusetsu* 草木図説 (An illustrated book of plants and trees), by the botanist Iinuma Yokusai 飯沼慾斎 (1783–1865).[59]

The table below is a comparison of the same entries indicated in the aforementioned texts and those listed in the LBT derived from the "Japanese list."[60] It is evident that numerous renderings included in the latter are taken from the three treatises, particularly the *Sōmoku zusetsu.*

TABLE 19 *Taisei honzō meiso, Shokugaku keigen, Sōmoku zusetsu,* and the "Japanese list"

Original term	*Taisei honzō meiso*	*Shokugaku keigen*	*Sōmoku zusetsu*	"Japanese list"
Decandria	十雄蘂	十鬚蕋	十雄蘂	十雄蕋
Diadelphia	雄蘂分為二體	鬚蕋根為二體	雄蘂分為二體	雄蕋分為二体
Diandria	二雄蘂	二鬚蕋	二雄蘂	二雄蕋
Icosandria	多雄蘂附萼	鬚蕋二十以上自萼生	多雄蘂附萼	多雄蕋附萼
Monadelphia	雄蘂上分下合為一體	鬚蕋之根合為一體	雄蘂上分下合為一體	雄蕋上分下合為一体
Tetrandria	四雄蘂	四鬚蕋	斉頭四雄蘂	齊頭四雄蕋
Pentandria	五雄蘂	五鬚蕋	五雄蘂	五雄蕋
Polyadelphia	雄蘂上分下合為三四以上數體	鬚蕋之根合為三四體	雄蘂上分下合為三四以上數體	雄蕋上分下合為三四体以上

56 "Educational Works," 141–42, and Fryer, "Scientific Terminology," 547–48.
57 Udagawa Yōan, *Shokugaku keigen* (Edo: Suharaya Ihachi, 1833).
58 Itō Keisuke, *Taisei honzō meiso* (Owari: Kajō Shooku zōhan, 1829).
59 Iinuma Yokusai, *Sōmoku zusetsu* (Edo: Suharaya Mohei, 1856).
60 Refer also to Endō Shōji, "*Sōmoku zusetsu* no kōzō—Shokubutsu yōgo no seikaku," *Yokusai kenkyūkai dayori,* no. 118 (2008) and particularly to the table included in Endō, Katō, Kōda, and Kiyoshi, *Udagawa Yōan,* 300–309.

By studying some entries included in the table above, we can better investigate the sources of terms within the "Japanese list" and the modality through which they were employed. For example, on page 49 of the LBT, "Decandria," denoted in the Linnaean taxonomy as a class of plants with ten styles, is rendered in the "Japanese list" as *shi xiongrui* 十雄蕋. "Diadelphia," a class of plants whose styles are united in two by their filaments, is translated as *xiongrui fen wei er ti* 雄蕋分為二体. Finally, "Diandria," a class of flowers with two styles, is rendered as *er xiongrui* 二雄蕋. The three entries are all included in the three aforesaid Japanese botanical texts; in particular, in the *Sōmoku zusetsu*, the first and third ones are slightly different, rendered as *jūsuzui* 十鬚蕋 and *nisuzui* 二鬚蕋. The solutions adopted for these two botanical classes in the "Japanese list" of the LBT are closer to the *Taisei honzō meiso* and the *Sōmoku zusetsu*. The only difference is represented by the last character, annotated as *zui* (*rui*) 蘂 in the two Japanese texts and as *rui* (*zui*) 蕋 in the LBT, both graphical variants of *rui* (*zui*) 蕊, indicating the stamen or pistil.[61]

Even though Fryer was integrating different sources in the compilation of the "Japanese list," he showed a penchant for the *Sōmoku zusetsu*. This supposition can be exemplified from the sixth item in table 19. In the *Sōmoku zusetsu*, the translation of "Tetrandria," a term denoting plants with four styles, is the same as listed in the LBT, *qitou si xiongrui* 齊頭四雄蕋, read as *seitō yon yūzui* in Japanese, while different solutions are included in the *Shokugaku keigen* and the *Taisei honzō meiso*. In any case, it should not be thought that Fryer only adopted the *Sōmoku zusetsu* as a reference. Texts employed are surely more than one since, for example, in the LBT more than one translation deriving from the "Japanese list" is provided for "Monoecia."

Furthermore, it is fundamental to note that within the lemmata of the "Japanese list," one of the most important innovations spread in Chinese scientific and technical lexicon in the second half of the nineteenth century is systematically present: the employment of the bound morpheme and right-hand component of words, *ke* 科, to denote the families of plants. This character had been used with this function in Chinese at least[62] since the eighth juan of *Zhiwuxue* 植物學 (Botany), composed by Alexander Williamson and published in 1858.[63] Instances of earlier attestations can also be found, such as *shancha ke* 山茶科, nowadays indicating the family of Theaceae and used for the first time

---

61 On the term *rui*, refer to Georges Métailié, "À propos du sexe des fleurs: le cas des 'rui,'" *Cahiers de linguistique—Asie Orientale* 23, no. 1 (1994).

62 Huang, *Jinxiandai ciyuan*, 441.

63 Alexander Williamson, Joseph Edkins and Li Shanlan, ed., trans., *Zhiwuxue* (Shanghai: Mohai shuguan, 1858). On the term *zhiwuxue*, refer to Pan Jixing, "Tan 'zhiwuxue' yi ci zai Zhongguo he Riben de youlai," *Da ziran tansuo*, no. 3 (1984).

TABLE 20 "Japanese list," the *TVM*, and contemporary Chinese terms

Plant family	"Japanese list"	*TVM*	Contemporary Chinese term
Arialiaceæ	五加科	阿拉里亞西依, 五加科	五加科
Campanulaceæ	桔梗科	看攀女辣西依, 鐘形科, 桔梗科	桔梗科
Convolvulaceæ	旋花科	康弗勿拉西依, 旋花科	旋花科
Aceraceae		阿西拉西依	槭樹科
Caryophyllaceae	石竹科, 馬齒莧科		石竹科
Asclepiadeæ	白前科	阿司可里比阿弟依, 白前科	马利筋族

in the *Yecai bolu* 野菜博錄 (Extensive records on wild vegetables), a treatise on plants composed by the scholar Bao Shan 鮑山 (n.d.) in 1622.[64] In the *TVM*, such use of *ke* is widespread; it also appears with this meaning in volume eight of the EAL manuscripts, more specifically in the entries denoting families of plants drawn from the "Japanese list."

Fryer frequently employs the character *ke* as a right-hand component in the *TVM*, particularly in the *XYDCYP*. In most cases, *ke* is used in translations that are meant as alternative renderings to phonemic loans, as clear from the first, second, third, and sixth items in the table above. Nevertheless, original terms translated in the *TVM* were frequently affected by inaccuracy within the botanical knowledge of that time. For example, the "Aceraceae," for which Fryer does not employ *ke* as a right-hand component, rendering it instead with a

64 In juan 4, 28, and 50 of the *BCGM*, it is possible to find the term *qie ke* 茄科, indicating on page 57 of the *XYDCYP* a tribe of the family Solanaceae, the "Solaneæ," which in turn was formerly considered a family. In any case, within the *BCGM*, *qie ke* presumably does not indicate the family of plants; refer also to Hua Linfu, Paul D. Buell, and Paul U. Unschuld, ed., *Dictionary of the Ben cao gang mu, Volume II: Geographical and Administrative Designations* (Oakland, CA: University of California Press, 2017), 419. The term is also included in juan 4 of *Cha Weng shi ji* 槎翁詩集 [Poetic collection of Cha Weng], by the scholar Liu Song 劉崧 (1321–1381), in juan 102 of *Zhengzhi zhunsheng* 證治準繩 [A criterion for diagnosing and treating illness], by the high official and physician Wang Kentang 王肯堂 (ca. 1552–ca. 1638), and juan 5 of *Wu bian* 武編 [Compilation on military affairs], by the mathematician Tang Shunzhi 唐順之 (1507–1560), all composed during the Ming dynasty. On the connection of the last one with the *BCGM*, refer to Zheng Jinsheng, Nalini Kirk, and Paul D. Buell, *Dictionary of the Ben cao gang mu, Volume III: Persons and Literary Sources* (Oakland, CA: University of California Press, 2018), 149.

phonemic loan, was recognised as a family only in a phase subsequent to the publication of the TVM.[65]

Other terms are listed exclusively in the "Japanese list," but not in the TVM; at times the opposite is true, as occurs in the fourth and fifth entries of table 20. The lemma "Asclepiadeæ" is a clear instance of how Fryer, on occasion, employs the character *ke* to indicate the wrong taxonomical level; the Asclepiadeae is nowadays recognised as a tribe within the family of the Apocynaceae. In this case as well, though, the incorrect choice derives from the mistaken knowledge of the time, which oscillated between considering the Asclepiadeae a family and a tribe.[66] Confirmation of the fact that Fryer most presumably used *ke* as a right-hand component, especially under the influence of the terms included in the "Japanese list," even more than those in the work by Williamson, can be derived exactly from these errors.

Finally, it is possible to locate, at least in the *Shokugaku keigen* and the *Sōmoku zusetsu*, another of the most interesting morphological peculiarities of the TVM: the use of the character *shu* 屬 as a right-hand component to indicate genera. It is plausible that Fryer, having certainly consulted at least the second of the two aforesaid texts, devised to abide by the translations comprised in it and also included this use of *shu* in the TVM. Moreover, it should be added that *shu* was also used as a right-hand component, albeit for different taxonomical levels, in the aforementioned manuscript *Hyakkō fu*, composed by Udagawa Yōan.

Let us now briefly analyse the three Japanese botanical treatises mentioned above. As for the first, *Taisei honzō meiso*, the version consulted is *Taisei honzō meiso oyobi furoku sankan* 泰西本草名疏及附録三巻 (Notes on Western names of herbs, with an appendix of three volumes); it comprises two juan and an appendix, *jō* 上, *ge* 下, and *zen* 全.[67] The text is said to have been composed by Itō Shunmin (Kinka) 伊藤舜民 (錦窠), Itō Shunmin being an alternative

---

65 Klaus Kubitzki, ed., *The Families and Genera of Vascular Plants—Volume X: Flowering Plants; Eudicots; Sapindales, Cucurbitales, Myrtaceae* (Berlin, Heidelberg: Springer-Verlag, 2011), 366.

66 An example of the first case can be found in John Michael Maisch, ed., *The American Journal of Pharmacy: Published by Authority of the Philadelphia College of Pharmacy; Volume XLIII; Fourth Series; Vol. I* (Philadelphia: Merriehew & Son, 1871), 501; William Darlington, *Agricultural Botany: An Enumeration and Description of Useful Plants and Weeds, which Merit the Notice, or Require the Attention, of American Agriculturists* (Philadelphia: J. W. Moore; New York: Mark. H. Newman, 1847), x, is an example of the second.

67 The author consulted the version stored in *Tōyō bunko* and the one housed at *Waseda daigaku toshokan* 早稲田大学図書館 (Waseda University library). A reproduction is included in Nagoya shi Hōka bunko, ed., *Nagoya sōsho sanpen* (Nagoya: Nagoya shi kyōiku iinkai, 1982–1990), XIX, 293–534.

name for Itō Keisuke and Kinka being his pen name. The text was published by Kajōshooku 花繞書屋 in Nagoya in the eleventh year of the Bunsei era (1818–1830), *bunsei jūni nen* 文政十二年, corresponding to 1829. The first juan also reports that the work was printed in Owari province, nowadays including the city of Nagoya.[68] It comprises a preface, *Taisei honzō meiso jo* 泰西本草名疏序; an introduction, *Taisei honzō meiso daigen* 泰西本草名疏題言; some explanatory notes, *hanrei* 凡例; a table of contents ordered according to the Japanese syllabary, *Taisei honzō meiso wamyō iroha mokuji* 泰西本草名疏和名以呂波目次; and finally the main points of the original work, *Taisei honzō meiso gensho kōmoku* 泰西本草名疏原書綱目.

In the last part of the text, the author of the original work, " 春別爾孤," pronounced as *Chunberugu* チュンベルグ, is mentioned: "春別爾孤原書二掲スル所ノ綱目ノ次第ナリ。今権二数字ノ符號ヲ作り毎目ノ上二標ス[...]" (The order [of this book] is made according to the original index annotated by Thunberg. For the time being, we use symbols with numbers, annotating them on the top of each entry [...]). The reference is to the Swedish botanist Carl Peter Thunberg (1743–1828), author of *Flora Japonica*[69]—the first text classifying Japanese plants according to the Linnaean system.[70]

The following table contains a comparison of the terminology for the classes of plants as expressed in the appendix of the first juan (main points of the original work of *Taisei honzō meiso*, 泰西本草名疏附録上，上，泰西本草名疏原書綱目) and in the "Japanese list" of the LBT:

TABLE 21 *Taisei honzō meiso* and the "Japanese list"

***Taisei honzō meiso*, first juan**	**"Japanese list"**
第一綱 (I. Monandria) 一雄蘂, 1A	一雄蕊
二 (Digynia) 二雌蘂, 1A	二雌蕋
第二綱 (II. Diandria) 二雄蘂, 1A	二雄蕋
第三綱 (III. Triandria) 三雄蘂, 1B	三雄蕋
十 (Tetragynia 四雌蘂), 1B	四雌蕋
第五綱 (IV. Pentandria) 五雄蘂, 2A	五雄蕋

68 Original text: "尾張, 錦窠伊藤先生撰, 藏版" (Owari, compiled by Mr. Kinka Ito).

69 Carl Peter Thunberg, *Flora Japonica: sistens plantas insularum Japonicarum secundum systema sexuale emendatum redactas ad XX classes, ordines, genera et species cum differentiis specificis, synonymis paucis, descriptionibus concinnis et XXXIX iconibus adiectis* (Leipzig: Müller, 1784).

70 As indicated in Federico Marcon, *The Knowledge of Nature and the Nature of Knowledge in Early Modern Japan* (Chicago: University of Chicago Press, 2015), 137.

Apart from the aforementioned variants for the character *zui* 蕊, the translations provided in the *Taisei honzō meiso* and the "Japanese list" of the LBT correspond.

Other valuable information concerning the textual interactions in the "Japanese list," and more generally in the LBT and the *TVM*, can be noticed in the third juan of the *Taisei honzō meiso*, where Chinese names expressing the nomenclature of the original text, *seiseki shosai kanmei shōroku* 西籍所載漢名抄錄, are annotated. They are thus introduced on page 1B: "[…] 直チ二漢名ノ義ヲ以テ譯スルモノアリ是ヲ舉ル益ナシトイヘドモ今初學ノ為二傍二譯字ヲ加へ，亦兹二贅ス唯異聞ヲ弘ムルノミ。ソノ西籍中引用ノ書ハ PUN TSAOU KANG MUH. 本草綱目 KWAN KEUN FANG POO 廣群芳譜ヲ初メ其佗諸書ヨリ抄出セリ" (Some [terms of plants] are translated using the meaning of Chinese names. Even if listing them is not much of help, for the sake of the beginners, we add here on the side their Japanese translations: however, they might convey to the reader unnecessary information. In Western books, the *Bencao gangmu* and the *Guangqun fangpu* are quoted; however, these quotations are maybe drawn from other texts). The author of the *Taisei honzō meiso* is therefore referring to the "KWAN KEUN FANG POO," to wit the *Guangqun fangpu* 廣群芳譜 (The extended collection of all perfumes), written by Wang Xiangjin 王象晉 (1561–1653), and to the "PUN TSAOU KANG MUH," namely the *BCGM*, two fundamental texts of Chinese botanical tradition. In the following table, some terms from the list of the third juan of *Taisei honzō meiso* mentioned above are reported, with the relevant accompanying notes:

TABLE 22 Third juan of *Taisei honzō meiso*

Latin term	Term in the third juan of *Taisei honzō meiso*	Notes
Aglaia odorata	五葉蘭花, 1A	
Aleurites	石栗花, 1A	
Amygdalus persica	飛桃花, 1B	
Epidendrum species 種	夷山吊蘭, 2A	
Iasminum officinale	素馨花, 2B	按ノ本草綱目群芳譜等二素馨一名耶悉茗花又野悉蜜花來自西域トアリ即千《ヤスミン》ノ音譯ナリ
Star aniseed	大茴,[a] 4B	TA HWUY
Tiger paw	虎掌, 5B	HOO CHANG

a The character *hui* 茴 presents the graphical variant with *hui* 囬 in place of *hui* 回.

What is important to again consider is the complicated connection between the lexicographical and, more generally, textual production at work in the TVM. Fryer consulted a consistent amount of texts for the second volume of the TVM, also drawing terms from the "Japanese list." The works constituting this list are in turn, at times, translations of original European works; at times they include references to Chinese autochthonous botanical tradition—these were exactly all the texts Fryer suggested consulting for the compilation of bilingual English-Chinese scientific and technical glossaries. Furthermore, it is important to stress that the second of the Chinese botanical treatises mentioned in the *Taisei honzō meiso*, the *Guangqun fangpu*, was quoted as one of the sources in different passages of *A Dictionary of the Chinese Language* by Morrison, which has in turn been indirectly used in other LBT sources.[71]

Addressing instead the other of the three Japanese sources adopted in the LBT, the *Sōmoku zusetsu*, two editions have been consulted: one, composed by Iinuma Yokusai and other authors, published by Izumoji Bunjirō 出雲寺文治郎 and twelve other authors in the third year of the Ansei era (1854–1860), corresponding to 1856. The other is the revised edition, *Shintei Sōmoku zusetsu sōbu nijū kan* 新訂草木図説草部二十巻 (New and revised edition of An illustrated book of plants and trees—Section herbs, twenty volumes), listed as composed by Iinuma Chōjun 飯沼長順, an alternative name for Iinuma Yokusai.[72] In the first page, the caption reported is "大日本飯沼慾斎翁著, 官許, 新訂艸木圖說, 明治七年十二月, 美濃大垣, 平林莊," with "安政三年原" on the top. As inferred from this information, this is the revised edition of the previous text. It was published in the seventh year of the Meiji era (1868–1912), corresponding to 1874, in Ogaki, by Hirabayashisō 平林莊, based on the original edition printed in the third year of the Ansei era. This is further confirmed by the Japanese preface, signed by the naturalist Tanaka Yoshio 田中芳男 (1838–1916), one of the revisers of the text: "此書世二行ハルルコト、既二久シト雖ドモ、慾斎翁物故シテ後，幾モナク嗣子長榮亦沒シテ、刷行セザルコト玆二數年。大二後進ヲシテ失望ノ嘆アラシム頃者、再ヒ刷行ノ擧アル ..." (Some time has passed since the publication of the book was accomplished; even so, after the death of Iinuma Yokusai, his son, Chōei, also died, and the book was not published for many years. In this time, where many scholars of the following generation were despairing and lamenting for this, we publish it again). Its French preface provides additional information:

71 By way of illustration, in Morrison, *A Dictionary of the Chinese Language*, I, part 1, 574 and 603.

72 The two texts are stored respectively in the Japanese National Diet Library, *Kokuritsu kokkai toshokan* 国立国会図書館, and in the *Tōyō bunko*.

> Préface de la deuxième édition du Sô-mokou-zoussets revue et augmentée": "Il y a longtemps que la première Edition de cet ouvrage a été publiée, et lue par les amateurs de Botanique; mais la mort a surpris, le Savant Yokoussaï avant qu'il ait terminé son œuvre, et peu après, son fils l'a suivi au tombeau. La publication de ce travail est ainsi restée inachevée, au grand regret des Botanistes. Aujourd'hui, je viens offrir une nouvelle Edition de la partie de cet ouvrage qui a déjà été publiée, en y joignant ce qui est resté trop longtemps ignoré des œuvres de cet illustre vieillard: Je me suis entendu pour cela avec son troisième fils, Monsieur Oudagawa Kôssaï, et son arrière petits-fils Iinouma Tchôzô.
>
> Some time has passed since the first edition of this book has been published and read by the lovers of botany; however, death has surprised the scholar Yokusai before he could complete his work. Shortly afterward, his son followed him. The publication of this book, therefore, remained uncompleted, for the regret of botanists. Today, I offer to you a new edition of the section of the book which has already been published, adding also the part of the works of this renowned old man which remained ignored for too long. For this, I made arrangements with his third son, Udagawa Kosai, and his great-grandson, Iinuma Chozo.

As further explained in the preface, within the *Shintei Sōmoku zusetsu*, Tanaka, together with the other reviser, the Japanese naturalist Ono Motoyoshi 小野職愨 (1837–1890), added to the Japanese names in *katakana* the European appellatives for plants, which in the first edition were only annotated in pencil. The *Shintei Sōmoku zusetsu* also included the name of the family, in Latin and Japanese, to which a certain plant belongs. Tanaka finally explains that he asked "Monsieur le Docteur Savetier résidant à Yokoska," i.e. the French botanist Paul Amedée Ludovic Savatier (1830–1891), to control the Latin name of plants.

The reference to the names of families is extremely important for the compilation history of the *TVM* and the LBT, in which terms from the *Shintei Sōmoku zusetsu* are adopted. To provide some examples, in juan 7, on page 34B and 35A, within the lemma Pentagynia, including a translation for this group of plants, *daigomoku* 第五目, and a definition for it, *tashizui* 多雌蕊, describing them as plants with many pistils, we can find the following information: "サジオモダカ (Saji-Omodaka), Alisma Plantago L. (Alismaceae)." The Alismaceae, a family nowadays known as Alismataceae, to which the Alisma Plantago belongs, is therefore translated as *omodakaka* 澤瀉科. The same term can be found on page 5 of the LBT, indicated with the colour purple. Analogously, in juan 8 of the *Shintei Sōmoku zusetsu*, on page 32B and 33A, we can read the term "Koba-no-tsumekusa (コバノツメクサ)" to indicate the "Alsine Verna Bart. Var Borealis Fr. (Caryophylleae)." The translation for Caryophyllaceae, denoted in the text as "Caryophylleae," a term today referring to a tribe within the family to which the plant belongs, *sekichikuka* 石竹科, is the same found on page 31 of the "Japanese list" of the LBT, designated with purple, together with *Umabiyuka* 馬齒莧科. The same applies to the "Hama—Benkeisō (ハマベンケイサウ)," indicating in the same juan, page 40B and 41A, the "Sedum populifolium L. (Crassulaceae)"; the rendering for the family to which the Sedum populifolium belongs, the Crassulaceae, corresponds to the one on page 43 of the "Japanese list" of the LBT, *Keitenka* 景天科.

Terms adopted in the examples above to indicate the families of plants designated by Fryer as drawn from the "Japanese list" are still employed in contemporary Chinese: *zexie ke* 澤瀉科 for the Alismataceae, indicated as "Alismaceae" in the *Shintei Sōmoku zusetsu*; *shizhu ke* 石竹科 for the Caryophyllaceae, denoted in the text as "Caryophylleae"; and finally *jingtian ke* 景天科 for the Crassulaceae. On the other hand, only the Alismataceae is referenced in the contemporary Japanese lexicon with the same term listed in the *Shintei Sōmoku zusetsu*, namely *omodakaka* オモダカ科; the use of katakana to indicate the reading of the two characters *ze* 澤 and *xie* 瀉 is today the most common way to represent them. Contemporary Japanese nomenclature for the Caryophyllaceae and the Crassulaceae is indeed different, *nadeshikoka* ナデシコ科 and *benkeisōka* ベンケイソウ科 respectively.

Furthermore, important chronological information for the compilation of the *TVM* and of the LBT should be highlighted. Since all terms analysed above were not included in the first edition of the *Sōmoku zusetsu*, but only added in the *Shintei Sōmoku zusetsu*, it can be ascertained that those consulted by Fryer for his "Japanese list" are exactly the terms drawn from the revised edition of the Japanese botanical treatise.

Finally, the third source of the "Japanese list" within the LBT, the *Shokugaku keigen*, is particularly important for better exhibiting the complicated connections between traditional Chinese nomenclature, Japanese terminology, and lexical innovations in the *TVM*, the LBT, and, more generally, late Qing textual production.[73] As pointed out by scholars, Udagawa Yōan, author of the *Shokugaku keigen*, drew extensively from the terminology of the *BCGM*, therefore adopting Chinese terms. Nevertheless, analogously to the work Fryer

73 The author consulted the version stored in the Japanese National Diet Library; there is also a reproduction of the text in Kokusho kankōkai, ed., *Bunmei genryū sōsho* (Tokyo: Kokusho kankōkai, 1914), II, 284–323.

carried out in the compilation of the TVM, when suitable terms could not be found, he chose to coin new terms by employing Chinese characters.[74] Similar to the *Sōmoku zusetsu* and the *Taisei honzō meiso*, numerous European sources of the *Shokugaku keigen* further add to the LBT's complicated terminological interaction.[75] It should be remembered that the latter was indeed compiled as an English-Chinese glossary, most likely to be included in a second volume of the *TVM*.

To sum up and conclude, by studying the LBT we are able to get a glimpse of the complex circumstances of Chinese scientific and technical nomenclature in late Qing China. Far from only being the result of sources compiled by Western missionaries and translators, the LBT, as well as the *TVM*, involves earlier Japanese botanical texts that frequently included specific terms of Chinese origin. This ready-made Chinese nomenclature, or *kanmei* 漢名, was a boon for foreign translators, Fryer included, who were striving for the creation of English-Chinese and Chinese-English scientific and technical glossaries. Conversely, Japanese treatises Fryer consulted for the LBT were frequently based on earlier Western language sources, at times referring to the sources composed by Western missionaries and translators as well, which in turn were the other inspiration, alongside the "Japanese list," for the terminology within the LBT. This circular connection at work in the LBT is also epitomised in texts composed at a later stage, such as the *Shokubutsu mei-i* 植物名彙 (A collection of names of plants). This treatise by the botanist Matsumura Jinzō 松村任三 (1856–1928) includes, among the sources consulted both in its original and revised edition, numerous works by foreign missionaries and translators of late Qing China mentioned in this section.[76] Through the analysis of the lexicographical production of late Qing China, including the *TVM* and its manuscripts, we are therefore able to clarify the lexical interactions that contributed to the creation and standardisation of Chinese scientific and technical nomenclature. Consequently, we can cast a brighter light on the transcultural flows of knowledge, textual production, and creation of ideas conveyed by said nomenclature.

---

74 Endō, Katō, Kōda, and Matsuda, *Udagawa Yōan*, viii.

75 These European sources are listed in ivi, 105.

76 Respectively Matsumura Jinzō, *Shokubutsu mei-i: Enumeration of Selected Scientific Names of Both Native and Foreign Plants, with Romanized Japanese Names, and in Many Cases Chinese Characters* (Tokyo: Maruzen kabushiki kaisha, 1897) and Idem, *Shokubutsu mei-i*, rev. ed. (Tokyo, Osaka, Kyoto, Fukuoka: Maruzen kabushiki kaisha, 1916).

5 "Vocabulary of Geological Terms"

Even though some manuscripts do not provide as much information as the LBT, they contribute to a better description of the lexicographical production and cultural interactions at work in late Qing China. This is the case of manuscript number seven of the EAL, the "Vocabulary of geological terms." It can be distinguished from the others by two peculiarities: the first is that numerous original terms, the great majority in English and scientific Latin and written on strips of paper glued onto the volume, are not combined with a Chinese translation. The most plausible hypothesis is that the compilation of the VGT started in a very late phase of Fryer's stay in China, remaining uncompleted due to his departure to Berkeley for the Agassiz Professorship of Oriental Languages and Literature at UC Berkeley. The second feature, making the aforementioned hypothesis more plausible, is that the VGT is the only volume among the manuscripts of the EAL for which the majority of original terms were written with a typewriter, even if Chinese renderings are handwritten.

As supposed by Ferdinand Dagenais, the manuscript was plausibly composed after Fryer brought the first typewriter to China, circa 1890.[77] The hypothesis is supported by a reference in secondary literature, according to which the first typewriter, a "Caligraph," was brought to China by Fryer in 1888.[78] Related information can also be derived from primary sources, such as the *Gezhi huibian*. In number 2 of the periodical, published in 1890, an article entitled "西國寫字機器圖說" (An illustrated exposition on Western typewriters) is found on page 48A to page 49A, introducing a typewriter, with pictures of two models reproduced on page 47A.[79] As all these dates are subsequent to the publication of the first volume of the *TVM*, namely 1888, it can be supposed that the VGT was planned for publication within a second volume of the text, which Fryer had already explicitly referenced. This hypothesis is confirmed by the fact that the VGT legend is not explicit. Analogously to other manuscripts, but with the exception of the LBT, a legend is indeed present in the VGT, listing the colours red, yellow, purple, blue, sepia, brown, light green, and light violet before the original terms; sources, however, are not indicated.

Pages in the volume are numbered in pencil, for a total of 114; the last page is empty and is followed by numerous blank pages. The strips of paper including

77 Personal communication to the author, who sincerely thanks Professor Dagenais.

78 Irwin and Dagenais, *John Fryer's Legacy*, 5.

79 Refer also to Li San-pao, "Letters to the Editor," 767. On the role of the *Gezhi huibian* in spreading knowledge about Western science and technology, refer to Wang Yangzong, "*Gezhi huibian* yu xifang jindai keji zhishi zai Qing mo de chuanbo," *Zhongguo keji shiliao—China Historical Materials of Science and Technology* 17, no. 1 (1996).

the original lemmata and their translations are divided into two columns, to be read from top to bottom, and from the column on the left to the one on the right. Two other features of the VGT are important for deriving more information on its compilation. The first is that a portion of the original lemmata is written in black ink, while others are annotated in blue ink, therefore denoting at least two different phases of writing. The second peculiarity is that some strips of paper are patently more yellowed than others; in this case as well, it can be inferred that the VGT was composed in at least two stages. Furthermore, on some strips of paper the part with the Chinese translation is separated from the original lemma by a red line and, above some colours, numbers can be read, particularly "(1)" and "(2)"; such indications can either refer to the sources from where terms or translations derived, or represent recommendations for the printing process.

As shown in table 23 and 24, drawn respectively from page 4 and page 17 of the VGT, the principal purpose of this manuscript was to collect scientific and technical nomenclature as expansively and accurately as possible. For the same original terms, various renditions, derived from different sources, are integrated together.

TABLE 23 VGT, page 4

Original term	Translation (colour)
Animal	肉性物 (blue)
Animal Kingdom	肉性物 (blue)
Animals	活物 (blue)
Animalcules	微虫 (blue)

TABLE 24 VGT, page 17

Original term	Translation (colour)
Carbon	炭瑛 (blue)
Carbonic acid	炭酸 (blue)
Carboniferous era	炭興時 (blue)
Carboniferous period	炭中期 (blue)
Carboniferous rocks	有炭石 (blue)
Carbuncle	夜明珠 (blue)

The sources indicated by the colours cannot be easily reconstructed, as the Chinese translations provided are extremely few. For example, on page 5 of the VGT, the entry "Arsenic" is translated as *xin shi* 信石, with blue colour, while on page 14 "Boron" is rendered with *peng jing* 硼精, with the same legend. The two Chinese terms are the same solutions provided for the two chemical elements in Martin's *Gewu rumen* 格物入門 (Introduction to the investigation of things). On the other hand, the same legend is attached to the character on page 20 of the VGT to indicate "Chromium," *lu* 鉻, adopted by Kerr and He Liaoran for the chemical element in the *Huaxue chujie*.[80]

From the point of view of translation choices adopted, in the VGT the percentage of phonemic loans is extremely low, analogous to the LBT; this solution, at times, goes against the renderings used in the *TVM*. For example, the term "Obsidian," translated in the *TVM* with the hybrid *Wabuxidi'en shi* 哇步西弟恩石 and followed by the descriptive labelling *huoshan boli* 火山玻璃, on page 73 of the VGT is rendered as *boli shi* 琉璃石.[81] In this case a descriptive labelling is used, indicated with blue colour.

In some circumstances, the renderings provided in the VGT are extremely close to contemporary Chinese nomenclature; an example is the lemma "Old red sand stone." The term, not included in the *TVM*, is translated with the calque *lao hongsha shi* 老紅沙石, marked with blue; it differs from the term employed in contemporary Chinese, *lao hongsha yan* 老紅砂岩, only on the last character. In any case, terms listed in the VGT are almost never coincident with the relevant ones included in the *TVM*—one of the few exceptions is "Channels," translated on page 19 of the VGT as *shui lu* 水路 and indicated with blue. The same rendering can be read on page 10 of the *QJ*, as the second translation for "Channe" [sic], and on page 9 of the same glossary, as the first rendering of the entry "Canal." It should be noted in this case that the *QJ* was published in 1889, one year after the first edition of the *TVM*. According to the hypothesis advanced in this section for dating the VGT, it is therefore possible that the *QJ* included terms common to the VGT.

To conclude, the fact that Chinese translations provided in the VGT diverge from those of the glossaries included in the *TVM*, and that even original lemmata are at times different, leads one to suppose that this glossary was also meant to be published in a second volume of the *TVM*. However, for the reasons mentioned earlier in this work, its destiny was to remain a manuscript.

80 Information on both can be derived from appendix 5 in Wright, *Translating Science*.

81 Page 24 of the *HXCL*; on page 24 of the *JS* the term "Obsidian, or Volcanic Glass" is rendered as *huoshan boli* 火山玻璃.

6 "Botanical Terms from William's Syllabic Dictionary of the Chinese Language"

At times, the entangled interactions of lexicological production analysed in this work can be seen in relation to texts composed by different translators and, at the same time, between and within the volumes compiled by Fryer himself. This is the case of volume number three of the manuscripts stored at the EAL, corresponding to the BTWDCL. Terms from three texts are included in the BTWDCL: those of the first part, representing the majority, are taken from Samuel Wells Williams' *SDCL*, described in section 5.3. Terms of "Vegetable kingdom ect [sic] of Shanghai" are derived from the *GDSE* by Walter Henry Medhurst. The text is indicated by Alexander Wylie as the fourth part of "The Chinese Miscellany; Designed to Illustrate the Government, Philosophy, Religion, Arts, Manufactures, Trade, Manners, Customs, History and Statistics of China."[82] Medhurst planned this work to be printed at regular intervals, but in the end only four issues were destined to be published. The last, composed of 168 pages, is the *GDSE*, described by Wylie as a "translation of the principal matters in the 上海縣志 *Sháng hae hëĕn che*, a topographical history of the city and district of Shanghae."[83] Finally, terms of the third part, listed in the "Sz-chiuen list of Plants,"[84] originally published in 1883 in number 6, volume 11 of the *China Review*,[85] are in reality divided into two lists, corresponding to the two sections of "Sz Ch'uan plants," from pages 301 to 308 of Parker's *UYT*.

An important *terminus ante quem* for the dating of the BTWDCL can therefore be identified. It is possible to infer that Fryer compiled the BTWDCL before 1892 since in the LBT, dated "April 25th., 1892" and marked white and red, the majority of terms included in the volume presented here are listed.

Pages in the BTWDCL are numbered; those of the first part, based on the *SDCL*, are divided into four columns. The first one includes numbers indicating

---

82 Alexander Wylie, *Memorials of Protestant Missionaries to the Chinese: Giving a List of Their Publications, and Obituary Notices of the Deceased* (Shanghai: American Presbyterian Mission Press, 1867), 38–39.

83 Wylie, *Memorials of Protestant Missionaries*, 39. Mention of the work can be located also in Paul Georg Von Möllendorff and Otto Franz von Möllendorff, *Manual of Chinese Bibliography, Being a List of Works and Essays Relating to China* (Shanghai, Kelly & Walsh; London: Trübner & Co., Görlitz, H. Tzschaschel, 1876), 37. Within the "List of Works which formed the foundation of the present compilation," in entry no. 305, among other translations without original Chinese text, it is possible to read: "General description of Shanghai and its environs. Shanghai, 1850, 8vo. Translation of the principal matters in the Shang-hai hsien-tshy, Ch. miscell, No. 4."

84 For more details, refer to Fan, *British Naturalists in Qing China*, 164.

85 In the text, the author presents a list of plants gathered during his travel to Sichuan identified by "Dr. Hance," namely the botanist Henry Fletcher Hance.

the page of Williams' dictionary in which the characters of the third column are annotated. The second column lists the original terms of plants, for the most part in English or scientific Latin; entries are presented according to alphabetical order, not on the basis of numbers from the first column. The third column is represented by Chinese translations as listed in the *SDCL*. Finally, in the fourth column are given definitions, descriptions, alternative names, and other information also drawn from Williams' dictionary but not included in the other three columns.

To sum up, the procedure put into practice within this part of the BTWDCL consisted in duplicating terms from Williams' dictionary, listing them according to the alphabetical order of the original lemma, and reporting the page numbers in the first column in which lemmata were included in the *SDCL*. In the latter, based on the pronunciation of "*Wu-Fang Yuen Yin*"—a reference to the rhyming dictionary *Wufang yuanyin* 五方元音 (Original sounds of the five directions) composed by the Chinese scholar Fan Tengfeng 樊騰鳳 (1601–1664)—characters are listed on the basis of their pronunciation; those terms employing that specific character as a head are reported below.[86]

It is important to remember here that, as already specified when describing Williams' dictionary, the *SDCL* adopts terms from other dictionaries as well: "I have followed [...] F. P. Smith's *Materia Medica*, and Hobson's *Medical Vocabulary*, for terms in those branches. [...] and Bridgman's *Chrestomathy*, have all furnished their quotas."[87] The texts mentioned by Williams, therefore, represent sources that Fryer employed for the compilation of his manuscripts and of the *TVM*, as in the case of Smith and Bridgman, or texts that he explicitly affirmed he was able to consult, as in the case of Hobson.[88] This represents further clarification of the link between flows of knowledge, lexicographical innovation, and adoption of existing terms in the *TVM*, in its manuscripts and, in a broader sense, in a crucial period for the formation and standardisation of the Chinese scientific and technical lexicon.

Hereafter some terms listed in the BTWDCL and in the *SDCL* are compared:[89]

86 For more information on the topic, refer to Léon Wieger, *Chinese Characters: Their Origin, Etymology, History, Classification and Signification; A Thorough Study from Chinese Documents*, translated by L. Davrout (Ho-kien-fu: Catholic Mission Press, 1915), 17, and Yong Heming and Jing Peng, *Chinese Lexicography: A History from 1046 BC to AD 1911* (Oxford: Oxford University Press, 2008), 358–59.

87 Williams, *A Syllabic Dictionary of the Chinese Language*, preface, ix.

88 Benjamin Hobson, *A Medical Vocabulary in English and Chinese* (Shanghai: Shanghae Mission Press, 1858). Refer to Yuan Yuan, "Zhongguo zaoqi bufen shenglixue mingci de fanyi ji yanbian de chubu tantao," *Ziran kexue shi yanjiu—Studies in the History of Natural Sciences* 25, no. 2 (2006): 171.

89 In table 25, dashes in the left column indicate the separation represented by the columns of the manuscript; in the right column they correspond to different parts of the

TABLE 25 BTWDCL and *SDCL*

BTWDCL	*SDCL*
188—Allium ascalonium[a]—薤—A spec of onion, the shallot or scallions (a: as:) with fistular leaves, com in Hu-kwang; it is forbidden to those who fast; (name of a mat).	薤—Composed of *plants, leeks and bad*—A species of onion, the shallot or scallions (*Allium ascalonicum*), with fistular leaves, common in Hu-kwang; it is forbidden to those who fast; name of a mat.
209—Almonds—杏—杏兒 name includes the sorts of Prunus genlly, almonds a plums	杏 [...]—The apricot fruit is 杏兒, but the name includes the sorts of *Prunus* generally, almonds and plums; [...][90]
539—Aleurites (nuts of)—石栗—Nuts of the Aleurites.	石栗—nuts of the *Aleurites*

a In the entry, in place of "Allium," the symbol """ is written, avoiding the repetition of the term of the line immediately above, corresponding precisely to "Allium."

As inferred from table 25, in the BTWDCL Fryer accurately reproduced all entries that, within the *SDCL*, referred to botany; the purpose was to publish them in a glossary of bigger proportions, presumably the second volume of the *TVM*. The work Fryer carried out is extremely accurate, having selected entries from a text of more than 1,200 pages and reorganised them in alphabetical order. The only slight differences are present in the fourth column, where the definitions supplied by Williams are reported. In the BTWDCL, elisions and abbreviations are frequently used, and the tendency to substitute italics when writing Latin names of plants with underlined text, being as the BTWDCL is handwritten, is evident.

After the first part of the BTWDCL analysed above, composed of 199 pages, another glossary follows. It is composed of eight unnumbered pages, and bound within the same volume of the BTWDCL, but with pages of different size and texture. On the first page of this smaller volume, the header "Shanghai Botany a. Sz-chiuen list of Plants," also pointing out the list of the subsequent

definitions within Williams' dictionary. Refer to table 68 in the appendices of this work for a reproduction of entries listed within the BTWDCL.

90 The character *xing* 杏 and the subsequent one, *li* 栗, are not actually written in Williams' dictionary, replaced instead by symbols to avoid repetition. In the definitions of the fourth column within the BTWDCL, these symbols are reproduced in a form similar to one or more exclamation marks.

part, can be read; the date "1883" is annotated below. The latter is a reference to the publication of the "Sz-chiuen list of Plants" within volume 11, number 6, of the *China Review*, printed precisely in 1883.

The first part of this glossary opens with the following description: "Vegetable kingdom ect of Shanghai. From a 'General description of Shanghae a its environs, extracted from native authorities'. S'hae L. M. S. printing press. 1850"; it refers to the *GDSE* by Medhurst, published in 1850.

The pages of this part of the BTWDCL are divided into four columns, analogous to the preceding vocabulary within the same volume. In the first one, the indication "Williams. Dic:" can be read. Below, numbers are reported, referring to the pages in which the respective characters are reported in Williams' dictionary, whenever the relevant term could be found in Medhurst's work. By way of illustration, the second entry of the table below, "Dendrobium ceraia," translated as *shihu* 石蔛, is listed below the second of the two characters, *hu* 蔛, on page 233 of the *SDCL*. However, descriptions in the *GDSE* only partially overlap with those by Williams, since Medhurst's work, strictly speaking, is not a dictionary, contrary to the *SDCL*.

Original terms are listed in alphabetical order in the second column, while in the third and fourth, the Chinese translation and other definitions, descriptions, and alternative names, all derived from the *GDSE* and related to a certain term, are indicated. However, numerous descriptions and supplementary information in the fourth column are completely lacking.

Some lemmata in manuscript number three and the corresponding ones listed in the *GDSE* are shown in table 26:

TABLE 26 "Vegetable kingdom ect of Shanghai" and *GDSE*

"Vegetable kingdom ect of Shanghai"	*GDSE*
Arrowheaded plant—箭鋒草—A med herb wh grows in marshy ground. The arrowheaded plt boiled in wine is said to be good for pains in the limbs.	"Of medicinal herbs, Shanghae abounds chiefly in those which grow in marshy grounds; [...] such as the 箭鋒草 tsëen fung ts'haòu, arrow-headed plant, which boiled in wine is said to be good for pains in the limbs [...]," 69
766, 233—Dendrobium ceraia—石蔛—One of the dry land herbs or flos wh are com here	"The dry land herbs and flowers, common in this district, are such as [...] 石斛 shĭh hŏ," 70
Fragrant boletus—香蕎—or wood ears	"[...] 香蕎 hëang joo, fragrant boletus, or wood-ears [...]," 70

It is clear again from the second example that Fryer integrated the *SDCL* with the *GDSE*; the latter lacked a rendering for the Chinese term "石斛 shĭh hŏ" and in this case Fryer adopted the solution of the former, Dendrobium ceraia, listed on page 233. Confirmation also comes from the fact that the second character, *hu* 斛, presents the radical *cao* 艹 only in the *SDCL* and not in the *GDSE*. This working modality is not surprising as Fryer always affirmed the need to gather a comprehensive nomenclature for the compilation of bilingual scientific and technical glossaries. This principle is strictly applied in his manuscripts, planned almost surely for the publication of a second volume of the *TVM*, notwithstanding the hypothesis advanced for the BTWDCL at the end of this section.

After the eight pages composing the glossary described above, the list annotated as "Sz-Ch'uan Plants. E. H. Parker 'China Review' May a June 1883" appears. The reference is to Parker's *UYT*, particularly to the section "Sz Ch'uan plants," originally published in the *China Review*. The pages of this list within the BTWDCL are divided into three parts: in the first column, numbers 1 to 90 for "List-no 1," coinciding with those of the "Sz Ch'uan plants" in Parker's text, are reported, albeit discontinuously. The second column instead comprises original terms, in English or scientific Latin. Finally, in the third one, Chinese translations can be read. Some examples from the first list are compared to terms in the text by Parker in table 27.

After this list, composed of ninety terms, corresponding to four and a half pages, in the lower part of page 5 "List. Nº ...2" commences; the latter is characterised by the same layout of the previous one. There are forty-four entries in this case, occupying exactly two pages, for a total of roughly seven pages for the two lists combined. The entries annotated in the second list correspond to the "Sz Ch'uan Plants," "List no. 2," reported on page 307 and 308 of the *UYT*.

TABLE 27 "Sz-Ch'uan plants—List-no 1" and *UYT*

"Sz-Ch'uan plants—List-no 1"	*UYT*
9—Cephalanthus naucleoides, D. C.—風箱 or 獺 (pro tᶜa') 狗耳	"9. *Cephalanthus naucleoïdes*, DC. 風箱 or 獺 (pron. 't'a) 狗耳," 304
17—Crepis japonica, Benth. 雀雀草 (or 菜) (or 包) (or 台)	"17. Crepis japonica, Benth. 雀雀草 (or 菜) (or 包) (or 台)," 304
Dichrocephalia latifolia, Dc.—山胡椒	"19. Dichrocephala latifolia, DC. 山胡椒," 304

TABLE 28 "Sz-Ch'uan plants—List. №…2" and *UYT*

"Sz-Ch'uan plants—List. №…2"	*UYT*
19—Iussiaea—對對草 or 過江藤	"19. Jussiaea 對對草 or 過江藤," 307
29—Polygonum—青蓼 or 蓼子草 (or 菜)	"29 Polygonum 青蓼 or 蓼子草 (or 菜)," 307
31—Ranunculus—貓脚草	"31. Ranunculus 貓脚草," 307

Finally, in the last section of the BTWDCL, the following annotation appears:

> Plants in N° 1 represent those out of a total of abt 300 collected by EHP in Sz Ch'uan. Identified by Dr Hance. List N° 2 illustrate alike the incompetence of the collector and the desperate efforts of the charitable botanist. Barely a quarter of my whole collection has been found susceptible of complete identification, whilst not more than a quarter of that quarter (those names wh are underlined or italicised) can be fitted with Chi written names at all coherent or certain. Of this happy few, again, not more than one half can be conscientiously pronounced as correct as to more than sound.

The words above have been almost literally copied by Fryer from those included in Parker's work.[91] The fact that Fryer annotated these lists and the circumstances of their composition in an extremely detailed manner, almost faithfully reproducing the various aforementioned texts as well, may also lead one to suppose that the BTWDCL was not conceived for publication, at least under its present aspect. This volume is indeed an accurate selection and copy of texts presented in the chapter, from which terms and translations pertaining to botany were drawn. It is conceivable that the volume, compiled before

91 The original text in Parker, *Up the Yang-tse*, 301–302, reads: "The plants enumerated in List No. 1 represent those, out of a total of about 300, collected by me in Sz Ch'uan, which my patient and obliging friend Dr. Hance has been kind enough to identify. […] The plants enumerated in List No. 2 speak for themselves. They illustrate alike the incompetence of the collector and the desperate efforts of the charitable botanist. […] Thus, barely a quarter of my whole collection has been found susceptible of complete identification, whilst not more than a quarter of that quarter (those names which are underlined or italicised) can be fitted with Chinese written names at all coherent or certain. Of this happy few, again, not more than one half can be conscientiously pronounced correct as to more than sound."

the LBT, was employed at a later phase to annotate terms within volume number eight.

7 "List of Geographical Names"

The LGN corresponds to volume one of the manuscripts Fryer composed for the publication of a second volume of the *TVM*. Analogously to other manuscripts, indications from the LGN can help to better describe its history and its implications on Fryer's work, and, consequently, further clarify the lexicographical circumstances of late Qing China and all their connections with earlier works. Information on the composition of the LGN is already available on the first page, where the annotation "L. D. Chapin" can be read—the reference is to Lyman Dwight Chapin, as reported in the aforementioned March and April 1880 issue of the *CRMJ*. Another reference to the LGN can be located in secondary literature, more specifically in the January 1892 issue of the *CRMJ*.[92] In the text, in which the history of the Educational Association of China is traced in broad terms, the LGN is mentioned as follows:

> The first session of the Publication Committee was held at the Mission Press, Shanghai, on November 16th, 1891, and following days. [...] The results of this session may be briefly summed up as follows:
>
> GEOGRAPHICAL AND BIOGRAPHICAL NAMES.
>
> A Committee was appointed to undertake the task of drawing up as complete a list of names as possible in English and Chinese. The members appointed are: [...] John Fryer, Shanghai (Chairman). The following suggestions were drawn up for the consideration of this Committee: [...]
>
> (4). Modify existing names where it may be done to advantage; as for instance by abreviating [sic] the number of characters, altering unsuitable characters, or having a fixed character to represent such common affixes as (堡) burgh, or prefixes as (聖) San or Saint. [...]
>
> (6). The lists of geographical names already drawn up by Rev. L. Chapin, Drs. Sheffield and Fryer and others, are recommended to be used as a basis for operations as far as they go.[93]

92 "Educational Notes," 30–36.

93 Ivi, 32–33. The points not quoted here all reflect the theories on translation endorsed in Fryer, "Scientific Terminology."

The list compiled by "Rev. L. Chapin" mentioned in the sixth point corresponds to the LGN; it is described as already completed by the *CRMJ* publication date, 1892, and the Committee suggested to use it as a reference for the lexicographical work to be accomplished. This is further confirmed by the fifth point of the text quoted: "The list to include all Scripture geographical names, as well as the principal names in ancient and modern history."[94] As shown in the preface to the LGN, reported below, one of its rules refers exactly to the anthroponyms annotated in the scriptures. However, the reference here could also hint, as stated further on, at the VPN, corresponding to volume ten of the EAL manuscripts: within the latter, names pertaining to ancient and modern history are indeed present.

On the other hand, the mention of the list of geographical terms Fryer compiled is ambiguous. It is not clear whether it is an allusion to the section "Names of places" within the *RMDM* or to volume ten of the EAL manuscripts, the VPN, in which anthroponyms, but also toponyms, are listed. If the mention is to the latter, it could be possible to ascertain that by 1892 the VPN was already accomplished, so much so to be indicated as a model by the Educational Association of China.

In any case Fryer, appointed president of the Committee in charge of the compilation of lists of English and Chinese terms, stored among his manuscripts the one Chapin provided. In the volume, the principles according to which the LGN has been compiled are briefly annotated below the name "L. D. Chapin":

> The following list of the most common foreign names has been prepared on the basis of the accompanying syllabary. In working out the form of the names in Chinese the author has endeavored to observe the following rules.
>
> (1) In Scripture names, and in those which have came into so general use as to render a change exceedingly difficult, the old form has been retained.
>
> (2) The sound of final r has usually been dropped, except when the name was very short and the sound of any other final consonant has often been dropped to avoid too great length. Less frequently, the sound of a single consonant has been dropped in the middle of a word.
>
> (3) The sound of final d or t after n or another consonant has been dropped.
>
> (4) The K. sound in x or Ks, and g in gs, has usually been dropped.

94 "Educational Notes," 33.

(5) The vague sound of all the vowels, as a, e, i, o, u, have usually been regarded as the same as ŭ.

Valuable criticisms have been received from Dr. Martin of Peking."

In the first point, it is made explicit that the LGN was composed on the basis of the "accompanying syllabary," presented in the next section. The first principle is in line with statements advanced by Fryer on the standardisation of scientific and technical nomenclature; those terms already in general use should be maintained, since it would be difficult to modify them. The subsequent four principles pertain instead to the phonetic rendering in Chinese of toponyms and anthroponyms, stating for the most part the necessity not to reproduce the pronunciation of every single consonant, particularly at the end of the word. The purpose was to obtain a more efficient rendering of original terms, creating phonemic loans composed of a reduced number of characters, again in line with precedent analogous statements Fryer gave. This can also be inferred from the second point, in which it is specified that this principle applies "except when the name was very short." In the last point, finally, the help provided in the compilation of the LGN by "Dr. Martin of Peking," namely William Alexander Parsons Martin, one of the most important translators active in the *Tongwenguan*, is referenced.

As in the case of the other manuscripts, compositional peculiarities of the LGN provide information on the history, compilation, and purpose of the text. In this case, these indications are mostly connected to the phonetic rendering of anthroponyms and toponyms, as highlighted in its preface. The LGN is composed of pages with pre-printed lines, divided into four columns, and meant to be read from top to bottom and left to right; the first and third line include the original terms, the second and fourth the relevant Chinese translations. Lemmata are written every other line, for a total in each line that usually amounts to twelve, corresponding to around twenty-four terms for each page. The text presents corrections, suggesting different stages of writing; for instance, additions in blank lines can be found, such as the entry "Hebron x 希伯崙."[95] Revisions are also present, as in the case of "Hyderabad," translated as *Haidelabade* 亥德拉巴德. The last character, *de* 德, is circled in pencil because, according to the third principle analysed in the header of the volume, it ought to be cancelled for a better phonetic representation of the original

95 Both "Hebron" and the "x" are written in pencil, thus indicating a different phase of drafting. On the implication of the symbol, also located in other glossaries, hypotheses have already been mentioned in this work.

term and, consequently, to increase its chance of being incorporated into Chinese lexicon.

At times, revisions pertain to the characters used, as for instance with "Caxias": in its Chinese translation, *Gashiyasi* 嘎施亞斯, the character *shi* 施 is circled in pencil and, below, *xi* 西 can be read, indicating the need to substitute the former with the latter, again for a better phonetic representation. However, one might rightly doubt whether the latter can justify the adoption of a common character bearing the meaning "West." Additions in the Chinese renderings can be read as well, as in the case of "King George 卓耳支王," in which the last character, *wang* 王, has been added in pencil, while *er* 耳 is circled in pencil, probably denoting the need to substitute it or eliminate it, again with a view to a more precise and terse phonetic representation. In the LGN, entries without the Chinese translation can also be found; in the same way, Chinese terms are listed at times without their corresponding original term.

The LGN is composed of ninety-seven pages. From the examples given in this section and from the relevant table in the appendices to this work, it is evident that its terms are, by its nature, prevalently represented by toponyms, even if rare exceptions can also be found.

Lemmata listed in the LGN do not overlap with those of the *TVM*, as with other manuscripts stored in the EAL. It is therefore possible to suppose that the manuscript, originally drafted by Chapin, was given to Fryer in order to assist his work on the standardisation of scientific and technical nomenclature and the publication of a second volume of the *TVM*.

8 "Syllabary for the Transfer of Foreign Names into Chinese"

The STFNC, already mentioned in the LGN, is included in volume two of the EAL manuscripts. The syllabary can be considered an important document from a linguistic perspective, with all its implications for its role as a facilitator in spreading knowledge, since it is a systematic attempt to standardise phonetic representation, through Chinese characters, of phonemic loans extensively used in the *TVM*. As expressed in the preface to the STFNC:

> In the following syllabary an attempt has been made to collect the sound and syllables in foreign names which can be represented either exactly or approximatively by single Chinese characters, and arrange them in alphabetical order in connection with selected corresponding Chinese characters. By always representing the same sound by the same character, it is hoped that much more of uniformity and system may be attained

> in the transfer of foreign geographical and biographical names. The diacritical marks employed are those used in Webster's Dictionary. Only the more important vowel sounds are given, as it is impossible to represent the lesser distinctions in Chinese.

The principles for the adoption of phonemic loans are explained, and the characters to be used for the transcription of a specific phone are listed in detail within the STFNC. As explained in the preface, the purpose was to ensure that, always associating the same character to the same phone, or groups of phones, of the original terms, the uniformity of toponyms and anthroponyms—which Fryer personally supported—could be attained. The STFNC, according to the LGN preface, was compiled by Chapin, even if his name is never explicitly mentioned in the volume. In any case, it is plausible that Fryer received, or copied, both texts acting as president of the Committee for the nomenclature of toponyms and anthroponyms of the Educational Association of China, and stored them among his manuscripts, presumably as a reference for a second volume of the *TVM*.

Pages in the STFNC, with the dimension of a small notebook, are divided into four columns; after the preface, a "Table of sounds" is annotated. Within the latter, the legend for the pronunciation of English terms is indicated. To offer a glimpse, it is possible to read "Ā = a in ale," "Ä = a in arm," "Ē = e in eve," "Ĕ = e in end," and so on. From page 3, the real syllabary begins: characters considered adequate for the representation of phonemes, or groups of phonemes, of the original term are carefully annotated.

The text is composed of a total of twenty-eight pages. After numerous pages without annotations, another version can be found, written in turn with different ink. The preface, the "Table of sounds," and the body text show much fewer corrections compared with the first version.[96] The only exception is represented by the presence of more entries in each column—that of lemmata written on each line. More specifically, modifications to the characters representing phonemes are extremely rare in the text, just as the elimination or

96 Hereafter, the preface to the second version of the manuscript is given: "In the following syllabary an attempt has been made to collect the elements and syllables in foreign names which can be represented either exactly or approximatively by Chinese characters, and arrange them in alphabetical order, in connection with selected corresponding Chinese characters. By always representing the same sound by the same character, it is hoped that much more of uniformity and system may be attained in the transfer of foreign geographical and biographical names. The diacritical marks employed are those used in Webster's Dictionary. Only the more important vowel sounds are given as it is impossible to represent the lesser distinctions in Chinese."

addition of entries after the first draft. It is worth analysing one correction, since it can provide information on the temporal connection between the two versions. On page 8 of the first notebook, the term "Fow 缶" is annotated; below *fou* 缶, *fa* "法," cancelled, can be read, with *fou* added on top in pencil. In the second version, "Fow 法," without corrections, can be found. It is possible to suppose that the first notebook includes a more recent version than the second, with modifications successively added on the basis of the latter. This hypothesis is also confirmed by the absence in the second version of other entries, such as "Dzĭn 盡" and "Dzĭng 靜." Furthermore, additions after a preliminary draft in the first notebook also frequently correspond to additions in the second. It is therefore conceivable to suppose they were both composed at the same time, and more numerous and recent modifications for the first of the two notebooks were done in distinct stages later on. Finally, another manuscript copy of the syllabary is annotated on a sheet of paper of remarkable dimensions within the FP, carton 3, folder 43.

The idea of composing a system for the phonetic rendering of foreign terms through the use of Chinese characters was of course not conceived for the first time by Fryer or Chapin. Without mentioning earlier attempts, in "Educational notes" there is already a reference to a system for the phonetic rendering of toponyms, different from the principles expressed in the STFNC.[97] The third point of the text actually reads: "Form a system of characters for phoneticising names not already fixed that shall be as much as possible of the same sound in all the various dialects. Several different characters should be used where practicable for each sound, so as to avoid the names being too much alike."[98] The principle exposed is exactly the opposite of that endorsed within the STFNC and by Fryer himself, who always stressed the need to adopt the same character for the same phone as much as possible. On the other hand, almost in the same time frame considered, there were radical attempts symmetrical to the one presented here, advocating the substitution of characters through a system of phonetic representation. An example is "A System of Phonetic Symbols for writing the Dialects of China" put forward by the American missionary Tarleton Perry Crawford (Gao Dipi 高第丕, 1821–1902).[99]

97 "Educational Notes," 30–36.

98 Ivi, 33. The issue would also be later addressed in Chen Duxiu, "Xiwen yiyin siyi," *Xin qingnian* 2, no. 4 (1916).

99 Tarleton Perry Crawford, "A System of Phonetic Symbols for Writing the Dialects of China," *The Chinese Recorder and Missionary Journal, Volume XIX*, no. 3 (Shanghai: Presbyterian Mission Press, 1888). The text would later be mentioned in *Records of the Triennial Meeting*, 56, and in Luo Changpei, *Guoyin zimu yanjin shi* (Shanghai: Shangwu yinshuguan, 1934), 3–4.

The last position was an expression of all foreign missionaries and scholars who believed in the inability of the Chinese language to maintain its writing system when facing the hurdles of scientific and technical lexicon; the STFNC is one of the lexicological solutions Fryer contributed in opposition to those voices.

9 "Vocabulary of Proper Names"

In volume number ten of the manuscripts housed in the EAL, the VPN is essentially a list of anthroponyms, annotated on strips of paper glued onto the volume in alphabetical order. Two different kinds of strips can be identified: the first is represented by strips of yellowed paper, with the original terms typewritten and their Chinese renderings handwritten or typewritten. The second type is a whiter strip with both original terms and relevant Chinese renderings handwritten.

Analogously to the VGT, a problem not easily resolved is the fact that a small portion of the Chinese translations is, in all likelihood, annotated with a typewriter. According to some secondary literature sources, the first typewriter for Chinese characters was invented in 1913.[100] It could be therefore conceivable that Fryer began the compilation of the volume after 1888, first annotating the strips of paper with handwritten Chinese renderings, then continuing the compilation at a later phase—based on the dating of the glossary mentioned above—after his arrival in Berkeley. According to other sources, though,[101] one of the first typewriter models for Chinese characters was devised much earlier[102] by Devello Zelotes Sheffield (Xie Weilou 謝衛樓, 1841–1913), an American missionary and author of the *Wanguo tongjian* 萬國通鑑 (A composite history of the ten thousand nations), published in 1882.[103] His purpose in creating the

100 William C. Hannas, *Asia's Orthographic Dilemma* (Honolulu: University of Hawaii Press, 1997), 266.

101 R. John Williams, *The Buddha in the Machine: Art, Technology, and the Meeting of East and West* (New Haven: Yale University Press, 2014), 133–35, and also confirmed in Jing Tsu, "Chinese Scripts, Codes, and Typewriting Machines," in *Science and Technology in Modern China, 1880s–1940s*, ed. Jing Tsu and Benjamin A. Elman (Leiden, Boston: Brill, 2014), 137, footnote 41.

102 The first prototype was created by Sheffield in 1886, as indicated in Thomas S. Mullaney, *The Chinese Typewriter: A History* (Cambridge, MA; London: The MIT Press, 2017), 127; refer in particular to 126–37.

103 In the aforesaid "Educational Notes," 32–33, it is possible to read a reference to a list of toponyms compiled by "Sheffield"; the reference is exactly to Devello Sheffield.

typewriter was precisely to help missionaries in composing their texts, and in this case it would seem possible that Fryer was able to use the one Sheffield devised, or a model derived from it, for his VPN.[104]

The VPN is composed of seventy-three pages. Original lemmata are extremely numerous and precede their Chinese translations; in turn, a huge quantity of alternative Chinese translations for the same original term is provided. Analogously to other manuscripts, distinct phases of corrections are evident in the VPN. For example, when an error in the alphabetical order of the lemmata occurs, Fryer indicates with numbers, written in pencil, the correct sequence to follow. An example can be read on page 5: the two entries "Ballard 把拉台" and "Ball 波辣" are followed respectively by the numbers "2" and "1," indicating the correct alphabetical order. Analogously, there are entries followed by the symbol "x," annotated in pencil, whose purpose is not easy to shed a light on, as already mentioned with other manuscripts. Moreover, in some of the strips of paper, the original term is written in pencil, while in other circumstances in dark brown ink. Corrections of other lemmata are annotated directly on the volume—all these confirm different writing phases of the VPN. Finally, a compositional feature connected to the question of the typewriter should be noted: when in the VPN lemmata composed of a name and surname are listed, if the Chinese rendering is handwritten, a graphical space is left between the name and surname, which is in turn missing when the Chinese translation is typewritten.

Original terms are mostly represented by anthroponyms and, at times, toponyms. Nevertheless, more generic names, such as those indicating the devotees of a certain religion or aristocratic titles, are annotated as well. Original terms are derived from different languages: apart from English, others include Arabic, Latin, Greek, Italian, Spanish, and Japanese.

From the perspective of translation choices, the vast majority of renderings in the VPN are represented by phonemic loans, given the nature of the glossary. Cases of hybrids, though, can also be located, such as the translation of "Bavaria," rendered on page 6 as *Bufeiliya bang* 不費利亞邦. However, titles and appellatives are not always rendered in the Chinese translation. For example, on page 15, "Count de Palma" is translated as *Balama jue* 把拉瑪爵, but on page 14 "Condé, Prince" is simply rendered as *Gandui* 干兌, thus not conveying the meaning "Prince." In this last example, it should be highlighted that the fourth point put forward in the January 1892 issue of the *CRMJ*, for the standardisation of scientific and technical nomenclature, reads: "Modify existing names

104 On the destiny of this typewriter, see Mullaney, *The Chinese Typewriter*, 16–37.

where it may be done to advantage; as for instance by [...] having a fixed character to represent such common affixes as (堡) burgh, or prefixes as (聖) San or Saint."[105] This method had already been extensively employed precisely in the *RMDM*, even if with a few exceptions. For example, on page 21 of the glossary, where "St. Domingo" is rendered as *Sangduomingge* 桑多明各, the character *sang* is used for its phonetic value instead of *sheng* 聖. In any case, the passage mirrors the influence that the *RMDM*, and the renderings Fryer adopted, had on the translation theories endorsed by the Educational Association of China.

In the VPN the legend is not annotated and, consequently, the sources of Chinese translations remain unknown. Terms included in the *RMDM*, published in the *TVM*, correspond with those of the VPN, as shown in table 29.

Regarding the dating of the VPN suggested in this section, given that some *RMDM* lemmata are reported in the VPN with the Chinese rendering typewritten, and that other lemmata are not included in the *RMDM*, it can be concluded that Fryer compiled the VPN after the *RMDM*. He presumably listed in the former those terms already published in the latter, integrating them with other sources, with the aim of a wider and more complete publication within a second volume of the *TVM*. Confirmation can also be derived from the fact that the great majority of both the original terms and their Chinese translations listed in the VPN and drawn from the *RMDM* are typewritten on strips of paper. Strips of paper with both an original term and a handwritten Chinese translation are not listed in the *RMDM*, with the exception of "Turin," rendered as *Dulin* 都林 in the VPN and annotated in the same way in the *RMDM*. Referring to the aforesaid hypothesis for the dating of the typewriter, it can be supposed that Fryer recorded these lemmata at a later stage and therefore, the manuscript

TABLE 29 *RMDM* and VPN

Original term	*RMDM*	VPN
Baron	巴侖	蘇的里巴侖, 巴侖納, 巴侖門, 巴倫, 波勞那
Fahrenheit	法倫海得	法倫海, 法倫海得
Pagan	巴干	巴干
Huguenots		胡基納

105 "Educational Notes," 33.

was drafted for another objective after the *RMDM* was printed—most probably, for a second volume of the *TVM*.

10 Conclusion

The vicissitudes of the second volume of the *TVM*, which in the end was never accomplished, reflect the complicated interactions between theoretical planning and actual lexicographical work and between lexicological innovations and the recovery of the relevant tradition, as well as the interplay between new ideas, new concepts coming to life, and the nomenclature expressing them. All these factors were at work in the scientific and technical lexicon of late Qing China, contributing differently to its shaping. The purpose of the manuscripts analysed above is analogous to that of the first bilingual glossaries compiled by Jesuit missionaries, starting from the late sixteenth century: to provide an exemplum for posterity. Somehow similar is their destiny: for both, the hope to contribute to uniformity remained on the drawing board.

On the contrary, the reasons why the great majority of the glossaries and dictionaries prepared by Jesuits and missionaries of other orders remained unpublished are technical, historical, and religious. In the case of the manuscripts presented above, other circumstances played a major role, particularly the fact that the translation process would later be led by Chinese scholars themselves, as Fryer already hypothesised. Translations from Western books were gradually replaced by those from Japanese sources, especially after 1895; consequently, the Chinese lexicon also began to be modernised accordingly. Fryer can be considered a pioneer of this process—due to his linguistic awareness that standardisation work required not only Chinese traditional texts and those compiled by Jesuits, earlier missionaries, and contemporary scholars, but also terms derived from Japanese.

The tradition of Dutch learning—*Rangaku* 蘭學, which started during the Tokugawa shogunate (*Tokugawa bakufu* 徳川幕府, 1603–1867) and allowed Japan to introduce and translate numerous books of Western technology and medicine—represented an important benchmark for translations during late Qing China.[106] As shown in this chapter, Japanese texts were rendered from

106 The translation methods adopted by experts of the *Rangaku* were not so different from those Fryer proposed, as indicated in Shen Guowei, "Xifang xin gainian de rongshou yu zao xinzi wei yici—Yi Riben lanxuejia yu lai Hua chuanjiaoshi wei li," *Zhejiang daxue xuebao (Renwen shehui kexue ban)—Journal of Zhejiang University (Humanities and Social Sciences)* 40, no. 1 (2010): 124.

original European works, therefore contributing not only new terms but also new concepts conveyed by "Western learning" in a manner easily accessible for Chinese translators. These interactions confirm that the analysis of the *TVM* contributes considerably to the clarification of historical lexicological exchanges between Chinese and Western languages. More importantly, it aided in the transmission of ideas and concepts in late Qing China, paving the way for the even more considerable vicissitudes of the twentieth century.

CHAPTER 6

Lexicological Features and Translation Choices: Conveying New Concepts with Old Terms and Lexical Innovations

1 Introduction

The suitability of a term within the lexical texture of a language is a prerequisite for its success. This principle also applies to the translation choices provided in the *TVM*; they are indeed closely connected to the lexicological peculiarities of the Chinese language. The importance of the linguistic suitability of these terms can be better shown in vivo, namely through the study of tendencies in the progress from the manuscript to the printed versions of the glossaries included in the *TVM*. The manuscripts are not only a useful tool for the reconstruction of the circulation of ideas and of the network of translators in late Qing China, but can also contribute to the evaluation of the working modality chosen by Fryer and his colleagues at the Jiangnan Arsenal. They adopted existing nomenclature as much as possible, which was in line with the standardisation objective of the scientific and technical lexicon and their theoretical guidelines. Their acute understanding of some peculiarities of the Chinese language contributed to the transmission of Western learning and the formation of a community of translators and scientists in late Qing China. They also heavily influenced the lexicographical production of the late nineteenth and early twentieth centuries: some terms coined by Fryer and his Chinese colleagues even permanently took root in the Chinese lexicon, as shown in chapter 7.

One of the most important aspects of the *TVM* is represented precisely by the lexical innovation generated in the modern Chinese scientific and technical lexicon. It has already been ascertained that a consistent part of the modern Chinese lexicon was influenced by the Japanese language, in different modalities and at various levels, and especially in the period after 1895.[1] Nevertheless, it is evident that the modern Chinese lexicon was born from the

1 Among the most authoritative works, Shen Guowei, *Kindai nicchū goi kōryūshi: shin kango no seisei to juyō* (Tokyo: Kasama shoin, 1994) and its revised Chinese version, Shen Guowei, *Jindai Zhong-Ri cihui jiaoliu yanjiu: hanzi xinci de chuangzhi, rongshou yu gongxiang* (Beijing:

 | DOI:10.1163/9789004443211_008

intricate interaction of various factors rooted well before 1895.[2] Pertaining to the Chinese scientific and technical lexicon, much attention has been particularly paid to the contribution provided by traditional Chinese works and those by Jesuit missionaries and Chinese and foreign translators, especially during the Ming dynasty; at a later phase, the focus was mostly on the influence of Japanese language.[3] Therefore, a description of the influence that the *TVM*, the first systematic collection of English-Chinese glossaries of scientific and technical terms, had on this specific sector of the lexicon and on other less researched lexicographical texts compiled at the end of nineteenth-century China, is necessary. In order to do so, the different modalities of translation adopted both in the *TVM* and its manuscripts, which represent the bases of its lexical innovations, will be discussed here.

2 Phonemic Loans

Terms listed within the *TVM* can be grouped into four major categories, the first one represented by phonemic loans.[4] It is evident that such translation modalities are not peculiar to this text. Nevertheless, the systematic way Fryer

Zhonghua shuju, 2010), can be mentioned; for the period before 1898, refer to Masini, *The Formation of Modern Chinese*.

2 Refer to Masini, *The Formation of Modern Chinese*, preface, iii–iv.

3 As for botany, one of the most important works on the influence exercised by the Japanese lexicon on Chinese lexicon is Shen Guowei, *Shokugaku keigen to Shokubutsugaku no goi: kindai nicchū shokubutsugaku yōgo no keisei to kōryū* (Osaka: Kansai daigaku shuppanbu, 2000). Influential contributions in the field are those by Georges Métailié, such as Georges Métailié, "La création lexicale dans le premier traité de botanique occidentale publié en chinois (1858)," *Documents pour l'histoire du vocabulaire scientifique*, no. 2 (1981); Idem, "Botanical Terminology of Li Shizhen in *Bencao gangmu*," in *Essays on Science*, ed. Hakim Mohammed Said (Karachi: Hamdard Foundation Pakistan, 1990); Idem, "Plantes et noms, plantes sans nom dans le *Zhiwu mingshi tukao*," *Extrême-Orient Extrême-Occident—Le Juste Nom*, no. 15 (1993); and Idem, "À propos du sexe des fleurs." Pertaining to chemistry, among the most important studies, apart from the already quoted Wang Yangzong, "A New Inquiry," it is possible to list Li Li, *Jindai huaxue*, and Wang Yangzong, "Guanyu '*Huaxue jianyuan*'." As for engineering terminology, one of the few studies focusing on the period examined in this work is Li Wen and Dai Wusan, "*Qiji Zhong-Xi mingmu biao* yu Zhongguo jindai jixie gongcheng shuyu de fanyi," in *Di san jie Zhong-Ri jixie jishu shi guoji xueshu huiyi lunwenji*, ed. CJICHMT—2002 bianji weiyuanhui CJICHMT—2002 (Zhongguo jixie gongcheng xuehui jixie shi fenhui, 2002). Finally, for geography, Arakawa Kiyohide, *Kindai nicchū gakujutsu yōgo no keisei to denpa: chirigaku yōgo wo chūshin ni* (Tokyo: Hakuteisha, 1997), can be consulted.

4 For a quantitative analysis of the categories pertaining to the *HXCL*, refer to Gabriele Tola, "Between Phonemic Loans."

and his colleagues at the Jiangnan Arsenal theorised and put their ideas into practice can help explain why neologisms coined in the *TVM* were successful from a lexicological perspective, greatly contributing to the diffusion of scientific and technical notions in late Qing China.

Fryer constantly reiterated that the use of phonemic loans when translating scientific and technical terms into Chinese should be avoided as much as possible. Fryer's acute linguistic awareness made it clear for him, well before the theorisation of the phenomenon, that the Chinese language was not prone to permanently incorporate these loans into its lexicon. Even if, on the one hand, the portion of disyllabic terms in Chinese was increasing, on the other, phonemic loans, particularly those composed of a conspicuous number of characters, were being rejected.[5] According to Fryer, apart from the intrinsic features of the Chinese language itself, the lack of a fixed system of phonetic rendering also contributed to the poor reception of phonemic loans in Chinese.

Fryer's attention to the issue is also reflected in his choice of characters for the graphical rendition of phonemic loans. For example, in the term "Acacia Karroo," translated on page 1 of the *XYDCYP* as *kailuo akaixiye* 楷羅阿揩西耶, he differentiates *kai* 楷 and *kai* 揩. Nevertheless, the two characters were both annotated as *kai* 楷 in the second version of the VCT section devoted to botany, the original translation therefore being *kailuo akaixiye* 楷羅阿楷西耶. However, Fryer was not completely immune from errors in this regard; an example is the use of the character *yi* 衣, bearer of its own specific meaning, within numerous phonemic loans of the *TVM*.

If phonemic loans should be avoided whenever possible, the only feasible path to follow consisted in the adoption of "descriptive terms" or, possibly, "combined terms."[6] Nevertheless, Fryer and his Chinese colleagues were not aprioristically against the casual adoption of phonemic loans, especially in the case of anthroponyms and toponyms. Hereafter, some examples from the *TVM* are listed:

5 On the increase of disyllabic terms in Chinese, refer to Shen Guowei, "Women weishenme xuyao erzi ci? Yuyan jiechu yu hanyu de jindai yanhua: xushuo," *Higashi Ajia bunka kōshō kenkyū—Journal of East Asian Cultural Interaction Studies* 10 (2017).

6 Fryer, "Scientific Terminology," 534.

TABLE 30 Phonemic loans in the *TVM*

Original term	*TVM*	Glossary	Contemporary Chinese term
Caprylic aldehyde	加布里里克阿勒弟海特	*HXCL*	辛醛
Cerite	昔來得	*HXCL*, *XYDCYP*[a]	铈硅石[b]
Laurite	羅耳愛得	*HXCL*	硫钌锇矿
Phillipsite	非勒白賽得	*JS*	钙十字沸石
Pyrene	貝里尼	*HXCL*	芘
Tarragon	太拉艮	*HXCL*, *XYDCYP*[c]	龙蒿

a Within the *XYDCYP* the descriptive labelling *han cuo zhi kuang* 含錯之礦 is listed as an alternative translation.

b On the character *gui* 硅, refer to Gong Yi, "Gui qu lai xi," *Zhongguo keji shuyu*, no. 5 (2008); Shao Jingyu, "Gui zi de laili he bianqian," *Zhongguo keji shuyu*, no. 1 (2008); and Wu Guoqing, "Gui zuowei yuansu ming yi kaishi jiu duzuo gui er bu duzuo xi," *Zhongguo keji shuyu*, no. 4 (2010).

c Within the *XYDCYP*, *qinghao* 青蒿 is listed as an alternative translation.

None of the phonemic loans listed in the table above and included in the *TVM* have been incorporated into the Chinese lexicon. This was something Fryer himself was aware of when he affirmed that phonemic loans should be substituted when more suitable terms could be found.[7] In any case, Fryer and his colleagues also tried to supply precise indications pertaining to the referent in cases where phonemic loans were adopted as alternative solutions. For example, "Zincite," on page 37 of both the *JS* and the *HXCL*, is rendered as *xin'gaide* 辛蓋得. As zincite is a mineral that predominantly includes zinc oxide, the first character, *xin* 辛, is the same found in the phonetic component in the term for zinc, *xin* 鋅. The latter was already annotated in the *Kangxi zidian* with the reading *zi*, third tone, with the meaning "strong," "solid" (*gang* 剛).[8] Nevertheless, Fryer and Xu Shou were the first to use that very character to indicate zinc, therefore exclusively adopting its phonetic value. The chemical element was until then known as *bai qian* 白鉛 or, alternatively, as *wo qian* 倭鉛.

7 Fryer, "Scientific Terminology," 536.

8 Wright, *Translating Science*, 209.

3 Hybrids

The second category of terms adopted by Fryer and his colleagues in both the printed version of the *TVM* and its manuscripts is represented by hybrids. It includes terms rendered with a phonemic loan in which one or more characters, providing indications on some of the peculiarities of the term, are added.[9] Fryer repeatedly affirmed the need to avoid the employment of phonemic loans as much as possible, as the Chinese language cannot easily incorporate

TABLE 31 Hybrids in the *TVM*

Original term	*TVM*	Glossary	Contemporary Chinese term
Dimethylamine	二迷以脫里阿美尼	*HXCL*	二甲胺
Chloride of thionyle	替哇內里綠[a]	*HXCL*	氯化亚砜
Nickel glance	鎳古闌司[b]	*HXCL*	辉砷镍矿
Nitriles	內特來里類	*HXCL*	腈
Aconitic acid	阿古尼低克酸	*XYDCYP*, *HXCL*	乌头酸
Adularia	阿度拉里阿石	*JS*, *HXCL*	冰长石
Alcohol amylicum	阿美里克醕	*XYDCYP*	戊醇
Whiskey	灰司記酒	*XYDCYP*, *HXCL*	威士忌酒
Palm oil	巴勒麻油[c]	*HXCL*	棕榈油
Amyle acetate, Amylacetic acid	阿埋里醋酸	*HXCL*	乙酸正戊酯
Carbotriamine	加耳波三阿美尼	*HXCL*	胍
Euchlorine	由綠氣	*HXCL*	优氯

a Within the *HXCL*, a second alternative rendering, *liu yang lü* 硫養綠, is provided; it adheres almost perfectly to the chemical structure of the compound $SOCl_2$.

b Within the *HXCL* a second alternative rendering, nie_{er} liu_{er} *shen* 鎳$_{二}$硫$_{二}$鉮, is provided; it partially adheres to the chemical structure of the mineral NiAsS.

c The adoption of this phonemic loan might have been influenced by Jesuit texts, such as *Shengjing zhijie* 聖經直解 [Direct explanation of the Bible], composed by the missionary Emmanuel Diaz (Yang Manuo 陽瑪諾, 1561–1639); in the last page of the third juan of *Shengjing zhijie*, the term *baerma shu* 巴爾瑪樹 indicates the palm tree. The author would like to thank Dr. Yu Yating 余雅婷 for kindly providing a personal copy of the text. On botanical terms in the Chinese translations of the Bible, see Yu Yating, "*Guxin shengjing* no kenkyū" (PhD diss., Kansai daigaku, 2018), 118–22.

9 At times, these characters are used as bound morphemes, as *quan* 權 and *dian* 電, as indicated in Masini, *The Formation of Modern Chinese*, 122–23.

them in its lexical texture. He suggested changing phonemic loans into hybrids, whenever feasible, through the addition of characters used for their semantic value. The vast majority of hybrids within the *TVM* is indeed composed of a phonemic loan with characters added, for the most part, as right-hand components of the word.

As inferred from table 31, characters adopted for their semantic value are frequently added within the hybrids as right-hand components of the term provided. They convey information pertaining to chemical structure, classes, acids, minerals, and other miscellaneous categories. In fewer cases they are used as left-hand components, as in "Dimethylamine," and in rarer circumstances they are embedded within the phonemic loan, as with "Carbotriamine." In any case, the tendency to change phonemic loans into hybrids whenever possible is constant and evident in the passage from the EAL manuscripts to the glossaries published in the *TVM*, as already shown in this work.

Finally, in the aforementioned connection between the *TVM* and Japanese sources, it is important to note that the hybrid *huisiji jiu* 灰司記酒, included in the table above, was already annotated with the pronunciation *uisukī* ウイスキー in volume 1, section number 41, of the *Hyakka kōgyō shinsho* 百科工業新書 (A new encyclopaedic book for the industry), edited by Miyazaki Ryūjō 宮崎柳条 (n.d.).[10] *Huisiji jiu* would be later listed on pages 58 and 67 of the *XYDCYP* as the translation of "Spiritus frumenti" and "Whiskey," and on page 37 of the *HXCL* for "Whiskey"; it will be eventually employed in juan 24 of the *Shiwu tongkao* 時務通攷 (Comprehensive examination of current affairs), analysed later.[11]

4 Calques

The third category of terms in the *TVM* and its manuscripts is represented by calques, the great majority of them compositional. This category includes all translations that adhere to the lexical structure of the original term.

In the examples listed in table 32, Fryer has tried to supply precise calques of the original terms; in some entries, the uncertainty about the exact referent of the translation is reflected. This can be seen, for example, in the second lemma. The peppermint of the "Oil of peppermint" is indicated with the contemporary Chinese correspondent, *hujiao bohe* 胡椒薄荷, but also appears in other translations referring to other plants of the family of Lamiaceae, or Labiate, as Mentha, *bohe* 薄荷, and the Stachys japonica, *shuisu* 水蘇.

10 Miyazaki Ryūjō, ed., *Hyakka kōgyō shinsho* (Tokyo: Seifūkaku, 1880).
11 Qilu zhuren, ed., *Shiwu tongkao* (Shanghai: Dianshizhai, 1897).

TABLE 32 Calques in the *TVM*

Original term	*TVM*	Glossary	Contemporary Chinese term
Abies nigra	黑杉[a]	*XYDCYP*	黑云杉
Peppermint oil, Oil of peppermint	薄荷油, 胡椒薄荷油, 水蘇	*XYDCYP*, *HXCL*[b]	胡椒薄荷油
Salt common	尋常食鹽	*XYDCYP*	食盐

a The term is the second translation provided after the hybrid *heise abisi* 黑色阿比司.

b On pages 42 and 45 of the *XYDCYP*, the entries "Oil of peppermint" and "Peppermint oil" are listed, both translated as *hujiao bohe you* 胡椒薄荷油 and *shuisu* 水蘇; in the *HXCL*, on page 26, the lemma "Peppermint, oil of" is in turn rendered only as *bohe you* 薄荷油.

This lexical referential ambiguity is further complicated if we temporally expand the research and consider earlier texts. For example, the *shuisu* can be found in the *Gezhi jingyuan* 格致鏡原 (Mirroring the origin for the investigation of things and extension of knowledge), a treatise published in 1735.[12] In juan 62 of the *Gezhi jingyuan*, corresponding to *shu lei* 蔬類 (vegetables), section *jie* 芥 (mustards), page 4A, it is possible to read: "《吳氏本草》芥菹一名水蘇[…]" (According to the *Wu shi bencao*, one of the names of *jie zu*—pickle of mustard—is *shuisu*—Stachys japonica—[…]). Therefore, in the *Gezhi jingyuan* passage where *shuisu* is mentioned, an older text is quoted, namely the *Wu shi bencao* 吳氏本草 (Materia medica of Mister Wu). The latter, also known as *Wu Pu bencao* 吳普本草 (Materia medica of Wu Pu], is a lost text composed by Wu Pu 吳普, whose dates of birth and death are unknown but are estimated to be circa the third century CE.[13]

In any case, it is not always easy to precisely circumscribe the boundaries of the categories of adopted choices for the terms listed in the *TVM* and its manuscripts. An example of this ambiguity is the lemma "Euchlorine," translated on page 14 of the *HXCL* as *youlüqi* 由綠氣. A portion of the term is conveyed by the phonetic loan *you* 由, while "-chlorine" is expressed through the correspondent term Fryer and Xu Shou used in the *Huaxue jianyuan* for chlorine, *lüqi* 綠氣. The latter also recalls in the Chinese translation the presence of the chemical elements within the gas. The resulting Chinese translation, *youlüqi*,

12 Chen Yuanlong, *Gezhi jingyuan* (1735). Refer to Ho Yi Kai, ed., *Science in China, 1600–1900: Essays by Benjamin A. Elman*. (Singapore: World Century, World Scientific, 2015), 36–37.

13 Zheng, Kirk, and Buell, *Dictionary of the Ben cao gang mu*, 518.

TABLE 33 Descriptive labelling in the TVM

Original term	*TVM*	Glossary	Contemporary Chinese term
Amadou	火絨[a]	*XYDCYP*	火绒
Diluents	沖淡薄之質	*XYDCYP*	稀释液
Millerite	毛形鎳礦	*JS, HXCL*[b]	针镍矿

a On page 2 of the HXCL, the translation is the second supplied after the phonemic loan *amadu* 阿瑪杜.

b On page 23 of the JS the term appears in the third column, "Terms in general use and chief elements," before its chief elements, *nie tie tong liu* 鎳鐵銅硫, while on page 23 of the HXCL it is the alternative translation of the phonemic loan *milelaide* 迷勒來得.

can therefore be placed at the intersection of the category of phonemic loans and that of calques.[14]

5 Descriptive Labelling

The last and fourth category in which the terms included in the TVM can be divided is represented by descriptive labelling. It includes all terms depicting certain peculiarities of the referent denoted.

As evidenced from table 33, descriptive labelling indicates specific characteristics of the referent. For example, in Fryer's translation, the rendering of "Millerite" simultaneously expresses the crystal habit composing it, shaped as a needle, or as "hair," *mao* 毛, and the presence of nickel *nie* 鎳 in the mineral, indicated with the character *kuang* 礦.

6 Multiple Translations

Finally, some examples of terms for which Fryer and his colleagues listed more than one translation are provided. The idea of adding alternative renderings was of course not invented by them—it has a long lexicographical tradition dating back to the first bilingual dictionaries, from the Ming and Qing dynasties.[15] In the TVM, more than one solution is given for numerous lemmata. As Fryer stated, "Where translation should end, and transliteration

14 Calques belonging to this typology can also be called "loan blends."

15 Refer to Yong and Peng, *Chinese Lexicography*, 273 and 292.

should begin, is a matter about which no one can lay down a hard and fast rule. Circumstances alone can determine what is best to do in each case."[16] It is exactly for this reason that Fryer frequently supplied the reader with more than one translation; the renderings provided were assumed to be equivalent. In the majority of cases, the first is a phonemic loan, while the second or third belong to one of the other categories analysed.

Lemmata for which multiple renderings are provided are precisely those for which Fryer considered the exclusive use of phonemic loans inadequate. The criticism he addressed in numerous circumstances in this regard led him to add other terms translated with different modalities that were more easily understandable and acceptable to the Chinese reader.

TABLE 34 Multiple translations in the *TVM*

Original term	Glossary	First translation	Second translation	Third translation
Acid radicals	*HXCL*	酸底子	配底子	
Amethyst	*HXCL*	阿迷替司得	黑晶	
Annealing	*QJ*	淬火	退火	
Bearing	QJ	軸頸	軸枕	
Cawk	*HXCL*	夸克	鋇養礦之類	
Garcinia Morella		磨來拉加西尼亞	產藤黃之樹	
Idocrase	*HXCL*	以度可拉司	雜形石	
Iron ore bog	*JX*	溼地鐵礦	無名異[a]	
Permanent white	*HXCL*	不退色之白	鋇養硫養$_{三}$	恆白
Tabasheer	*XYDCYP*	他罷西耶	竹節內結成矽養$_{三}$粉	
Tetradymite	*HXCL*	低特拉代埋得	四角礦	鉍碲硫礦

a In the case of the *JS*, the alternative terms indicated in table 34 do not include those listed within the second column, "Terms used in Dr. Macgowan's translation," namely terms used in the translation of *Manual of Mineralogy* by Dana, neither the "chief elements" of the third column. For example, for "Idocrase," annotated on page 17 of the *JS* as "Idocrase, or Vesuvian," Fryer supplies the term employed in the *Jinshi shibie* by Hua Hengfang and Macgowan, *aidukelaisi* 愛度客來斯, and the "Terms in general use and chief elements," respectively *zaxing shi* 雜形石 and *lügaixiyang*$_{er}$ 鋁鈣矽養$_{二}$. The latter, though, represents only the elements composing the mineral, or its chemical formula as expressed at that time. In any case, *JS* terms for which two alternative translations are provided within the third column are rare.

16 Fryer, "Scientific Terminology," 540.

To illustrate, the entry "Idocrase," on page 18 of the *HXCL* is first translated with a phonemic loan, *yidukelasi* 以度可拉司. Considering this solution inadequate, though, Fryer and his Chinese colleagues added a second rendering, *zaxing shi* 雜形石. The latter belongs to the group of descriptive labelling, indicating the multiple habits of the crystals of the idocrase, or vesuvianite. The term *zaxing* 雜形, moreover, retraces the Greek etymology of the mineral, from εἶδος, "form," and χρᾶσις, "mixture," with the addition of *shi* 石 as a right-hand component indicating the nature of the referent.

It is not always the case that the first translation provided for a lemma is a phonemic loan. For example, on page 2 of the *QJ*, in which phonemic loans are extremely rare, the first rendering of "Annealing" is *cuihuo* 淬火, while the second is *tuihuo* 退火. In this case, Fryer and his colleagues simply realised that the first term, indicating quenching, was not the precise rendering of the term, therefore adding the second, which in turn refers exactly to the process of annealing.

There are also cases of three translations for a single lemma, as with "Permanent white." Listed on page 26 of the *HXCL*, the lemma is first translated with a descriptive labelling, *bu tuise zhi bai* 不退色之白, describing it as a kind of white that does not lose colour. The second translation is the chemical formula *beiyang liuyang$_{san}$* 鋇養硫養$_{三}$, which is the exact reproduction,[17] according to the Berzelian dualistic formulae, of the chemical formula of the sulphate of baryta, also known as permanent white.[18] Finally, the third translation is, according to Fryer, the vernacular name of the term, *hengbai* 恆白, belonging to the group of calques.

It should be noted that multiple translations were also essential for disambiguating Chinese terms involving phonetic representation of the original lemma. Fryer and his colleagues frequently provided two lemmata for the same referent, one in English and the other in scientific Latin, and this could sometimes result in two different Chinese renderings when phonemic loans or hybrids were provided. Page 63 of the *XYDCYP* serves as an example: the Latin lemma "Tinctura veratris viridis" is rendered in Chinese both as *lüse feiladeluomo jiu* 綠色非辣得羅末酒 and *suan lilu jiu* 蒜藜蘆酒. In the same page, the correspondent English lemma, "Tincture green hellebore," is again translated as *suan lilu jiu* 蒜藜蘆酒," with the only difference represented by

17 Refer for example to John Eddowes Bowman, *An Introduction to Practical Chemistry, Including Analysis*, 2nd ed. (London: John Churchill, 1854), 199.

18 Erwin Andres, *A Practical Treatise on the Fabrication of Volatile and Fat Varnishes, Lacquers, Siccatives, and Sealing-waxes* (Philadelphia: Henry Carey Baird & Co.; London: Sampson Low, Marston, Searle & Rivington, 1882), 247.

the addition of the radical *cao* 艹 in the second character, and a hybrid, *lüse helepuer jiu* 綠色赫勒蒲爾酒; the latter, the original lemmata being different, diverges from the one listed above. However, this was not always the case, particularly when the Latin and English original lemmata shared the same root. Furthermore, Fryer and his colleagues were aware of the problem, as evident from "Unguentum cetacei," translated on page 65 of the *XYDCYP* as *sibamaxide yougao* 司巴瑪息的油膏. The hybrid is not based on the Latin lemma, but on the corresponding English one, reported as "Ointment spermaceti" and "Spermaceti ointment" and rendered as *sibamaxide yougao* on pages 42 and 57 of the glossary.

In any case, the attention that Fryer and his colleagues dedicated to locating the exact scientific term of a certain referent, whenever possible, can be appreciated in the *TVM*. For instance, in the Chinese renderings of the concepts of "Air" and "Steam," they always took care to distinguish *qi* 氣, in most entries listed in the form of *kongqi* 空氣, and *qi* 汽. This is clear when reading the two lemmata on page 1 of the *QJ*, "Adiabatic line for air," rendered as *qi zhi bu chuan li quxian* 氣之不傳力曲綫 and "Adiabatic line for steam," translated as *qi zhi bu chuan li quxian* 汽之不傳力曲綫.

Nevertheless, notwithstanding their strenuous efforts on compiling glossaries in which only suitable terms, both from a scientific and linguistic point of view, were included, at times a referential ambiguity can be noticed. In the *QJ*, for example, the same referent for "engine" is rendered at least six different ways: *jiqi* 機器, as in "Blowing engine," *chuifeng zhi jiqi* 吹風之機器; *qiji* 汽機, in "Agricultural engine," rendered as *nong qiji* 農汽機; *ji* 機, as can be inferred from "Air engine," *kongqi ji* 空氣機; *qiche* 汽車, for example in "Aveling's traction engine," translated as *Afeiling malu qiche* 阿非令馬路汽車; *qi* 器, as in "Adamson's vertical blowing engine," *Adansheng zhili jinfeng qi* 阿担生直立進風器; and finally, as a "zero translation," for instance in "Cylinder of steam engine," rendered as *qi tong* 汽筩, or "Engine beams," translated as *ganggan* 槓桿.[19]

7 Using Existing Nomenclature

It is evident that the production of bilingual glossaries requires the consultation of existing relevant materials; these materials, in turn, can include other references to earlier texts, as we have already seen in the case of the *TVM* and

19 Respectively on pages 6, 1, 1, 3, 1, 14, and 18. In the entry on page 18 of the *QJ*, the lemma "Engine" only includes the translations *jiqi* 機器 and *qiji* 汽機.

its manuscripts. When starting their work, Fryer and his colleagues would consult an enormous corpus of texts, particularly pertaining to native Chinese nomenclature. It is therefore not surprising that a consistently large part of Chinese terms included in the TVM, especially in the field of botany and mineralogy, already appeared in earlier texts.

Fryer actually referred to different lexicographical works compiled by coeval Protestant missionaries and translators, who certainly used native books or those compiled by Jesuits; the Jesuits also referred to traditional Chinese books and worked in cooperation with Chinese literati when producing their compilations. Protestant missionaries and translators active in China in the first part, and particularly in the second half, of the nineteenth century, were operating with the help of Chinese colleagues as well. The latter, in turn, frequently were literati who received a classical education in view of the imperial examinations, the *keju*, and therefore were able to masterfully make use of traditional sectoral literature.

The examples in tables 35 and 36 show that numerous terms listed in the TVM pertaining to botany, such as parts and names of flowers or plants, plus mineralogy and other scientific branches, were already present in representative traditional Chinese texts, such as the *Tong ya* 通雅 (The general guide) and the *Gezhi jingyuan*.[20] These two works have been selected as a basis for comparison since they represent veritable lexicographical milestones, gathering numerous existing Chinese terms. The first, composed of fifty-five volumes, conducts detailed textual research on various aspects of terminology, including meanings and definitions, and represents a "marvellous reference" for lexicological studies.[21] The second is a "post-Jesuit collection of practical knowledge" in which an enormous amount of texts is included, frequently quoting works already lost.[22]

In some circumstances, terms included in the TVM were only partially listed in traditional Chinese works, at times with different meanings, or in different forms from those adopted within the TVM. As an example, *boluomi* 波羅蜜, Artocarpus heterophyllus, commonly known as jackfruit, was already annotated in the *Gezhi jingyuan*, within juan 66, *mu lei* 木類 (trees), section *zhu mu* 諸木 (various kinds of trees), on page 22A, quoting the relevant definition drawn from the *Fei xue lu* 霏雪錄 (Records of the swirling snow): "《霏雪錄》波羅蜜樹如荔枝樹 [...]" (According to the *Fei xue lu*, the jackfruit tree is similar to the lychee tree). The *Fei xue lu* is a text composed during the early Ming dynasty

20 Fang Yizhi, *Tong ya* (Fang shi fushan ci cang xuan keben, 1666).

21 Yong and Peng, *Chinese Lexicography*, 306.

22 Ho Yi Kai, *Science in China, 1600–1900*, 36.

TABLE 35 *Tong ya* and the *TVM*

Chinese term	*Tong ya*	*TVM*
澤蘭	Juan 41 (植物: 草), 1A: "澤蘭又名虎蘭"	Iris: *XYDCYP*, 31
花莖	Juan 41 (植物: 草), 1B: "丫蘭最貴。 華莖生岐枝也"	*XYDCYP*, 60[a]
白芷	Juan 41 (植物: 草), 5A: "白芷一名莞"	Iris florentina: *XYDCYP*, 31
蜜陀僧	Juan 46 (金石), 3A: "炒為黃丹。其渣為蜜陀僧"	Lead protoxyde, 34; Litharge, lithargyrum, 36; Massicot, 38: *XYDCYP*. Litharge: *HXCL*, 21. Litharge: *JS*, 20[b]
阿魏	Juan 46 (金石), 5A: "猶走馬射阿魏"	Assafœtida: *XYDCYP*, 7. Asafoetida: *HXCL*, 4
寶石	Juan 46 (金石), 5B: "九玉與寶石"	Spinelle: *HXCL*, 32[c]

a *Huajing* 花莖 is listed in two lemmata: "Quercus pedunculata," translated firstly as *you huajing kuerkusi* 有花莖苦爾苦司, and "Quercus sessiflora," rendered with *wu huajing kuerkusi* 無花莖苦爾苦司. In Huang, *Jinxiandai ciyuan*, 334, the first attestation of *huajing* 花莖 is reported in 1890, in "Taixi bencao cuoyao 泰西本草撮要" [Synopsis of Western herbs], *Gezhi huibian*.

b *Mituoseng* 蜜陀僧 is always annotated in the *TVM* as *mituoseng* 密佗僧. In the first two occurrences within the *XYDCYP* it is the translation alternative to *qianyang* 鉛養, while in the third it is annotated before it. In the fourth, finally, it is the rendering alternative to the phonemic loan *maxigete* 瑪西各特. In the *HXCL*, it is listed before *qianyang*; the same can be said within the *JS*. Moreover, *mituoseng* 密佗僧 is included as part of numerous other lemmata within the *XYDCYP*.

c *Baoshi* 寶石 is listed as a translation alternative to the hybrid *sipineile shi* 司批內勒石, in addition to being attested as part of translations in other lemmata of the *HXCL* and the *JS*.

by Liu Ji 鎦績 (n.d.) and it is included in the *Qinding siku quanshu* 欽定四庫全書 (Imperial Edition of the Complete Collection of the Four Treasuries), *zi bu* 子部 number 10, *zajia lei* 雜家類 number 3, among the "miscellaneous anecdotes" (*zashuo zhi shu* 雜說之屬).[23] This example shows how compilers of the *TVM* drew fully from the boundless tradition of Chinese nomenclature, while sometimes applying new meanings to existing terms; *Boluomi* 波羅蜜 is

23 On the *Fei xue lu*, refer also to Zheng, Kirk, and Buell, *Dictionary of the Ben cao gang mu*, 131.

TABLE 36 *Gezhi jingyuan* and the TVM

Chinese term	*Gezhi jingyuan*	*TVM*
石英	Juan 33 (珍寶類, section 水晶), 11A: "《庶物異名疏》[…] 廣雅謂之石英"	Quartz: *HXCL*, 29. Quartz: *JS*, 28
蜜陀僧	Juan 34 (珍寶類, section 銀), 15B: "《格古要論》[…] 又次之松紋假金花以蜜陀僧為之"	Lead protoxyde, 34; Litharge, lithargyrum, 36; Massicot, 38: *XYDCYP*. Litharge: *HXCL*, 21. Litharge: *JS*, 20
樟腦	Juan 64 (木類, section 樟), 17A: "《彙苑詳註》 其汁可為腦。《 胡演升鍊方》 樟腦用樟木新者切片井水浸三日夜 […]"	Camphora officinarum: *XYDCYP*, 11. Camphor: *HXCL*, 8
不灰木	Juan 66 (木類, section 諸木), 12A–12B: "《太平寰宇記》 不灰木俗多為鋌子燒之成炭 […]"	Asbestos: *XYDCYP*, 7. Asbestos: *HXCL*, 4. Asbestus: *JS*, 4. Asbestos: *QJ*, 3

listed in the *HXCL*, on page 27, within the lemma "Pine apple flavouring," translated as *renzao boluomi xiangshui* 人造波羅蜜香水.

Another instance of the use, adaptation or partial modification of existing nomenclature is the lemma "Acacia serissa," translated on page 1 of the *XYDCYP* with the hybrid *selisa akaixiye* 色里薩阿揩西耶. Within the *Gezhi jingyuan*, juan 67, *mu lei* 木類 (trees), section *yi mu* 異木 (other trees), on page 4B, it is possible to read: "《 物類相感志》尸利沙樹生西域 […]" (According to the *Wulei xianggan zhi*, the Acacia sirissa grows in the Western regions). In this case, the translation provided, drawn from the *Wulei xianggan zhi* 物類相感志 (Record of the mutual responses of the ten thousand things), a lost book composed by the Northern Song dynasty (960–1127) monk Zan Ning 贊宁 (919–1001), is a hybrid, *shilisha shu* 尸利沙樹, with the phonetic part presenting a different rendition compared with *selisa* 色里薩.[24]

24 On the *Wulei xianggan zhi*, refer also to ivi, 516.

8 Neologisms: Dating Back through Etymological Dictionaries

When compiling a scientific and technical glossary, the first step is the consultation of existing relevant lexicographical production in the field and its adoption, with the necessary modifications, as seen in the previous section. When confronted with referents that have never been translated, or for which a suitable Chinese equivalent could not be found, Fryer and his colleagues coined new terms, or simply used others that were not recorded before, or whose first attestation had not yet been discovered. Sometimes, the impossibility of finding such equivalents can be attributed to personal inability, such as a lack of understanding of the translated texts. It can also be due to practical conditions, such as the unavailability of a certain treatise, or again to problems in the co-operation between foreign and Chinese translators. In any case, the results of their work contributed to the enrichment of Chinese scientific and technical lexicon, either through coining new words or annotating existing terms on the paper. As already stated, the TVM's importance is not only the lexical innovation itself, but the fact that it made available a set of sectoral nomenclatures published through an influential institution in a condensed form.

In table 44 in the appendices, some of the first attestations listed in the most authoritative etymological dictionary for scientific and technical nomenclature of modern Chinese, the *Jinxiandai ciyuan* 近现代辞源 (Modern etymology), are backdated to the TVM and compared with the database *Zhongguo jiben guji ku*.[25] The main purpose of the table is to demonstrate that the TVM represents an innovative work from a lexicological point of view, including numerous neologisms until now considered to have appeared at a later phase. Nevertheless, by consulting other texts or through the use of more advanced technological means, some terms might be further backdated in the future. Every accepted scientific theory is considered true as long as another might prove more suitable to explain a certain phenomenon; the same applies to lexicological research.

It is evident that only some of the terms Fryer and his Chinese colleagues coined or adopted were destined to be included in the TVM; some of them were adopted in other texts published before 1888, to wit the publishing year of the first edition of the TVM. As an example, the term *renzao* 人造, with the meaning "artificial," is considered to appear for the first time in *Huaxue qiushu* 化學求數 (Seeking numbers in chemistry), translated by Fryer and Xu Shou

25 Huang, *Jinxiandai ciyuan*. See also Xianggang Zhongguo yuwen xuehui, *Jinxiandai hanyu xinci ciyuan cidian* (Shanghai: Hanyu da cidian chubanshe, 2001).

and printed in 1883.[26] Among the neologisms adopted by Fryer and his Chinese colleagues there are also toponyms. For example, *Balagui* 巴拉圭, a phonemic loan indicating Paraguay, is indicated as first attested in *Dizhi xuzhi* 地志須知 (Essentials of topography), composed by Fryer and published in 1882.[27] Nevertheless, the term will not be listed in the *RMDM*, but will be a component of the translation for the lemmata "Paraguay tea," rendered as *Balagui cha* 巴拉圭茶 on page 25 of the *HXCL*, and "Ilex oak Paraguayensis," translated as *Balagui yilikesi* 巴拉圭衣里克司 on page 30 of the *XYDCYP*.[28]

It should be noted that the analysis of the neologisms conducted in this section has not included the manuscripts housed in the EAL; because those were presumably compiled by Fryer himself, published versions could not be located. Their adoption in later works has not yet been ascertained and the influence they could have in manuscript form must be considered less relevant than within the *TVM*. On the other hand, manuscripts whose sources have been explicitly located were not directly composed by Fryer, and are worthy of distinct research. A specific analysis of the manuscripts housed in the EAL can bring other neologisms to light, as demonstrated with one example: on page 95 of the VGT, the entry "Specific gravity," translated as *bizhong* 比重, is listed. Earlier studies traced the first attestation of *bizhong* with the meaning in the *Xin erya* 新爾雅 (The new guide),[29] section *Shi gezhi* 釋格致 (Explaining the investigation of things and extension of knowledge).[30] As it is impossible to identify a published version of the VGT, therefore making the diffusion of the manuscript untraceable, the term has been excluded from neologisms annotated in table 44.

---

26 Huang, *Jinxiandai ciyuan*, 628.

27 Huang, *Jinxiandai ciyuan*, 11. The term can actually already be traced to juan 68 of the renowned *Haiguo tuzhi* 海國圖志 [Illustrated treatise on the maritime countries], by the Chinese scholar Wei Yuan 魏源 (1794–1857), and in juan 10 of the *Yinghuan zhilüe* 瀛環志畧 [A short account of the maritime circuit], by the geographer Xu Jiyu 徐繼畬 (1795–1873). Another phonemic loan, *yalanmi* 呀蘭米, listed in the *XYDCYP* on page 16 as the translation of "Coccus" and "Cochineal," as well as on page 11 of the *HXCL* to indicate again "Cochineal," was already included in juan 64 of *Haiguo tuzhi*.

28 In table 44 all the neologisms coined by Fryer and his colleagues and already included in Huang, *Jinxiandai ciyuan*, were specifically not included. To give a glimpse of the fields covered, some among them can be mentioned, such as *haili* "海裡" (nautical mile), whose first attestation can be traced to the *Gezhi huibian*, vol. 1, 1876, or also *guolu* "鍋爐" (furnace), which appeared for the first time in 1871 in the *Huaxue jingyuan*, juan 3.

29 Wang Rongbao and Ye Lan, *Xin erya* (Shanghai: Mingquanshe, 1903). For the text, refer to Yong and Peng, *Chinese Lexicography*, 217–18.

30 Huang, *Jinxiandai ciyuan*, 41.

9 Neologisms: Dating Back through Databases

Lexicographical studies can now be assisted by databases and corpora; terms that are not considered in *Jinxiandai ciyuan* can instead be traced through the MCST database, compared with the *Zhongguo jiben guji ku*.

From the enormous amount of information included in table 45 in the appendices, two main conclusions can be drawn. The first one is that numerous terms listed in the *TVM* and considered to be neologisms coined at a later stage in reality existed in the Chinese language, at least since the publication of the text. Even though some relevant considerations pertaining to other fields have been proposed in secondary literature, this hypothesis was never put forward pertaining to the *TVM*.[31] Lexicological research can never aspire to completeness; furthermore, a digitalised version of the *TVM* does not exist, so the tables used here were compiled manually. Nevertheless, the attestations provided point out that more detailed research on native Chinese texts and bilingual glossaries compiled during late Qing China can shed a brighter light on the structure, formation, and evolution of the Chinese lexicon, and on the relevant linguistic context. This principle can be applied not only to the scientific and technical lexicon, but also to the core lexicon, since some terms listed in table 45 can be included in this category. Moreover, such an analysis is instrumental in showing how the concepts and ideas behind the neologisms have circulated since late Qing China, and how translation theories were constantly shaped by practice. Finally, it should be specified that those terms already annotated in the MCST database as Fryer's neologisms are not listed. For example, *aotuli* 凹凸力, included on page 18 of the *QJ* as a rendering of "Elasticity,"[32] is signalled in the MCST as a neologism attested for the first time in *Shengxue* (Acoustics), a text translated by Fryer and Xu Jianyin.[33]

The second conclusion that can be reached is that the *TVM* has been, directly or indirectly, extensively used by numerous works published after 1888.[34]

31 Wright, *Translating Science*, 183–247.

32 The term is also part of other translations in numerous lemmata within the glossary. From here on in this work, such cases will not be specified and only examples of lemmata listed independently will be considered, unless otherwise specified.

33 Fryer and Xu, *Shengxue*.

34 The contrary must also be considered true. When comparing the *TVM*'s entries with those of the MCST and *Zhongguo jiben guji ku* databases, it is actually possible to infer that Fryer consulted and adopted other sources, in addition to those already observed in this work. Among the most representative ones, Lobscheid, *English and Chinese Dictionary*; the second part of Gabriel Lemaire and Prosper Marie Giquel, *Dictionnaire de poche français-chinois: suivi d'un dictionnaire technique des mots usités à l'arsenal de Fou-Tcheou* (Shanghai: American Presbyterian Mission Press, 1874), to wit the *Dictionnaire Technique*

Among the remarkable number of examples, two can be mentioned. The first is the *TT* by Mateer, published sixteen years after the *TVM* and fifteen years after the compilation of the *QJ*.[35] Mateer drew extensively from the glossaries compiled by Fryer and his colleagues in discussing the steam engine, which is covered in the *QJ*. However, the work within the *TT* was not limited to copying terms from the *TVM*—it included the integration and correction of the latter, adopting solutions supplied by other translators as well. As an example, the term "Boiler maker" is rendered on page 47 of the *TT* as *guolu jiang* 鍋爐匠, while on page 6 of the *QJ* it is translated as *zao guolu jia* 造鍋爐家; in this case, the solution provided by Fryer and his Chinese colleagues was rejected. In other circumstances, renditions of the *TVM* are integrated with those supplied by other translators, as with the term "Return flue boiler," first rendered on page 47 of the *TT* as *hui huolu zhi guolu* 回火路之鍋爐, followed by the alternative translation *huitong guolu* 回筩鍋爐; Fryer adopted the same one on page 42 of the *QJ*.

The same can be said, even more considerably, for the *ECSD*, edited by Yan Huiqing, also known as Wei Ching Williams. Yan was an important Chinese diplomat as well as a prime minister and foreign minister during the first half of the twentieth century, but also a renowned linguist.[36] There are two editions of the dictionary in the MCST database: one published in 1910[37] and one printed two years later.[38] Analysing table 45, it is possible to see that numerous terms considered neologisms, attested for the first time in the *ECSD* by Yan Huiqing, were in reality directly adopted from the *TVM*. These terms, contrary to the *TT*, which particularly employed the *QJ*, are homogeneously drawn from all the glossaries Fryer and his Chinese colleagues compiled, comprising different translation choices, phonemic loans included, which are basically missing in the *QJ*. The *ECSD*, as illustrated in section 7.4, has been published in different

Français-Chinois; comprenant les terms professionnels et scientifiques employés à l'arsenal de Fou-Tcheou; and Kuang Qizhao, *Hua-Ying zidian jicheng—English and Chinese Dictionary* (Hong Kong, 1882) can be mentioned. To delve further into this strand of research would exceed the boundaries of this work; the author hopes to be able to devote another monograph to it in the future.

35 Calvin Wilson Mateer, *Technical Terms, English and Chinese Prepared by the Committee of the Educational Association of China* (Shanghai: Presbyterian Mission Press, 1904).

36 Refer also to his autobiography, Yan Huiqing, *East-West Kaleidoscope, 1877–1946: An Autobiography* (Jamaica, NY: St. John's University Press, 1974).

37 Yan Huiqing, *An English and Chinese Standard Dictionary*, 3rd ed. (Shanghai: Shangwu yinshuguan, 1910).

38 Yan Huiqing, *An English and Chinese Standard Dictionary*, 4th ed. (Shanghai: Shangwu yinshuguan, 1912).

editions, collecting the neologisms included in the *TVM* and extensively contributing to their diffusion.

10 Beyond Lexical Innovations

The *TVM* greatly influenced the *ECSD*, as well as other dictionaries and works. This was exercised not only by means of lexical innovations, but also through partial calques of the morphological structures of the starting terms that the *TVM* lemmata were based on. It indirectly provided a model for later textual production: for example, the term "Amyloid" is translated on page 4 of the *XYDCYP* as *si ameili* 似阿美里; in the latter, *si* traces the bound morpheme -oid, with the meaning "similar to." The term was not listed in Yan Huiqing's dictionary, but five other lemmata would be included, in which the bound morpheme -oid is rendered with the character *si* by analogy to the *TVM* usage: "alkaloid," *si jian de* 似鹼的; "amianthoid," *si shima de* 似石麻的; "amoeboid," *si bianxing shuichong de* 似變形水蟲的; "antheroid," *si huafen nang de* 似花粉囊的; and "anthropoid," *si ren de* 似人的. Particularly pertaining to "alkaloid," the fact that the lemma "Alkaloids" has been rendered by Fryer on page 2 of the *HXCL* as *si jian lei* 似鹻類 has certainly influenced Yan Huiqing's rendering; the others were thus translated by analogy.

Other morphological features, which in the MCST database are considered as developed in a phase after the publication of the *TVM*, actually existed before the *TVM* was published; they were annotated, perhaps for the first time, precisely in Fryer's work. An example is the use of the character *jia* 假 as a left-hand component in compounds to convey the meaning of the element "pseudo-." In the *JS*, the terms "Pseudomalachite," rendered on page 27 as *jia malakaide* 假瑪拉開得 and "Pseudomorph, steatitic," *jia feizao shi* 假肥皂石, and on page 32 "Steatitic pseudomorphs," translated exactly like the latter, all show this usage of the character *jia*.[39] This feature spread in later texts, presumably due to the *TVM*'s influence, as demonstrated by the terms *jiaxing qiangyingzheng* 假性強硬症, translated as "pseudo-catalepsy," and *jiaxing mabi bing* 假性麻痺病, rendered as "pseudo-paralysis," both indicated in the

39 The translations quoted above are listed in the third column of the *JS*, namely the "Terms in general use and chief elements." In the second column, corresponding to the "Terms used in Dr. Macgowan's translation," the first lemma is translated as *jia mailaigaituo* 假麥來蓋脫, the second and third as *jia sidi'aideaituo* 假斯底哀得愛脫.

MCST database as neologisms attested for the first time in *A Dictionary of Philosophical Terms*.[40]

It should be noted that the employment of the character *jia* as a left-hand component with the meaning "pseudo-" is also widely attested in other glossaries within the *TVM*.[41] Furthermore, the character *jia* is used as a left-hand component with the meaning "artificial," "false," and at times with a zero translation in numerous other lemmata of the glossaries quoted, and of the *QJ*.

All these considerations clearly show how the contribution of the *TVM* is not limited to lexical innovation. It is not surprising that, due to the nature of lexicological works, even if the neologisms listed in the tables in the appendices are numerous, many other terms were already attested in previous works. To give some examples, *bai zhi* 白芷 or *ma huang* 麻黃[42] can be found in the dictionary *Pian ya* 駢雅 (The rhythmical guide),[43] while others, such as *shi zhuyu* 食茱萸 and *changpu* 菖蒲,[44] annotated in the farming treatise *Qimin yaoshu* 齊民要術 (Essential techniques for common people), date back to even earlier texts.[45] The use of some characters as right-hand components of words as well, such as *shi* 石 and *lei* 類, is attested in precedent works, at

40 Timothy Richard and Donald MacGillivray, ed., *A Dictionary of Philosophical Terms: Chiefly from the Japanese* (Shanghai: Christian Literature Society for China, 1913).

41 Respectively, on page 28 of the *HXCL*, in "Pseudomorphine," rendered as *jia moerfeini* 假莫爾非尼; on page 7 of the *XYDCYP*, in "Asclepias pseudo-sarsa," *jia shasha asikelibia* 假沙沙阿司可里比阿; on page 45, in "Panax pseudo-ginseng," *jia renshen banakesi* 假人參巴拿克司; on page 49 in "Pseudaconitine," *jia agunidini* 假阿古尼低尼 and "Pseudo-toxine," *jia dukexini* 假毒克西尼; and finally on page 59 in "Strychnos pseudo quina," *jia ji'na situoligeni* 假雞那司脫立格尼.

42 On pages 31 and 12 of the *XYDCYP* they indicate "Iris florentina" and "Cannabis sativa," respectively. According to the *Zhongguo jiben guji ku*, the first attestation of both *bai zhi* and *ma huang* dates back to the anonymous text *Jiran wanwu lu* 計然萬物錄 [Jiran's records of the ten thousand things], purportedly composed during the Zhou dynasty (1046–256 BCE) by the scholar Jiran 計然 (n.d.), also known as Wenzi 文子.

43 Completed by the scholar Zhu Mouhan 朱謀垏 (1552–1624) in 1587; the edition consulted by the author is Zhu Mouhan, *Pian ya* (Beijing: Zhonghua shuju, 1985), 83.

44 Respectively given, within the *XYDCYP*, on page 60 as a translation for "Sumach" and on pages 2 and 60 as a rendering for "Acorus calamus" and "Sweet flag." According to the *Zhongguo jiben guji ku*, the first attestation of *changpu* dates back to the sixteenth juan of *Hanfeizi* 韓非子, composed by the renowned philosopher Han Fei 韓非 (?–233 BCE) in the third century BCE, and that of *shi zhuyu*, to the *Jinkui yaolüe fanglun* 金匱要略方論 [On the essential prescriptions of the gold casket], a treatise composed during the Eastern Han dynasty (25–220) by the physician Zhang Zhongjing 張仲景 (ca. 150–ca. 219), also known as Zhang Ji 張機.

45 The text was completed by the scholar Jia Sixie 賈思勰 (n.d.) during the Eastern Wei dynasty (534–550); the edition consulted by the author is Miao Qiyu and Miao Guilong, ed., *Qimin yaoshu yizhu* (Jinan: Qilu shushe, 2009), 315 and 813.

times together. For example, the combination of the two characters can be read within the lemma "Granitic rocks," *huagang shi lei* 花剛石類, on page 16 of the *HXCL*, but it already appears since the *Gezhi jingyuan*: juan 33, *zhenbao lei* 珍寶類 (precious stones), section *ma'nao* 瑪瑙 (agate), page 6B. As the *Gezhi jingyuan* passage reads: "《拾遺記》瑪瑙石類也一說云馬腦者是惡鬼之血[...]" (According to the *Shiyi ji*, agates can also be called *ma'nao* and are the blood of demons). Analogously to precedent analyses in this work, in the *Gezhi jingyuan* an earlier text is mentioned, in this case the *Shiyi ji* 拾遺記 (Record of retrieved anecdotes), composed by the alchemist Wang Jia 王嘉 (?–ca. 385)[46] during the Eastern Jin dynasty (317–420).[47]

The same also applies to the character *shu* 屬; it is already listed in the *Gezhi jingyuan*, juan 62, *shu lei* 蔬類 (vegetables), section *manjing* 蔓菁 (turnip), on page 17B. The original text reads: "《本草》李時珍曰別錄以蔓菁蘆菔同條致諸說猜度或以二物為一種或謂在南為萊菔在北為蔓菁今按二物根葉花子俱別非一類也蔓菁是芥屬蘆菔是菘屬" (In the *Bencao*, Li Shizhen states that on the basis of other records, [the turnip] is also called *manjing* or *lufu*. According to other authors, it can be stated that the two are varieties of the same species, or that in the south it is called *laifu*, in the north *manjing*. Since the roots, leaves, and flowers of the two are different, we can affirm they do not belong to the same kind. The *manjing* belongs to the *jie*, while the *lufu* belongs to the *song*). In this case, the reference is to the well-known botanical treatise *BCGM*, already mentioned in this work.

One of the most remarkable achievements ascribable to the *TVM* lies in the stabilisation of the use of the aforementioned characters as right-hand components of words, even if the uncertainty in some of their usage and meaning is also evident within the *TVM*, and in the passage from the manuscripts to the printed version. One example of this doubt is the cancellation of entries from the *XYDCYP* manuscript that were destined to finally be included in the *TVM*. Within these entries, the character *ke* was used in the form of a bound morpheme to indicate families of plants. As an example, on page L 5 of the *XYDCYP* manuscript, Fryer has cancelled the entry "Linaceæ," translated with the phonemic loan *li'naxiyi* 里那西依 and as *ma ke* 麻科; however, the lemma was later included on page 35 of the *XYDCYP*.

Slightly different is the case of "Valerianaceæ," on page V 1 of the aforesaid manuscript. As in the previous example, the lemma was first translated with a phonemic loan, *falili'a'naxiyi* 發里里阿那西依, to which Fryer added another

46 Zheng, Kirk and Buell, *Dictionary of the Ben cao gang mu*, 498.

47 On the text, refer also to Erik Zürcher, *The Buddhist Conquest of China: The Spread and Adaptation of Buddhism in Early Medieval China*, 3rd ed. (Leiden: Brill, 2007), 276.

rendition, *baijiang ke* 敗醬科, the same found in contemporary Chinese dictionaries. The entry "Valerianaceæ," though, has not been cancelled, but is instead preceded by the symbol "x." It is precisely here that the uncertainty on the employment of such a character is more clear; in any case, the term will be published on page 66 of the *XYDCYP*. Such uncertainty, analysed further on in this work, will also persist in texts published years after the *TVM*. For example, in the *Bowu da cidian* (The great dictionary of natural history), in the Chinese term for the family of "Cruciferae," *shizihua shu* 十字花屬, the character *shu* is used, nowadays employed for genera.[48]

At times, more than the uncertainty in the use of characters as right-hand components of words, Fryer and his colleagues were dealing with problems inherent to the original scientific nomenclature itself. For example, the entry after the family of "Valerianaceæ," on page 66 of the *XYDCYP*, the "Valerianeæ," is a tribe within the Valerianaceæ. The "Valerianeæ" is first translated with the phonemic loan *falili'a'nayi* 發里里阿那依, followed by *baijiang ke* 敗醬科, the same term used to translate "Valerianaceæ." The name for the tribe is therefore merely treated as a variant of "Valerianaceæ." It is highly plausible that Fryer hesitated on the translation for the two lemmata due to the fact that in numerous botanical treatises of the nineteenth century, the Valerianeae was still considered a family.[49] This hypothesis could also explain the aforesaid symbol "x" added next to the entry within the *XYDCYP* manuscript, and the fact that the first of the two has been written in pencil.

The same applies to the lemma "Urticaceæ," rendered on page 65 of the *XYDCYP* with both the phonemic loan *erdikaixiyi* 爾低楷西依 and the term still used in contemporary Chinese, *qianma ke* 蕁麻科. Analogously to the previous example, Fryer and his colleagues translated the lemma "Urticeæ," a tribe of the Urticaceae listed in the *XYDCYP*, with the phonemic loan *erdixiyi* 爾低西依, accompanied by the same rendition employed for the "Urticaceæ," *qianma ke* 蕁麻科. In this case as well, numerous botany and pharmacology texts considered the Urticeae a family.[50] Nevertheless, in this circumstance, there are no

48 Zeng Pu and Xu Nianci, *Bowu da cidian* (Shanghai: Hongwenguan, 1907), 8.

49 Refer to Jacob Bell, ed., *Pharmaceutical Journal and Transactions: Volume I; 1841–2* (London: John Churchill; Edinburgh: Maclachlan & Stewart; Dublin: Fannin & Co., 1842), 575, and *The Asiatic Journal and Monthly Register for British and Foreign India, China and Australasia: Vol. XXXVII—New Series* (London: Wm. H. Allen and Co., 1842), 251.

50 As an example, refer to Henri Milne-Edwards and Pierre Dominique Vavasseur, *A Manual of Materia Medica and Pharmacy, Comprising a Concise Description of the Articles Used in Medicine; Their Physical and Chemical Properties; The Botanical Characters of the Medicinal Plants; And the Formulæ for the Principal Officinal Preparations of the American, Parisian, London, Dublin, Edinburgh, &c. Pharmacopœiæ; With Observations on the Proper*

indications in the manuscript of doubts in the translation choices; the printed version is congruent with the manuscript. Such uncertainty is evident in other manuscripts as well, such as the "Japanese list" within the LBT: on page 19, both the family of Berberidaceae and the tribe of Berberideae, belonging to the former, are translated as *funiu ke* 伏牛科.

While the use of some characters as right-hand components of words for taxonomical categories was already listed in earlier texts, one of the first attestations of the character *zu* 族 to indicate tribes of plants can be located in the *TVM*. The character, listed on page 40 of the VGT in the lemma "Family tribe or genus," marked with blue, is not indicated with this meaning and usage in either the MCST database or the *Jinxiandai ciyuan*. I could not locate it in the most relevant texts printed before the *TVM* that are not considered in the MCST and the *Jinxiandai ciyuan*.

This employment of the character *zu* can be considered a semantic loan based on the English correspondent, "tribe." In the *TVM* there are forty-eight occurrences of *zu* as a right-hand component of words, all within the *XYDCYP*, distributed across forty-three lemmata. In some lemmata the character is used as a right-hand component in alternative translations supplied for the same term, the first of which is usually a hybrid. Among the forty-eight recurrences, in three circumstances the character appears within a short syntagma in which the original correspondent entry is indicated as a tribe within a higher taxonomic level, therefore not falling within the aforesaid cases of usage in the strictest sense of the word. Furthermore, it should be specified that among the forty-eight occurrences, six denote taxonomic levels of the kingdom Animalia, nowadays reclassified as belonging to other levels of the zoological nomenclature. *Zu* is indeed still used in contemporary Chinese to indicate tribes of animals, an intermediate level between subfamily and subtribe.

This morphological innovation has to be considered even more important than the lexical ones. Its replicability permits its usage in numerous terms, without distinction among hybrids, calques or descriptive labelling; it therefore transcends the boundaries of neologisms, which in turn can be only incorporated, adapted, or discarded in a language. In any case, in line with different statements by Fryer, when hybrids incorporated the character *zu*, the phonetic portion was rendered with the same characters used for the hybrid indicating the family of plants to which the tribe in question belonged, or was at least believed to belong, sometimes only partially due to phonetic reasons. To clarify this statement with an example, on page 53 of the *XYDCYP*, the "Pyrolaceæ,"

Mode of Combining and Administering Remedies (Philadelphia: Carey, Lea & Carey, 1829), 143.

formerly considered a family and later reclassified as a subfamily within the Ericaceae, has been rendered by Fryer with the phonemic loan *boluolaxiyi* 伯羅拉西依 and *luticao ke* 鹿蹄草科, using in the latter the character *ke* for the families of plants. The subsequent lemma, "Pyroleæ," indicating a tribe of plants belonging to the Ericaceae, is translated as *Boluoliyi* 伯羅里依 and *luticao zu* 鹿蹄草族. In this case, a phonemic loan employing the same first two characters of the precedent one, *boluo* 伯羅, is used; the second rendering is instead the same as "Pyrolaceæ" in which the character *zu*, indicating tribes, is appended in place of *ke*.

The uncertainty seen in the manuscripts on the use of right-hand components of words also confirms that the *TVM* was the result of a long process with different phases. By way of illustration, on page 18 of the VCT section devoted to botany, the entry "Camellieæ" is first translated with the phonemic loan *kaimiliyi* 楷米里依, to which Fryer added *cha ke* 茶科 at a later phase. In the third version Fryer will change the term for this tribe, also known as Theeae, to *cha zu* 茶族. Here, the error simply derives from doubt in the use of the character for the adequate taxonomic level, since within numerous treatises of the time the "Camellieæ" was already considered a tribe.[51]

Finally, a mention of the usage in the *TVM* of the character *bu* 部 as a right-hand component of words is necessary. In the VGT, *bu* is annotated on page 74 as the rendering of the lemma "Orders,"[52] while in contemporary Chinese it is used to convey "cohort," a taxonomic level between the order and the class.[53]

In the *XYDCYP* seven occurrences of *bu* as a right-hand component of words can be located, but the referents all belong to taxonomic levels different from the cohort: the class of "Articulata,"[54] page 7, translated as *jiesheng bu* 節生部; the "Cryptogameæ," page 18, rendered as *wuhua bu* 無花部, plants with reproductive organs not completely visible or simply hidden; "Diploneura," indicating a suborder of the order of Tricladida, translated on page 20 as *jiesheng bu* 節生部; "Mollusca," *jie bu* 介部,[55] listed on page 39; "Pisces," generically referring

51 Refer to John Forbes Royle, *Essay on the Productive Resources of India* (London: Wm. H. Allen and Co., 1840), 260.

52 In contemporary Chinese, the character used for orders is *mu* 目.

53 Robert C. King and William D. Stansfield, *A Dictionary of Genetics*, 5th ed. (New York, Oxford: Oxford University Press, 1997), 371.

54 Ulrich Lehmann and Gero Hillmer, *Fossil Invertebrates*, trans. Janine Lettau (Cambridge: Cambridge University Press, 1983), 245.

55 As indicated in the *Zhongguo jiben guji ku*, the term is already attested as *linjie bu* 鱗介部 in the text *Huang lan* 皇覽 [The imperial conspectus], composed during the Three Kingdoms period (220–280) by different authors; on the *Huang lan*, refer also to Rafe De Crespigny, *A Biographical Dictionary of Later Han to the Three Kingdoms (23–220 AD)* (Leiden, Boston: Brill, 2007), 45.

to fish, translated on page 47 as *yu bu* 魚部;[56] "Spina cerebrata," an alternative form of "spini-cerebrata," a term in which vertebrates were formerly known, rendered on page 58 with *jigu bu* 脊骨部, and finally "Vertebrata," annotated on page 66 as *jigu bu* 脊骨部.[57]

It is difficult to affirm whether the choice of this character to indicate the taxonomic level of the cohort might have been influenced by the classification of the *BCGM*. There are indeed examples of terms attested for the first time in the latter, according to the *Zhongguo jiben guji ku*, and present in the same form in the *TVM*, such as *suomumian* 莎木麪, included in juan 3 and 31 in Li Shizhen's work, and listed on page 53 of the *XYDCYP* as a translation for "Sago palm," as well as on page 30 of the *HXCL* for "Sago."[58] One entry of the *XYDCYP* in particular, "Saffron," rendered on page 53 as *safa ji fanhonghua* 撒法即番紅花, is reported in exactly the same way it was attested for the first time in juan 52 of the *BCGM*. The two alternative translations were later included in analogous forms in juan 9 of the *Lu shu* 露書 (Book of dew), composed by the Ming dynasty scholar Yao Lü 姚旅 (?–1622) in the first part of the seventeenth century, and in juan 29 of the *Shiwu yiming lu* 事物異名錄 (Record of different names for things), written by Li Quan 厲荃 (n.d.) during the Qing dynasty.[59] In any case, notwithstanding the fact that the employment of *bu* in the *TVM* presents various uncertainties analogous to those of *ke*, the pattern supplied is the

56 The term was already included, with a different meaning, in the classification of the *Shuowen jiezi*, juan 12.

57 On the term "spini-cerebrata," see James Elwick, *Styles of Reasoning in the British Life Sciences: Shared Assumptions, 1820–1858* (London: Pickering and Chatto, 2007), 172, footnote no. 17.

58 Refer also to Hua, Buell, and Unschuld, *Dictionary of the Ben cao gang mu*, 426, and to Zheng, Kirk, and Buell, *Dictionary of the Ben cao gang mu*, 424.

59 There are also instances of terms attested in earlier texts and later employed in both the *BCGM* and the *TVM*. For example, *wa'naqi jiu* 膃肭臍酒 was used for the first time in juan 8 of the *Leibian Zhu shi jiyan yifang* 類編朱氏集驗醫方 [A categorised compilation of the collected and tested medical prescriptions of Mister Zhu], a text comprising medical recipes composed during the Song dynasty (960–1279) by the medical scientist Zhu Zuo 朱佐 (n.d.), and later in juan 3 of the *Yinshan zhengyao* 飲膳正要 [Necessary knowledge for drinking and feasting], a dietary manual written during the Yuan dynasty (1279–1368) by the doctor Hu Sihui 忽思慧 (n.d.). *Wa'naqi jiu* was later included in juan 25 of the *BCGM* and in juan 219 of the *Puji fang* 普濟方 [Prescriptions for universal benefit], a treatise including prescriptions compiled in 1406 by various authors. It is true that the *Leibian Zhu shi jiyan yifang* was frequently quoted in the *Puji fang*, which in turn was employed as a source of quotations by Li Shizhen, as indicated in Zheng, Kirk, and Buell, *Dictionary of the Ben cao gang mu*, 212. *Wa'naqi jiu* was finally adopted, presumably quoting from the *BCGM* and not from other earlier works, in the *XYDCYP* on page 12 for "Castor tincture of" and "Castorei tincture," on page 62 for "Tinctura castorei," and on page 63 for "Tincture castor."

same of *zu*, which proved to be adequate, permanently entering the scientific and technical lexicon of contemporary Chinese.

11 Conclusion

The work Fryer and his Chinese colleagues accomplished in the compilation of the TVM is remarkable from several different points of view. By adopting existing nomenclature, creating neologisms whenever believed necessary, and substituting them in case their lexicographical research led them to find a more suitable term, the translations carried out at the Jiangnan Arsenal and crowned with the TVM all contributed to enriching the Chinese lexicon in a crucial phase of its development. Their work also influenced later textual, and more specifically, lexicographical production for a considerable length of time, as the next chapters demonstrate.

The information provided in the tables in the appendices should not let the reader believe that the TVM is only worth researching for its lexicological achievements. The TVM is a mirror reflecting the status quo of the scientific and technical lexicon at the end of nineteenth-century China. It can let us infer, in a condensed form, both contemporary and earlier texts available for foreign translators and their Chinese colleagues to adopt as a reference for their work.

There were also examples of institutions or translators that intentionally chose not to adopt John Fryer's vocabularies, instead developing their own nomenclature. This was the case, for example, of the Publication Committee of the Educational Association of China, which Fryer himself was a member of. Even though Fryer was not re-elected at a later phase by the committee, he still used personal funds for the publication of the glossaries edited in cooperation with his Chinese colleagues. Among the reasons for the refusal of Fryer's choices, it is possible to include linguistic considerations: in an untitled document in the FP, signed by Mateer and Hayes, Fryer's nomenclature for chemical elements was disapproved because his "list has six elements all called 'shi'."[60] Despite the criticisms, we have already seen that it was, on the contrary, Fryer and Xu Shou's adopted method that survived even in contemporary Chinese.

Personal hostilities were at work as well. In the same document, citing the need to republish a huge quantity of texts if the revisions required by the

60 FP, Banc Mss C—B 968, carton 3, folder 34, "Miscellaneous Materials—Material of Chinese Scientific terminology." Courtesy of The Bancroft Library, University of California, Berkeley.

committee were implemented, Mateer and Hayes resentfully affirmed: "If changes generally admitted to be desirable are ruled out, not on the real merits of the case, but simply because of an aversion to making so many alterations, or interfering with the work of this or that author, we may find it difficult to show that we have been faithful to the trust committed on us."[61] The reference here is most probably to Fryer. Despite such harsh considerations, we will see many years later that Mateer would change his ideas and incorporate, voluntarily or not, numerous neologisms and terms derived from the TVM. Life and work decisions by Fryer himself also contributed to the partial rejection of his nomenclature; he left China in 1896 for California, abandoning his project for the second volume of the TVM and for the improvement of the nomenclature.

In any case, the most relevant factor for the rejection of the terms listed in the TVM, other than power relations between different missionaries and purely linguistic considerations, was represented by the historical and political circumstances of late Qing China. China's request of texts pertaining to science and technology was all the more pressing after its defeat in the First Sino-Japanese War, as stated by Fryer and analogously by Mateer and Hayes: "The present time is opportune—more so than any that is likely to occur again. The opportunity is upon us, and must be embraced or it will be lost forever. The Japanese war has excited a great interest in western learning—an interest which is just now taking shape and seeking the means for its development."[62] It was exactly in this context that a portion of the nomenclature proposed in the TVM was discarded, while other terms were incorporated into the lexicon. Chinese scholars were more and more eager to assimilate Western learning, especially after 1895. In order to fulfil this thirst for knowledge, one of the quickest ways was to address Japanese texts, which had the undeniable advantage of sharing, at least partially, written forms with the Chinese language.

61 *Ibidem.*
62 *Ibidem.*

CHAPTER 7

The Translator's Vade-mecum: Role, Diffusion, and Influence on Later Texts

1 Introduction

Had the lexicological and morphological innovations presented in the TVM remained confined to it, they would have been unlikely spread into the Chinese lexicon and, consequently, the TVM's contribution would be extremely limited. This is not the case, however: the TVM circulated extensively and its nomenclature was adopted within other works, printed contemporaneously to and after 1888 and 1889, respectively the publishing year of the first edition of the TVM and the QJ.

The traces of its lexicological innovations, and consequently of the circulation of ideas behind them, lingered in Chinese lexicon for a long time after its publication. The TVM's success was connected to the historical circumstances and the demand for new kinds of publications in late Qing China. Such texts, divided by categories, are listed here and analysed according to publication year, demonstrating with various degrees of certainty the use of neologisms included in the TVM and their contribution to its objectives. The lexicological legacy of the TVM extended chronologically, reaching the first decades of the twentieth century. After contributing to the formation of scientific and technical knowledge in the last part of the nineteenth and the beginning of the twentieth century, its influence gradually started to fade, yet it left its marks on the contemporary Chinese lexicon.

2 Encyclopaedias, Collectanea, and Periodicals

One of the first texts to adopt the TVM as a reference for technical nomenclature, extensively employing its terms, was the *Shiwu tongkao*. This encyclopaedic work was compiled by Qilu zhuren 杞廬主人 (n.d.) and published in Shanghai in 1897.[1] Numerous terms adopted in the TVM, generally considered neologisms that appeared in a later phase, were actually included in the *Shiwu*

1 Refer to the relevant entry in Rudolf G. Wagner, "The Formation of Encyclopaedic Commonplaces during the Late Qing: Entries on the Newspaper," in *Chinese Encyclopaedias of New*

 DOI:10.1163/9789004443211_009

tongkao. For example, *Pulushi lan* 普魯士藍, listed in juan 24 and 25 of the text, respectively, dedicated to *huaxue* 化學 (chemistry) and *dianxue* 電學 (electricity), was already annotated in the *XYDCYP*, on page 24, as the translation for "Ferri sesquiferrocyanidum," and in the *HXCL*, on page 6 for "Blue Prussian," page 19 for "Iron sesquiferrocyanide," and page 28 for "Prussian blue." Analogously, *Menzi tong* 們子銅 can be read in juan 27 of the *Shiwu tongkao*, devoted to *qixue* 汽學 (pneumatics), recovering the term employed on page 34 of the *QJ* for "Muntz metal." At times, the author of the *Shiwu tongkao* also modified the solutions adopted in the *TVM*. A typical example is *jinlei he yang zhi zhi* 金類合養之質 (The substances [composed] of metals combining with oxygen), listed in four passages of juan 13 of the work, devoted to *kuangwu* 礦務 (mineralogy). The locution is reported instead as *jinlei he yangqi zhi zhi* 金類合養氣之質 on page 22 of the *HXCL*, to indicate "Metallic oxides"; the change consists in the elision of the character *qi* 氣 within the term referring to oxygen. These examples only represent a small portion of those included in the *Shiwu tongkao*, covering the various scientific and technical fields touched upon in the *TVM*'s glossaries. The fact that the *Shiwu tongkao*—a representative work the scientist Ren Hongjun 任鴻雋 (1886–1961) indicated was one of the texts that late Qing Chinese scholars used for exam preparations[2]—extensively and systematically adopted the *TVM*'s terminology confirms that the latter was considered and proved to be an efficient means of conveying scientific and technical knowledge, being adequate from the point of view of both form and content.

Among the texts inspired by the *TVM* and adopting its nomenclature is one of the editions of the *Huangchao jingshi wenbian* 皇朝經世文編 (Collection of Qing dynasty writings on statecraft). The compilation of the *Huangchao jingshi wenbian*, an encyclopaedic collection of texts on government affairs, was started by two eminent Qing scholars, Wei Yuan and He Changling 賀長齡 (1785–1848), in close connection with the *Jingshi zhi xue* 經世之學 (statecraft learning) movement; the work was printed in 1828.[3] The editions derived from

Global Knowledge (1870–1930): Changing Ways of Thought, ed. Milena Doleželová-Velingerová and Rudolf G. Wagner (Heidelberg: Springer, 2013), 107.

2 The inclusion of the *Shiwu tongkao* among texts for exam preparations is indicated in Iwo Amelung, "*The Complete Compilation of New Knowledge*, Xinxue beizuan 新學備纂 (1902): Its Classification Scheme and its Sources," in *Chinese Encyclopaedias of New Global Knowledge (1870–1930): Changing Ways of Thought*, ed. Milena Doleželová-Velingerová and Rudolf G. Wagner (Heidelberg: Springer, 2013), 90.

3 For more specific details, consult Zhong Shaohua, *Renlei zhishi de xin gongju—Zhong-Ri jindai baike quanshu yanjiu* (Beijing: Beijing tushuguan chubanshe, 1996); Andrea Janku, "Translating Genre: How the 'Leading Article' became the *Shelun*," in *Mapping Meanings: The Field of New Learning in Late Qing China*, ed. Michael Lackner and Natascha Vittinghoff

the *Huangchao jingshi wenbian* are numerous. Among them, the *Huangchao jingshi wen sanbian* 皇朝經世文三編 (Third collection of Qing dynasty writings on statecraft), edited by Chen Zhongyi 陳忠倚 (n.d.) in 1898, can be mentioned as one that surely adopted the terms included in the TVM.[4]

More specifically, the twelfth juan of the *Huangchao jingshi wen sanbian* is devoted to chemistry, and within the juan, a text attributed to Yang Yuhui 楊毓輝 (n.d.), "Zhong wai huaxue mingmu yitong kao 中外化學名目異同考" (Considerations on the differences and similarities of Chinese and foreign chemical terms), can be read.[5] As indicated by the title, in the pages written by Yang Yuhui the question of discrepancies existing in the chemical nomenclature between Chinese and Western languages is debated. Numerous technical terms adopted in the paper are drawn from exactly those employed by Fryer and his colleagues in their translations, and later gathered within the TVM. Terms specifically pertaining to chemistry can be located, such as *cuoyang caosuan* 錯養草酸, a translation of "Cerii oxalas" on page 14 of the XYDCYP, and a rendering on page 9 of the HXCL of "Cerium oxalate."[6] Confirmation of the use within the *Huangchao jingshi wenbian* of the TVM's nomenclature can be specifically derived from the term *gong*$_{er}$*lü* 汞二綠, for which in the former, another rendering, *qingfen* 輕粉, is also annotated.[7] The two translations are listed, with the same order, as an alternative to each other on page 39 of the XYDCYP within the lemma "Mercury subchloride." It should be noted that *qingfen* is still widely used in contemporary Chinese to indicate mercury(I) chloride, or mercurous chloride.

As to the adoption in the *Huangchao jingshi wenbian* of the TVM's nomenclature, the relationship between some *Jingshi wenbian* 經世文編 (Collection of writings on statecraft) style publications and the *Gezhi shuyuan*, established

(Leiden, Boston: Brill, 2004), 343–44; and especially Benjamin A. Elman, "The Relevance of Sung Learning in the Late Ch'ing: Wei Yuan and the *Huang-ch'ao ching-shih wen-pien*," *Late Imperial China* 9, no. 2 (1988).

4 Chen Zhongyi, ed., *Huangchao jingshi wen sanbian* (Shanghai: Baowen shuju, 1898). The version referred to here is the one republished in Idem, *Huangchao jingshi wen sanbian* (Taipei: Guofeng chubanshe, 1965).

5 Yang Yuhui was also the author of two other essays on chemistry: "Huaren jiangqiu xixue yong huawen yong xiwen libi ruohe lun 華人講求西學用華文用西文利弊若何論 [On the pros and cons of using Chinese or Western languages for Chinese people striving for Western learning]" and "Zhongguo liandan huaxue yitong shuo 中國煉丹化學異同說 [An exposition of the similarities and differences of Chinese alchemy and chemistry]," as reported within the bibliographical section of Wright, *Translating Science*, 498.

6 Chen, *Huangchao jingshi wen sanbian*, 222, corresponding to page 3B of the original edition.

7 Ivi, 226, corresponding to page 5B of the original edition.

and managed for a long period of time by, among others, Fryer,[8] has already been noted.[9] Other than the lexicological link between the two texts, an indirect hint at the connection that existed between the *Huangchao jingshi wenbian* and Fryer's texts, not least the *TVM*, can be derived from the file "Shanghai shi jiaoyuju guanyu shili tushuguan jieshou *Gezhi shuyuan* shuji qingce 上海市教育局关于市立图书馆接收格致书院书籍清册" (A list by the Shanghai Bureau of Education of books donated by the Polytechnic institution to the municipal library).[10] In the document, among the texts supplied by the *Gezhi shuyuan* for the creation of a library in the city of Shanghai,[11] the *Huangchao jingshi wen xinbian* 皇朝經世文新編 (New collection of Qing dynasty writings on statecraft), the *Huangchao jingshi wenbian* and the *Huangchao jingshi wen sanbian* are all mentioned.[12] Even more important to highlight is that in the *Huangchao jingshi wen xinbian*,[13] composed by Mai Zhonghua 麥仲華 (1876–1956), one of Kang Youwei's 康有為 (1858–1927) students, the glossaries analysed in this work are explicitly mentioned: "傅蘭雅在制造局所譯化學汽機各書皆列中西名目表[...]" (All texts on chemistry and the steam engine translated by Fryer at the Jiangnan Arsenal included glossaries with parallel translation [...]).[14]

---

8 Refer to John Fryer, ed., *Gezhi shuyuan xixue kecheng* (Shanghai: 1895). On the connection between Fryer and the *Gezhi shuyuan*, secondary literature is abundant. Consult in particular Biggerstaff, "Shanghai Polytechnic Institution," and Wang Ermin, *Shanghai gezhi shuyuan zhilüe* (Hong Kong: Xianggang zhongwen daxue chubanshe, 1980).

9 A reference can be found in Xiong, *Xixue Dongjian*, 293; it should be noted, though, that the *Huangchao jingshi wen sanbian* is not expressly mentioned.

10 Archive number Q235 1 2133, file number 2133, dated twelfth month of the twentieth year of the Chinese Republican period, stored in the Shanghai dang'anguan 上海档案馆 (Shanghai Municipal Archives).

11 For the role of the *Gezhi shuyuan* in education, refer to Hao Bingjian, "Shanghai Gezhi shuyuan ji qi jiaoyu chuangxin," *Qing shi yanjiu—Studies in Qing History*, no. 3 (2003) and Li Zhijun, "Gezhi shuyuan yu shixue jiaoyu," *Qing shi yanjiu*, no. 3 (1999). On the influence the *Gezhi shuyuan* had on Chinese intellectuals, see Shang Zhicong, "1886–1894 nian jian jindai kexue zai wan Qing zhishi fenzi zhong de yingxiang—Shanghai Gezhi shuyuan gezhi lei keyi fenxi," *Qing shi yanjiu—Studies in Qing History*, no. 3 (2001). Pertaining in particular to its connection with the Chinese reformer and translator Wang Tao 王韜 (1828–1897), consult Wang Tao, ed., *Gezhi shuyuan keyi,* fifteen volumes (Shanghai: Shanghai tushu jicheng yinshuju, 1887–1894) and Paul A. Cohen, *Between Tradition and Modernity: Wang T'ao and Reform in Late Ch'ing China* (Cambridge, MA: Harvard University Press, 1974), 181–84.

12 Refer also to Wang Yangzong, "Shanghai gezhi shuyuan de yi fen yishu qingdan," *Zhongguo keji shi zazhi—The Chinese Journal for the History of Science and Technology* 27, no. 1 (2006).

13 Mai Zhonghua, ed., *Huangchao jingshi wen xinbian* (publication place and year unknown).

14 The text also includes Luan Xueqian, "Gezhi shuyuan jiaoyan huaxue ji," *Xinxue bao*, no. 1 (1897), later also reported in the journal *Kexue* 科學 [Science]; see Liu Guangding,

Moving to other works, proof of the diffusion of the *TVM* is also recognisable within the *Gezhi xinbao* 格致新報; some instances from the periodical, published in a limited period of time from January to August 1898, are self-explanatory.[15] In number 9, on page 16A, questions on how to resolve, with the help of chemical notions, problems encountered in everyday life are reported: question number 109 examined the way to remove stains on fabric, and the reply suggested was to use water with "Oxalate of potash." The term, also reported in English in the text, is rendered as *jiayang caosuan* 鉀養草酸, a neologism attested for the first time in the *XYDCYP* on page 44 as the rendering of the namesake entry, and on page 48 of "Potash oxalate."[16]

Similarly, in number 5, 1898, on page 6A, it is possible to read "[…] 則煤與養氣之愛力顯，故煤被燒之後，即於空中與養氣相合而成一種氣，名為炭酸。 炭酸亦名炭氣即炭養二 […]" (The affinity between carbon and oxygen shows and therefore, after the carbon is burnt, it mixes in the air with oxygen, creating a gas, whose name is carbonic acid. Carbonic acid is also known as *tan qi* and *tanyang er*). In the passage, the term *tan suan* 炭酸 is annotated. It was listed as the translation of "Acid carbonic" on page 2 of the *JS*, specifically in the second column, referring to "Terms used in Dr. Macgowan's translation"; *tan suan* was, therefore, used originally within the *Jinshi shibie*, the translation by Macgowan and Hua Hengfang of Dana's *Manual of Mineralogy*. In the passage within the aforementioned number of the *Gezhi xinbao*, though, two other alternative translations can also be read, respectively *tan qi* 炭氣 and *tanyang*$_{er}$ 炭養$_{二}$, attested in the *XYDCYP*, the *HXCL*, and the *JS*.[17]

Nevertheless, in some circumstances, ambiguities at the referent level between the translation choices of the *TVM* and those of the *Gezhi xinbao* are

---

Zhongguo kexue shi lunji (Taipei: Guoli Taiwan daxue chuban zhongxin, 2002), 508, and Zhao Zhongya, "Huaren bianji Luan Xueqian yu *Gezhi huibian*," *Shilin—Historical Review*, no. 2 (2011).

15 For a presentation of the periodical, refer to Xiong, *Xixue Dongjian*, 354–66, in which a useful comparison with the contents of the *Gezhi huibian* is also supplied; Elman, *On Their Own Terms*, 386; and Wright, *Translating Science*, 100–27. On the influence the text had on scientific divulgation in China at the end of the nineteenth century, consult Wang Fengyun and Gu Jiashan, "*Gezhi xinbao* ji qi dui huaxue zhishi de chuanbo," *Huaxue jiaoyu*, no. 10 (2012).

16 The first attestation, according to the *Zhongguo jiben guji ku* is indeed in John Fryer and Zhao Yuanyi, trans., *Yejin lu* (Shanghai: Jiangnan zhizaoju, 1873), a translation of *Moulder's and Founder's Pocket Guide*, composed by Frederick Overman (1803–1852) and published for the first time in 1851. This confirms once more that numerous terms analysed in this work were included for the first time in the texts composed by Fryer and his Chinese colleagues at the Jiangnan Arsenal, later merging into the *TVM*.

17 In *tanyang*$_{er}$, though, the symbol corresponding to the number of atoms is not reported in subscript, due to typographical reasons.

evident. An example can be located in number 6 of the periodical, within the section *Da wen* 答問 (Questions and answers), page 13B, in which the term *jiabulike suan* 加布里克酸 is used to render "Carboric Acid," whereas the same phonemic loan was employed on page 8 of the *HXCL* to indicate "Capric acid." In any case, translation choices converging to or diverging from those adopted in the *TVM* are traceable in many passages within the periodical, as for example again in number 6, page 21A: "Alizarine" is rendered with the phonemic loan *ailiselin* 愛理色林, not *alisalini* 阿里撒里尼, listed on page 2 of the *HXCL* as a translation of the namesake entry.

These examples confirm the hypothesis that terms in the *Gezhi xinbao* drawn from the *TVM* represent in their entirety a consistent number, but also that different translation choices and modifications were adopted in the periodical. The examples supplied aim to demonstrate the importance that a periodical published in a limited time span, but with national circulation, played in the spread of the *TVM*'s scientific and technical nomenclature.[18]

Another text extensively adopting the *TVM* sectoral nomenclature was the *Fuqiangzhai congshu xuji* 富強齋叢書續集 (A sequel to the Collectanea of the prosperous and strong studio).[19] In the latter, a considerable quantity of books published as first editions by the Jiangnan Arsenal is included. Among them, those translated by Fryer in collaboration with his Chinese colleagues, such as *Yidali canshu* 意大利蠶書 (Italian silkworms), *Qiukuang zhinan* 求礦指南 (The guide of prospecting for minerals),[20] *Bingchuan qiji*, and *Huaxue qiushu*, are numerous.[21] In order to demonstrate that the *Fuqiangzhai congshu xuji* contributed to spreading the *TVM*'s terminology, the fact that it comprises two glossaries already published in the *TVM*, the *QJ* and the *XYDCYP*, within volumes 34 and 11 respectively, is even more relevant.[22]

The role the *Fuqiangzhai congshu xuji* played in divulging the *TVM*'s scientific and technical nomenclature is not limited to the reproduction of entire texts Fryer and his Chinese colleagues rendered in Chinese, in which their translation choices were adopted. This also happened with numerous other

18 On the circulation of the text, see Xiong Yuezhi, *Xixue Dongjian yu wan Qing shehui*, rev. ed. (Beijing: Zhongguo Renmin daxue chubanshe, 2011), 356.

19 Yuan, *Fuqiangzhai congshu xuji*. When referring to this text, "volume" is meant as a translation of the term *ce* 冊. The author consulted the edition stored in the library of rare books of Fudan University.

20 John Fryer and Pan Song, trans., *Qiukuang zhinan* (Shanghai: Jiangnan zhizaoju, 1899).

21 For other details on the publication year, the authors of the original texts, and the foreign and Chinese translators of these and other works published by the Jiangnan Arsenal, refer to Wang Yangzong, "Jiangnan zhizaoju."

22 The inclusion of the two glossaries is also confirmed in Xia Jing, "Wan Qing keji shuyu de fanyi—Yi Fu Lanya wei zhongxin" (PhD diss., Wuhan daxue, 2011).

texts: in the *Cefu tongzong* 策府統宗 (A summary of the imperial library), works such as *Dili xuzhi* 地理須知 (Essentials of geography), *Dianxue xuzhi* 電學須知 (Essentials of electricity),[23] and *Tianwen xuzhi* 天文須知 (Essentials of astronomy), edited by Fryer for the "Science Outline Series," *Gezhi xuzhi*, or *Shuishi caolian* 水師操練 (Training for naval forces), translated by Fryer and Xu Jianyin and published by the Jiangnan Arsenal, are included.[24] The same can be said for the *Xixue fuqiang congshu* 西學富強叢書 (Collectanea of Western learning, wealth, and power),[25] *Xizheng congshu* 西政叢書 (Collectanea on Western governance),[26] and *Tongxuezhai congshu* 通學齋叢書 (Collectanea of the profound knowledge studio).[27] The importance of the *Fuqiangzhai congshu xuji* was not limited to the inclusion of some of the glossaries published in the TVM; other texts in the *Fuqiangzhai congshu xuji* in which the TVM's terminology is adopted are also included, taking into consideration that this collectaneous work gathered numerous publications first printed by the Jiangnan Arsenal.

Only two examples are quoted here: the *Neike lifa qianbian* 內科理法前編 (First collection of the principles and methods of internal medicine) and the *Neike lifa houbian* 內科理法後編 (Second collection of the principles and methods of internal medicine), whose original text corresponds to *Hooper's Physician's Vade Mecum: A Manual of the Principles and Practice of Physic; With an Outline of General Pathology, Therapeutics, and Hygiene*, by the English physician Robert Hooper (1773–1835).[28] The two texts are included in the eighth and ninth volume of the *Fuqiangzhai congshu xuji*, translated by Shu Gaodi 舒高第 (1844–1919), who spent more than ten years in the United States and became one of the most important translators active at the Jiangnan Arsenal, and annotated by Zhao Yuanyi; they both list numerous lemmata included in the TVM.[29] In juan 6, pages 2A and 2B, for example, terms such as *gaiyang linyang*$_{wu}$

23 John Fryer, *Dianxue xuzhi*, collected in *Gezhi xuzhi yi jiben*, ed. John Fryer (Shanghai: Gezhi shushi 1887).

24 Cai Meian, *Cefu tongzong* (Shanghai: Hongwen shuju, 1889).

25 Zhang Yinhuan, *Xixue fuqiang congshu* (Shanghai: Hongwen shuju, 1896), as also indicated in Kokusai Kirisutokyō daigaku Ajia bunka kenkyū iinkai, ed., *Nihon Kirisutokyō bunken mokuroku: Meiji ki* (Tokyo: Kokusai Kirisutokyō daigaku, 1965), 275. On the text, refer to Doleželová-Velingerová and Wagner, *Chinese Encyclopaedias*, 11–12.

26 Qiu ziqiangzhai zhuren, *Xizheng congshu* (Shanghai: Shenji shuzhuang, 1897).

27 Zou Lingyuan, *Tongxuezhai congshu* (publication place and year unknown).

28 As mentioned in Bridie Andrews, "Blood in the History of Modern Chinese Medicine," in *Historical Epistemology and the Making of Modern Chinese Medicine*, ed. Chiang Howard (Manchester: Manchester University Press, 2015), 122.

29 The information on Shu Gaodi comes from Biggerstaff, *The Earliest Modern Government Schools*, 178–79.

鈣養燐養$_{\text{五}}$, *bai shifen* 白石粉, and *danqing*$_{si}$*yang linyang*$_{wu}$ 淡輕$_{\text{四}}$養燐養$_{\text{五}}$ can be read. The original text is as follows: "補其骨之所少土類質用鈣養燐養$_{\text{五}}$五釐至十釐合於白石粉[…]如嬰孩身力軟弱淡輕$_{\text{四}}$養燐養$_{\text{五}}$可代鈣養燐養$_{\text{五}}$" (In order to obviate to the lack of earth substances in bones, use from five to ten *li* of phosphate of lime to combine in chalk […] If children are weak, the ammonia phosphate can substitute the phosphate of lime). Phosphate of lime was indeed used to treat rickets and its Chinese equivalent is listed in the *TVM* as *gaiyang linyang*$_{wu}$ 鈣養燐養$_{\text{五}}$.[30] In the *XYDCYP*, it is annotated on page 11 as the rendering of "Calcis phosphas," on page 35 as "Lime phosphate" and on page 46 as "Phosphate lime"; in the *JS*, it is attested on page 4 as the rendering of "Apatite." *Bai shifen* 白石粉 is annotated within the *XYDCYP* on page 11 as the translation of "Calcis carbonas friabilis," on page 14 as "Chalk" and on page 18 as "Creta"; in the *HXCL* it is annotated on page 9 as the rendering of "Chalk"; on page 8 of the *JS* it can be found as the translation of "Chalk, (Calcite)."[31] Finally, *danqing*$_{si}$*yang linyang*$_{wu}$ 淡輕$_{\text{四}}$養燐養$_{\text{五}}$ is annotated within the *XYDCYP* on page 4 as a rendering of "Ammonia phosphate" and "Ammoniæ phosphas," and on page 46 as the equivalent of "Phosphate ammonia." It is also used as part of the translation in another lemma of the *XYDCYP* and in the *JS* it can be found on page 3 as the rendering of "Ammonia phosphate."

Phonemic loans can also be found further on in the *Neike lifa qianbian* and the *Neike lifa houbian*, such as *feibulini* 非布里尼, within the same text, on page 3A: "血有病形非布里尼之多 […]" (If there is a disease of the blood, there is too much fibrin). The term indicating fibrin is attested in the *XYDCYP* and in the *HXCL*, and the same also applies to hybrids on page 4B, as *yibige zasan* 叱嗶喀雜散: "在臨卧時用叱嗶喀雜散十釐至十二釐為一服" (Before sleeping, take a dose of around ten to twelve *li* of compound powder of ipecacuanha). *Yibige zasan* is annotated within the *XYDCYP* on page 31 as the translation of "Ipecacuanha pulvis comp:," on page 49 as "Powder ipecacuanha compound," and on page 50 as "Pulvis ipecacuanha compositus."

Another example can finally be mentioned: the nomenclature adopted in the *Nongxue chuji*, a text included in volume 25 of *Fuqiangzhai congshu xuji*, also derives from the *TVM*, as demonstrated by numerous examples.[32] One of them is on page 2A, where *lü*$_{er}$*yang*$_{san}$ *xiyang*$_{er}$ 鋁$_{\text{二}}$養$_{\text{三}}$矽養$_{\text{二}}$ can be read, attested

---

30 On the use of phosphate of lime, see *Encyclopaedia Britannica: Or a Dictionary of Arts, Sciences, and Miscellaneous Literature; Enlarged and Improved*, vol. 12, 5th ed. (Edinburgh: Printed for Archibald Constable and Company, 1817), 737.

31 On the term *bai shifen*, refer also to Shellen Xiao Wu, *Empires of Coal: Fueling China's Entry into the Modern World Order, 1860–1920* (Stanford: Stanford University Press, 2015), 85.

32 Francis Huberty James and Fan Xiyong, trans., *Nongxue chuji* (Shanghai: Jiangnan zhizaoju, 1898).

on page 2 of the *HXCL* as a translation of "Alumina silicate" and on page 7 of the *JS* as indicating the principal element of "Bucholzite, (Sillimanite)," within the third column of the glossary, referring to the "Terms in general use and chief elements."

In conclusion, as already pointed out in secondary literature, the *Fuqiangzhai congshu xuji* played an important role in China in the circulation of knowledge, particularly scientific and technical, in the last part of the nineteenth century and in the first part of the twentieth century, thanks to the diffusion of lithography.[33] Therefore, the fact that a considerable number of terms coined or employed by Fryer and his Chinese colleagues is included within such an important work, which also gathers entire glossaries drawn from the *TVM*, confirms their indisputable contribution to the diffusion of the terminology included in the latter.[34]

Proof of the *TVM* 's influence on later textual production can be found within many other works, for example in the *Taixi yixue tongkao* 泰西藝學通攷 (A comprehensive examination of Western technical learning), edited by He Liangdong 何良棟 (n.d.) and published in Shanghai by Hongbao shuju 鴻寶書局 in 1901. The *Taixi yixue tongkao* is an encyclopaedic work composed during the late Qing dynasty and divided into sixteen juan, among which the seventh is devoted to chemistry; some terms from the text can provide confirmation of the previous statement.[35] Analysing, for example, juan 7, devoted to the description of chemical compounds, on page 27A *danqing*$_{si}$*yang tanyang*$_{er}$ 淡輕$_{四}$養炭養$_{二}$ is listed. The term was annotated on page 4 of the *XYDCYP* as a rendering of "Ammonia carbonate" and "Ammoniæ carbonas," and on page 12 of "Carbonate of ammonia." It also appeared in the *HXCL* on page 3 as a rendering of "Ammonia carbonate," and in the *JS* on page 3 as a translation of the namesake entry.

Analogously to the texts previously analysed in this section, more proof of the consultation of the *TVM* by *Taixi yixue tongkao* authors can be obtained

---

33 As indicated in Han Qi, "Le début de la diffusion des techniques d'imprimerie occidentales en Chine à la fin des Qing: l'exemple de l'introduction du procédé lithographique," *Histoire et civilisation du livre: Revue internationale III* (Geneva: Librairie Droz S. A., 2007): 145.

34 The importance of the *Fuqiangzhai congshu* in the diffusion of texts translated at the Jiangnan Arsenal has been hinted at in Bai Limin, "Practicality in Curriculum Building: A Historical Perspective on the Mission of Chinese Education," *Frontiers of Education in China*, 8 (4): 527–28.

35 More specific references to the text can be located in Zhong Shaohua, "Studies on the Characteristics of Late Qing Encyclopaedia Entries," in *Chinese Encyclopaedias of New Global Knowledge (1870–1930): Changing Ways of Thought*, ed. Milena Doleželová-Velingerová and Rudolf G. Wagner (Heidelberg: Springer, 2013).

from phonemic loans. An example is a term listed on page 32A, in juan 6, consecrated to mineralogy: *beilumofeide* 貝路莫非得 is precisely the same phonemic loan adopted on page 29 of the *HXCL* to indicate "Pyromorphite." In this case as well, the adoption of the terminology of the *TVM*'s glossaries within a highly specialised text such as the *Taixi yixue tongkao*[36] further confirms the *TVM*'s influence on the Chinese scientific and technical lexicon.

In chronological order, the *Fenlei shiwu tongzuan* 分類時務通纂 (Comprehensive classified compilation of current affairs) follows in this list. The text is an encyclopaedic work compiled by Chen Changshen 陳昌紳 (n.d.) and was printed by Wenlan shuju 文瀾書局 in 1902. The *Fenlei shiwu tongzuan* is composed of different sections, among which the last, number 5, *Gewu lei* 格物類 (On the investigation of things), is further divided into various subsections, one specifically dedicated to chemistry. A notable number of terms is drawn from the *TVM*. For example, on page 2B of juan 240 of the section, *qianyang geyang*$_{san}$ 鉛養鉻養$_{三}$ is annotated. The term was adopted within the *HXCL*, on page 10, as the translation of "Chrome yellow," on page 20 as "Lead basic chromate" and "Lead chromate," and on page 37 as "Yellow, chrome." The term is also attested in the *JS* for "Chromate of lead" on page 8, on page 11 as the rendering of "Crocoisite, (Lehmannite)," and on page 19 as "Lead chromate"; in the latter case, it represents the principal element of the mineral within the third column.

In the case of the *Fenlei shiwu tongzuan*, there is also confirmation of the hypothesis that the *TVM*'s nomenclature was used, based on an extremely high number of phonemic loans attested within the former. For example, in juan 239, on page 18A, *xieliulusi* 寫留路司 can be found; *xieliulusi* was employed on page 9 of the *HXCL* as the rendering of "Cellulose." The substantial presence of terms derived from the glossaries Fryer and his Chinese colleagues edited in an encyclopaedic work such as the *Fenlei shiwu tongzuan* certifies the *TVM*'s importance and, at the same time, the role played by the former in spreading the lexicological innovations of the latter.

Continuing the list in chronological order, one of the subsequent texts of encyclopaedic nature in which a consistent number of terms drawn from the *TVM* is used, is the *Xixue santong* 西學三通 (Three compilations on Western knowledge).[37] The *Xixue santong* was supervised by Xie Ruochao 謝若潮 (n.d.)

36 Li Xiaoti, "Jianli xin shiye: wan Qing de baike quanshu jia," *Dongwu xueshu*, no. 2 (2012): 50. Refer also to Idem, "Jianli xin shiye: wan Qing de baike quanshu jia (xu)," *Dongwu xueshu*, no. 3 (2012).

37 A more detailed description of the text can be found in Wagner, "The Formation of Encyclopaedic Commonplaces," 107.

and edited by Yuan Zonglian 袁宗濂 (n.d.) and Yan Zhiqing 晏志清 (n.d.).[38] The examples attesting the adoption of the nomenclature of the TVM's glossaries, fourteen years after its first edition, are drawn from juan 134 of the last of the three sections in which the *Xixue santong* is divided, namely *Xiyi tongkao* 西藝通考 (Comprehensive investigation of Western techniques); juan 127 to 136 are all invariably dedicated to the description of flora and fauna.

Juan 134 opens with the description of *ehua zhiwu* 萼花植物, identifiable from their Chinese name as plants provided with calyx. Within the TVM, the closest term to *ehua zhiwu* is *ehua shu* 萼花屬, indicating the "Calycifloræ" on page 11 of the XYDCYP; "Calycifloræ" are indeed plants with stamina inserted within the calyx.[39]

In the introductory part of juan 134, numerous other families of plants are also described, in which the TVM's nomenclature, phonemic loans included, is employed. One of the families described is indicated by the term *tuolipingtayi ke* 脫里平他以科; the hybrid identifies Terebinthaceae, a term indicating various families of plants later reclassified separately, as the Chinese text also explains: "此科植物近時植物學家分為數科" (This family of plants has been recently classified by botanists in various families). Below the entry, other terms follow, printed in smaller characters—they all refer at various degrees to Terebinthaceae. To cite two examples: *bisitaxiya* 比司他西亞 indicates the genus Pistacia of the Anacardiaceae family, the latter an alternative name for Terebinthaceae; "Pistacia" had already been translated as *bisitaxiya* on page 47 of the XYDCYP. The second term considered is *duteng rusi* 毒籐如司, listed on page 52 of the XYDCYP beside the lemma "Rhus toxicodendron." It is exactly from this example that another important confirmation of the adoption of the TVM's nomenclature by *Xixue santong* authors can be derived: in the phonemic portion of the hybrid, both in the XYDCYP and in the *Xixue santong*, within the character *teng* 籐 the radical *zhu* 竹 (bamboo) is used, not *cao* 艹 (grass). These examples, analogously to the other works presented, demonstrate that the authors of the *Xixue santong*, an influential text composed during the late Qing dynasty, made full use of the TVM's nomenclature, phonemic loans included, therefore providing an important contribution to its diffusion.[40]

---

38 Yuan Zonglian and Yan Zhiqing, ed., *Xixue santong* (Shanghai: Cuixin shuguan, 1902).

39 Refer, among others, to Colin Milne, *A Botanical Dictionary: Or, Elements of Systematic and Philosophical Botany*, 3rd ed. (London: Printed for H. D. Symonds by Bye and Law, 1805) and *Encyclopaedia Britannica; Or a Dictionary of Arts, Sciences, and Miscellaneous Literature*, vol. 4, 3rd ed. (Edinburgh: Printed for A. Bell and C. Macfarquhar, 1797), 52.

40 On the role of the *Xixue santong*, refer to Zhang Qing, "'Cai xixue': xueke cidi zhi lunbian ji qi yiyi—Lüe lun wan Qing dui 'Xixue menjing' de tantao," *Lishi yanjiu*, no. 3 (2007): 126.

The subsequent and last item in this list of works instrumental in the diffusion of the *TVM*'s nomenclature, demonstrating its importance and influence, is another text derived from the *Huangchao jingshi wenbian*, namely *Huangchao jingshi wen xinbian xuji* 皇朝經世文新編續集 (A sequel to the New collection of Qing dynasty writings on statecraft), compiled by da Gan Han 甘韓 (n.d.) and published in Shanghai in 1902.[41] Within juan 7 of *Huangchao jingshi wen xinbian xuji*, in the section *Nongxue lun* 農學論 (On agriculture), the modalities to cultivate and take care of plants are described; a notable portion of terms in the chemical field is derived from the glossaries of the *TVM*. For example, *jiayang er guosuan* 鉀養二果酸,[42] *jiayang liuyang*$_{san}$ 鉀養硫養$_{三}$,[43] and many others[44] are mentioned.[45]

---

41 The edition consulted is the one republished in Lai Xinxia, ed., *Qingdai jingshi wen quanbian*, vol. 129 (Beijing: Xueyuan chubanshe 2010).

42 Lai Xinxia, *Qingdai jingshi wen quanbian*, 21 and 27, corresponding to pages 11A and 14A of the original edition. *Jiayang er guosuan* appears on page 18 of the *XYDCYP* as the translation of "Cream of Tartar," on page 48 as the rendering of "Potash acid tartate," "Potash tartrate acid" and "Potassæ bitartras," and finally on page 49 as the equivalent of "Potassæ tartras acida." Within the *HXCL*, *jiayang er guosuan* is annotated on page 12 as the translation for "Cream of tartar," on page 28 for "Potash bitartrate," and on page 34 for "Tartar."

43 Lai Xinxia, *Qingdai jingshi wen quanbian*, 21, corresponding to page 11A of the original edition. *Jiayang liuyang*$_{san}$ appears on page 48 of the *XYDCYP* as the equivalent of "Potash sulphate" and on page 49 as the rendering of "Potassæ sulphas." *Jiayang liuyang*$_{san}$ is also annotated within the *HXCL* on page 28 to translate "Potash sulphate," and "Sal polychrest" on page 30.

44 Both terms are reported as neologisms in table 45. According to the *Zhongguo jiben guji ku*, the first attestation of the former is indeed within Fryer and Zhao, *Yejin lu*.

45 Other texts can be noted here; first is Zhu Dawen and Ling Gengyang, *Wanguo zhengzhi yixue quanshu* (Shanghai: Hongwen shuju, 1902). Numerous terms within the work are drawn from the *TVM*, for example, in juan 103 of the section *Yixue congkao* 藝學叢攷 (Collectaneous examination of technical learning), *ningshui yao* 甯睡藥 can be found. *Ningshui yao* is attested for the first time on page 41 of the *XYDCYP* as the translation of "Narcotics," and was later also employed in juan 31 of the *Shiwu tongkao*. The second is Duan Fang, ed., *Da Qing Guangxu xin faling* (Shanghai: Shangwu yinshuguan, Xuantong period). By way of illustration, on page 1681 *han danyang*$_{wu}$ *zhi zhi* 含淡養$_{五}$之質 can be read, a translation on page 41 of the *XYDCYP* and page 24 of the *HXCL* of the lemma "Nitrates." In the renowned periodical *Dongfang zazhi* 東方雜誌 [The Eastern Miscellany], founded in 1904 and printed until 1948, many terms were also derived from the *TVM*, for example *danqing*$_{san}$ 淡輕$_{三}$. The latter, included in numerous Chinese lemmata of the *TVM*, is the translation of "Ammonia" on page 3 of the *JS*, the *XYDCYP*, and the *HXCL*, and can be found in different juan and issues of the *Dongfang zazhi*, such as juan 12, issue number 1, p. 7, with the number of atoms of hydrogen in the compound written as a subscript, analogously to the *TVM*. Finally, one of the most important newspapers of late Qing China, the *Shenbao* 申報 [Shanghai News], also included neologisms attested for the first time in the *TVM*. For example, *axiduoni* 阿西多尼, a rendering of "Acetone" on page 1 of the *XYDCYP*, is reported, together with the corresponding English lemma,

3 Dictionaries, Vocabularies, and Glossaries

In this section, attention is directed to the dictionaries, vocabularies, and glossaries that extensively employed the *TVM*'s nomenclature. One of the texts that certainly used the glossaries gathered in the *TVM* is the *TT* by Calvin Wilson Mateer, published in 1904. The dictionary adopted the *TVM*'s terminology especially for the field of engineering; numerous terms included in the *TT* are attested for the first time within the *QJ*, even if terms drawn from other glossaries are also present. It should be again pointed out that Mateer's dictionary, notwithstanding some problems pertaining to the creation of characters,[46] represents a fundamental step in the process of standardisation of the scientific and technical lexicon of modern Chinese,[47] as well as a vehicle for the neologisms coined in the *TVM*.

Pertaining to the link between the *TVM* and the *TT*, some of the terms signalled in the MCST database as neologisms attested for the first time in Mateer's dictionary were later used in the *De-Ying-Han wen kexue zidian* 德英華文科學字典—*Deutsch-Englisch-Chinesisches Fachwörterbuch*, compiled by the German missionary Richard Wilhelm (Wei Lixian 尉禮賢, 1873–1930) in 1911.[48] The *DECF* was published before the *ECD*, analysed further on, and its compilation is tightly connected to Mateer's *TT*. Examples of terms listed within the *DECF* and derived from the *TVM* are numerous; by way of illustration, two can be mentioned. The first is *laiwang dong* 來往動, a translation of "Motion reciprocating" on page 34 of the *QJ*, later used with the same meaning both by Mateer and, seven years later, in the dictionary by the German sinologist. The second is *chuqi men* 出汽門, a rendering of "Eduction valve" on page 18 and of "Exhaust valve" on page 20 of the *QJ*, later employed both in the *TT*, as a translation of "Exhaust valve," and in the *DECF*, for "exhaust-valve" and "head valve."

The complex interaction between the *TVM* and later lexicographical production is even more evident when considering other dictionaries and their

---

in issue number 21219, published on May 1, 1932, while *tulunzhou* 凸輪軸, translation of "Cam shaft" and "Cam wheel shaft" on page 9 of the *QJ*, is adopted in numerous issues of the newspaper, in some cases also with the first of the two relevant English lemmata, as in numbers 18270, 19003, 19569, and 21571.

46 Shen Guowei, *Jindai Zhong-Ri cihui*, 304.

47 Ji Meng and Ukai Atsuko, ed., *Translation, History and Arts; New Horizons in Asian Interdisciplinary Humanities Research* (Cambridge: Cambridge Scholars Publishing House, 2013), 19.

48 Richard Wilhelm, *De-Ying-Han wen kexue zidian—Deutsch-Englisch-Chinesisches Fachwörterbuch* (Qingdao: Deutsch-Chinesische Hochschule, 1911).

relevant connections. For example, the *ECD*, composed in 1916 by the German sinologist Karl Ernst Georg Hemeling (He Meiling 赫美玲, 1878–1925), is tightly connected to the employment of the *TVM*'s nomenclature within the *TT*. Numerous terms considered neologisms in the MCST database, attested for the first time in the *TT*, in reality originally appeared in the *TVM*; a portion of these neologisms, backdated in table 45, reappeared in Hemeling's dictionary, published in 1916, twelve years after Mateer's *TT*. Confirmation that the *TT* was used for the compilation of the *ECD*, at least indirectly, comes from the inclusion, among the dictionaries consulted within Hemeling's text, of *Technical Terms, English and Chinese*, composed in 1910 by the American Methodist missionary George A. Stuart (Shi Tuer 師圖爾, 1859–1911).[49] The proof can be located in the preface: "While all standard works on the Chinese language have been freely consulted, the following books have been more extensively drawn from: [...] 15. Geo. A. Stuart: English-Chinese technical terms,"[50] the latter indeed representing the enlarged version of Mateer's *TT*.[51] The connection between the nomenclature of the *TVM* and that of later glossaries and dictionaries is further complicated by the fact that Hemeling also mentions the *DECF*, "[...] 16. R. Wilhelm: German-English-Chinese dictionary of technical terms," among the texts used for the compilation of his *ECD*.[52]

Some examples, drawn from the different dictionaries cited, can help clarify the links between them. The term *huoyi* 活翼, annotated in the MCST database as a neologism attested in the *TT*, actually appeared for the first time as the translation of "Feathering paddle" on page 21 of the *QJ*; *huoyi* would later enter the *ECD* with the same meaning. The same applies to *minglun huoyi* 明輪活翼; attested for the first time as the rendering of "Reefing paddle" on page 42 of the *QJ*, it would appear again in the *ECD* through the *TT*. The term *juzhong luosiqi* 舉重螺絲器, a translation of "Screw jack" on page 44 of the *QJ*, would also be later used, with the same meaning, in the dictionaries composed by Mateer and Hemeling.

It is evident that not all *QJ* neologisms, later used in the *TT*, finally merged into the *ECD*. For instance, terms such as *qiyan tuilu* 汽罨推路, "Slide valve travel," or *Menzi tong* 們子銅, "Muntz metal," already seen in the first text,

49 George A. Stuart, *Technical Terms, English and Chinese, Prepared by the Committee of the Educational Association of China* (Shanghai: Methodist Publishing House, 1910).

50 Karl Ernst Georg Hemeling, *English-Chinese Dictionary of the Standard Chinese Spoken Language* (官話) *and Handbook for Translators, Including Scientific, Technical, Modern and Documentary Terms* (Shanghai: Statistical Department of the Inspectorate General of Customs, 1916), preface, iii.

51 Shen Guowei, *Jindai Zhong-Ri cihui*, 436.

52 Hemeling, *English-Chinese Dictionary*, preface, iii.

would later be given in the second, but not the third.[53] In any case, it is possible to confirm that, analogously to the *TT* and the *DECF*, the *ECD* also served as a platform for spreading the neologisms attested for the first time in the *TVM*, even twenty-eight years after the publication of the latter.

Momentarily setting aside the *ECSD*, to which the next section is specifically devoted, another work that adopted the terms listed in the *TVM* is the *Putong baike xin da cidian* 普通百科新大詞典 (New general encyclopaedic dictionary), compiled in 1911 by the literary critic and writer Huang Moxi 黃摩西 (1866–1913).[54] In this case as well, the examples chosen to demonstrate this hypothesis are phonemic loans. Other examples can also be consulted in tables 44 and 45, even though they are fewer than previous works considered in this section. The first example is the neologism *miyituoli chun* 迷以脫里醕, indicating "Alcohol methylated" and "Methylated spirit," attested on pages 2 and 22 of the *HXCL*; *miyituoli chun* is employed within the *PBXDC*, published twenty-six years after the *TVM*, by substituting the character *chun* 醕 with its variant *chun* 醇. The same applies to another version of this phonemic loan, *miyituoli chun* 米以脫里醕, a translation of "Spirit methylated" on page 58 of the *XYDCYP* and later adopted in the *PBXDC*, again with the replacement of the character *chun* 醕 with the homophone variant *chun* 醇.

4 Yan Huiqing's *Ying-Hua da cidian*

A specific section is devoted to the *ECSD*, compiled by Yan Huiqing, because it is one of the dictionaries that more extensively adopted the *TVM*'s nomenclature, and because of the role the text played in late Qing China and in the first half of the twentieth century. I directly consulted four editions of the dictionary: the first is the one published in 1908;[55] the second is the fourth edition, published in 1916;[56] and the third and fourth are editions in small type published in 1920[57] and 1925 respectively.[58] Through the MCST database, two other

53 In Hemeling, *English-Chinese Dictionary*, 906, "Muntz metal" is rendered as "捫子銅 *mên tzŭ t'ung*."

54 Huang Moxi, *Putong baike xin da cidian* (Shanghai: Zhongguo cidian gongsi, 1911).

55 Yan Huiqing, *An English and Chinese Standard Dictionary*, 1st ed. (Shanghai: Shangwu yinshuguan, 1908).

56 Yan Huiqing, *An English and Chinese Standard Dictionary*, 4th ed. (Shanghai: Shangwu yinshuguan, 1916).

57 Yan Huiqing, *English and Chinese Standard Dictionary: Small Type Edition* (Shanghai: Shangwu yinshuguan, 1920).

58 Yan Huiqing, *English and Chinese Standard Dictionary: Small Type Edition*, 5th ed. (Shanghai: Shangwu yinshuguan 1925). The author would like to thank Professor Zou Zhenhuan for kindly providing a personal copy of the dictionary.

editions of the *ECSD* could be indirectly accessed: the one published in 1910[59] and the edition printed two years later.[60]

In this case as well, pertaining to the diffusion of the *TVM* and its interactions with later lexicographical production, it is necessary to stress that in the *ECD*, among the texts consulted, the first is precisely "[...] 1. Commercial Press. English-Chinese Standard Dictionary."[61] As of June 1916, seven editions of the *ECSD* were already published.[62]

On the other hand, the sources Yan Huiqing consulted for the compilation of the *ECSD* are not explicitly expressed, but some indirect indications can be found in the preface to the text.[63] Immediately after specific information on the translation of scientific and technical texts, the standardisation of relevant nomenclature, and the necessity of consistently printing glossaries and bilingual dictionaries, a reference to Fryer's work can be found on pages II and III of the preface: "So many works have been consulted that it is not possible to acknowledge our indebtedness to all, but we desire to mention in particular the publications of the Educational Association of China, and of the Medical Association, to the glossaries published by the Kiangnan Arsenal, to the text-books of the Commercial Press [...]." This passage alludes to the glossaries published by the Jiangnan Arsenal. The analysis of lemmata listed in the two texts provides evidence that the *TVM* can be included among glossaries used by Yan Huiqing and other *ECSD* editors.

The choice falls again on those entries considered more disambiguating—for example, lemmata that could not be, with the highest possible probability, drawn from traditional Chinese texts. More specifically, the terms listed here are different kinds of translations pertaining to a relatively new branch of knowledge, such as chemistry, and once more, phonemic loans. As for the first one, consider the term "Acetate of lead," annotated in the entry "Acetate" on page 19 of the 1908 and 1916 editions of the *ECSD*, and on page 9 of the 1920 and 1925 editions. In both cases, providing an example of a term including the word "acetate," Yan Huiqing supplies the rendering of lead acetate, *qianyang cusuan* 鉛養醋酸. The neologism, as inferred from table 45, was already included in numerous lemmata within the *TVM*; it is annotated on pages 1, 34, 48,

59 Yan, *An English and Chinese Standard Dictionary* (1910).

60 Yan, *An English and Chinese Standard Dictionary* (1912).

61 Hemeling, *English-Chinese Dictionary*, preface, iii. On the Commercial Press, refer to Wang Zhongchen, "Xinshi yinshua, zujie dushi yu jindai chuban ziben de xingcheng—Shangwu yinshuguan chuangli de qianqian houhou," *Zhongguo xiandai wenxue yanjiu congkan*, no. 1 (1999).

62 Zou Zhenhuan, "Guangxu huangdi de yingyu xuexi yu jinru Qing mo gongting de yingyu duben," *Qing shi yanjiu—The Qing History Journal*, no. 3 (2009): 111.

63 The reference is here to the 1908, 1916, 1920, and 1925 editions.

and 59 of the *XYDCYP* for the entries "Acetate lead," "Lead acetate" and "Lead sugar of," "Plumbi acetas," and "Sugar of lead," as well as on page 20 of the *HXCL* for the lemma "Lead acetate."[64] In "Lead sugar of" and "Plumbi acetas," *qianyang cusuan* precedes *qian san* 鉛散, and in "Sugar of lead" it follows *qian tang* 鉛糖; Fryer and his Chinese colleagues actually slightly confused some of these terms. The term "plumbi acetas" is simply the translation in scientific Latin of the English equivalents within the *XYDCYP*, to wit "Acetate lead" and "Lead acetate," in which alternative translations are not provided. The rendering *qian san* is listed in the lemma "Lead sugar of" as the second translation, an alternative to *qianyang cusuan*; sugar of lead is in reality one of the appellatives that lead acetate, or diacetate, is known as. Moreover, the term *qian tang*, which in the *XYDCYP* is listed in the entry "Sugar of lead," but not in "Lead, sugar of," was later employed by Yan Huiqing on page 2258 of the 1908 edition and on page 969 of the 1920 and 1925 editions. In the lemma "Sugar," Yan actually lists the term "*Sugar of lead*," therefore supplying the alternative scientific appellative, "acetate of lead," providing *qian tang* as the first of the Chinese translations.

Another confirmation of the use of the *TVM* within the *ECSD* can be obtained from phonemic loans, because the quantity of this typology of terms drawn from the *TVM* is very consistent. The first instance is "Aconitine," translated as *agunidini* 阿古尼低尼 on page 20 of the 1908 and 1916 *ECSD* editions, and on page 9 of the 1920 and 1925 *ECSD* editions; the phonemic loan *agunidini* is the same neologism attested on page 1 of *HXCL*. The second, *mi'erbeini* 迷耳貝尼, is annotated on page 1449 of the 1908 edition, and on page 634 of the 1920 and 1925 editions as the third translation for the lemma "Mirbane," an artificial oil; the phonemic loan is the same one listed on page 23 of the *HXCL*.

5 Other Works

After the analysis of the encyclopaedic works, glossaries, and dictionaries that adopted the *TVM*'s nomenclature, culminating in the *ECSD*, this section will consider other miscellaneous works that served as vehicles for translations of

64 According to the *Zhongguo jiben guji ku*, the term appeared for the first time in the *Gewu Zhongfa* 格物中法 [Chinese methods for the investigation of things], by the politician and scholar Liu Yueyun 劉岳雲 (1849–1917). The compilation of the text started around 1870 but, as indicated in Ho Yi Kai, *Science in China, 1600–1900*, 218, it was only published in 1900.

the *TVM*. However, this fact is less certain for the second and third titles considered in the section.

The first one is *Zhong-Xi binglüe zhizhang* 中西兵略指掌 (An easy explanation of the military strategies of China and the West), by Chen Longchang 陳龍昌 (n.d.).[65] The text particularly adopted the *QJ*'s terminology, and numerous terms considered attested for the first time in the *Zhong-Xi binglüe zhizhang* were in reality drawn from the glossary. For example, *kongqi guan* 空氣管, listed in juan 17 of the work by Chen Longchang, was already annotated as the translation of "Air pipe" and "Air tube," respectively, on page 1 and 2 of the *QJ*. The same can be said of *hui tang* 灰膛, listed in juan 21 of *Zhong-Xi binglüe zhizhang* and later also recovered in different passages of the *Shiwu tongkao*. *Hui tang* was already annotated on page 3 of the *QJ* as a rendering for both "Ash box" and "Ash pit," in addition to being included in other Chinese lemmata of the glossary.

The second work considered is *The Principal Articles of Chinese Commerce* (*Import and Export*), compiled by Ernest Watson (n.d.)[66] in 1923.[67] In the first part of the *PACC* various fibres are analysed; in the second oils, fats, and waxes; in the third gums and resins; in the fourth dyes, colours, pigments, and paints; in the fifth metals; and in the sixth and last part, miscellaneous products are presented. Even though on pages X and XI of the *PACC* the *TVM* is not mentioned among the works consulted, from some terms listed within the text, particularly phonemic loans, it is possible to suppose that Watson consulted, perhaps indirectly, the nomenclature Fryer and his Chinese colleagues adopted.

---

65 Chen Longchang, *Zhong-Xi binglüe zhizhang*.

66 For more information on Ernest Watson, refer to Zhongguo haiguan zong shuiwu sishu, *Documents Illustrative of the Origin, Development and Activities of the Chinese Customs Service* (Vol. 3. Shanghai: Statistical Department of the Inspectorate General of Customs, 1938), 701. See also Charles Alfred Speed Williams, *Hai kuan yü yen pi hsü—An Anglo-Chinese Glossary for Customs and Commercial Use* (Shanghai: Commercial Press Limited, 1914).

67 Watson, *The Principal Articles*. The version consulted is the one housed in the library of Ricci Institute for Chinese-Western Cultural History at the University of San Francisco. As described in the first page of the *PACC*, the text includes "a description of the origin, appearance, characteristics, and general properties of each commodity; an account of the methods of preparation or manufacture […]." The *PACC* was "Published by the Order of the Inspector General of Customs" and sold by Kelly & Walsh in Shanghai, Hong Kong, Yokohama, and Singapore; by Commercial Press in Shanghai; by Tientsin Press in Tianjin and Beijing; finally, by P. S. King & Son in London and in all the "Maritime Custom Houses" on the Chinese territory.

Consider the first example, the entry "Bdellium (假沒藥, *Chia-mu-yao*),"[68] which Watson specifies is "sometimes called 'false myrrh' […]": on page 41 of the *XYDCYP* the lemma "Myrrh, false" was rendered exactly as *jia moyao* 假沒藥.[69] The second example from the *PACC*, "Copaiba, Copaiva, Capivi, or Copaiba balsam (Copaiba 哥拜把, *Kê-pai-pa*),"[70] presents a translation very similar to the one listed on page 17 of the *XYDCYP*, in which "Copaiba" and "Copaiva" are both rendered as *gebaiba* 哥拜把.[71] The only difference is the addition of the radical *kou* 口 in the character *ge* 哥, in line with Fryer's statements on the use of phonemic loans. The term *gebaiba* is also listed in the *XYDCYP* as part of other translations of correlated lemmata. In the same way, describing a derivative of the copaiba, namely the oil produced from the plant, Watson affirms that "This volatile oil, known as oil of copaiba (哥拜把油, *kê-pai-pa-yu*), may be separated by means of distillation […]."[72] In the passage a term present in two lemmata of the *XYDCYP* is annotated: "Copaibeæ oleum" and "Oleum copaibæ," respectively, on page 17 and 43, are both rendered as *gebaiba you* 哥拜把油. The only discrepancy between the solutions provided in the *XYDCYP* and the *PACC* lies in the aforementioned insertion of the radical *kou* in the first character.

In order to better clarify the connection between the *PACC* and the *TVM*, it is important to note that in the former, the entry "Gamboge (籐黃, *T'êng-huang*)" adopts the radical *zhu* 竹, not the expected *cao* 艹, in the character *teng* 籐.[73] This uncertainty was already noticeable in the namesake entries on page 15 of the *HXCL* and on page 25 of the *XYDCYP*, as well as in the numerous correlated lemmata listed in the latter.

In this circumstance as well, the uncertainty on the hypothesis of the use of the *TVM* within the *PACC* can be partially clarified with a phonemic loan. In the *PACC*, the entry "Mastic (洋乳香, *Yang-ju-hsiang*, or 瑪司的克, *Ma-ssŭ-ti-k'ê*)" shows the same phonemic loan that is translated as the lemma "Mastiche" on page 38 of the *XYDCYP*.[74] Not coincidentally, *masidike* can also be located in juan 24 of the *Shiwu tongkao*, devoted to *huaxue* 化學 (chemistry).

68 Ivi, 154.

69 The term "Myrrh" was translated as "Moot yauk 沒藥" in Thomas T. Devan, *The Beginner's First Book in the Chinese Language, (Canton Vernacular): Prepared for the Use of the House-Keeper, Merchant, Physician, and Missionary* (Hong Kong: China Mail, 1847), 125.

70 Watson, *The Principal Articles*, 156.

71 On page 11 of the *HXCL*, "Copaiva" is rendered instead as *kepaiyifa* 可派以發.

72 Watson, *The Principal Articles*, 157.

73 Ivi, 162.

74 Ivi, 169.

Even if the words included in the preface to the *PACC* are not completely disambiguating, as is the case of the *ECSD*, the examples shown make for a plausible hypothesis that terms attested for the first time in the *TVM* merged indirectly into the *PACC*; for example, through the consulted texts in its preface. It should be finally noted that the *PACC* preface, dated August 13, 1920, is signed by "C. A. V. Bowra," employed at Chinese maritime customs, and son of Edward Charles Macintosh Bowra. The latter was indeed indicated by Fryer as the author of one of the sources consulted in volume eight of the EAL manuscripts, offering another glimpse into the complicated interaction between the *TVM*, its manuscripts, and later lexicographical production.

The third and last text considered in this section is the dictionary compiled in 1935 by the Royal Engineer James Vivian Davidson-Houston (1901–1965) and the lieutenant-colonel Raymond Varley Dewar-Durie (1905–1999), *Chinese and English Modern Military Dictionary*.[75] In this case as well, the *TVM*'s influence on this dictionary can be affirmed with less certainty, compared with the other works considered; the examples, analogously to the *PACC*, are also fewer. The first is *yougao* 油膏, considered in the MCST database as a neologism, and attested for the first time in the *CEMMD* as an equivalent of grease, "*yu*^{2} *kao*^{1}."[76] In reality, it was already listed as part of numerous translations within the *TVM*, invariably with the meanings "ointment," "unguentum," "plaster," "emplastrum," and "amalgam."[77] Another term adopted in the *CEMMD* can be derived from the *QJ*: on page 28, *jinqi men* 進汽門 represents the translation of "Induction valve" and was later used in the *CEMMD* as the rendering of "inlet valve," "*chin*^{4} *ch'i*^{4}-*mên*^{2}."[78] Given the scarce quantity of terms coincident with those annotated in the *TVM*,[79] it is difficult to establish with absolute certainty the relation

75 James Vivian Davidson-Houston and Raymond Varley Dewar-Durie, *Han-Ying junshi cidian—Chinese and English Modern Military Dictionary* (Peiping: Henri Vetch, The French Bookstore, 1935).

76 Idem, 72.

77 According to the *Zhongguo jiben guji ku*, the term already appeared in three Buddhist texts of the Northern and Southern dynasties (420–589): the *Za a han jing* 雜阿含經 [Saṃyuktāgama or Miscellaneous Agama], translated by the Indian monk Guṇabhadra (394–468), the *Da ban nie pan jing* 大般涅槃經 [Mahāparinirvāna-sūtra], translated by the Indian monk Dharmakṣema (385–433), and the *Si fen lü* 四分律 [Dharmaguptaka Vinaya], rendered by the monk Buddhayaśas (4th to 5th century) and the Chinese translator Zhu Fonian 竺佛念 (4th to 5th century). *Yougao* was also included in the *Jing-Chu suishi ji* 荊楚歲時記 [Record of the times and seasons of Jing-Chu], by the scholar and minister Liang Zonglin 梁宗懍 (ca. 501–ca. 565). In all four texts, however, the meaning is slightly less technically connoted than in the *CEMMD* and the *TVM*.

78 Davidson-Houston and Dewar-Durie, *Chinese and English Modern Military*, 166.

79 Others can be found, such as "buoyancy," translated as *fuli* 浮力 both on page 8 of the *QJ* and in Davidson-Houston and Dewar-Durie, *Chinese and English Modern Military*, 22,

between the *TVM* and the *CEMMD*. Nevertheless, in the latter, the terminology representing the gaseous elements chlorine, hydrogen, and oxygen includes the disyllabic terms Fryer and Xu Shou used in their translations.[80]

6 Conclusion

The fact that the kinds of texts analysed in the chapter, particularly encyclopaedias and journals, adopted Fryer's terms is of course no coincidence. The lithographic printing technologies greatly spurred the publication of texts, especially in the turning point represented by China's defeat in the First Sino-Japanese War; among the texts published, those intended for the diffusion of scientific knowledge were in great demand.[81] Late Qing journals and encyclopaedias particularly benefited from the rise in printing production, and the latter were especially successful from the perspective of sales for two basic reasons: their classification was similar to that of traditional Chinese texts and their purpose was not purely commercial, as they had a "civilizing mission."[82] Another peculiarity of late Qing encyclopaedias is that they were meant for an extremely broad readership, even wider than that targeted by the Jiangnan Arsenal translations.

The success of these texts was based on the fact that they conveyed scientific and technical knowledge, required in everyday life, using the Chinese language.[83] Such new notions and knowledge had already been carefully prepared and rendered into Chinese with a suitable linguistic perspective by, among others, Fryer and his Chinese colleagues. The same applies to the journals presented in this chapter; even if some of them were later destined to be superseded by others, they contributed to the creation of a solid awareness of science in late Qing China.[84]

"cast iron" as *sheng tie* 生鐵 on page 9 of the former and page 25 of the latter, "exhaust pipe" and "exhaust valve," respectively as *chuqiguan* 出汽管 and *chuqimen* 出汽門, on page 20 of the *QJ* and 56 of the dictionary by Davidson-Houston and Dewar-Durie, "fly wheel" as *feilun* 飛輪, respectively on pages 23 and 64, and "horsepower" as *ma li* 馬力, on pages 27 and 78 of the two texts.

80 Davidson-Houston and Dewar-Durie, *Chinese and English Modern Military*, 28, 79, and 111.

81 Han, "Le début," 145.

82 Barbara Mittler, "China's 'New' Encyclopaedias and Their Readers," in *Chinese Encyclopaedias of New Global Knowledge (1870–1930): Changing Ways of Thought*, ed. Milena Doleželová-Velingerová and Rudolf G. Wagner (Heidelberg: Springer, 2013), 399–400.

83 Ivi, 422.

84 Elman, *On Their Own Terms*, 386.

The great majority of these texts, such as the *Huangchao jingshi wenbian*, became extremely popular in the historical period considered,[85] helping to spread the ideas, terms, and scientific knowledge included in the TVM and its glossaries. The encyclopaedic texts illustrated in this chapter were indeed very successful and popular—they were used as teaching material by Chinese students for exam preparations and in schooling, based on *Xinzheng* educational reforms, carried out between 1901 and 1904.[86]

Analogous considerations can be proposed for the dictionaries and miscellaneous texts presented. Scholars and translators who wanted to access Western knowledge and incorporate it in their work had the advantage of having the ready-made lexicological forms of the TVM. Referring to the dictionary compiled by Yan Huiqing in particular, its importance in the history of Chinese bilingual glossaries has been widely recognised.[87] Eminent contemporary personalities expressed their appreciation for the ECSD, such as the scholar Dai Liuling 戴鎦齡 (1913–1998), who especially credited the technical terminology adopted in the ECSD, and the very well-known translator Yan Fu 嚴復 (1854–1921), who always gave considerable prominence to the question of Chinese scientific and technical nomenclature.[88] Yan Fu's appreciation for the ECSD can also be inferred from the preface he wrote for the dictionary.[89] Such positive judgements by important Chinese personalities on a text extensively using the TVM's nomenclature, and the fact that the last edition of the ECSD dates back to 1935, twenty-seven years after the first one, and forty-seven years after the publication of the TVM, confirms and highlights the importance it had as a vehicle for the linguistic innovations by Fryer and his Chinese colleagues.[90]

The evidence presented confirms once more that the TVM's influence was not limited to the chronological and spatial context of the Jiangnan Arsenal, nor to an inner circle of users. Instead, it constantly expanded both in time and space, further proving the hypothesis that its lexical innovations played

85 Janku, "Translating Genre," 343.

86 Elman, *On Their Own Terms*, 430–31.

87 Refer for example to Zou, "Guangxu huangdi de yingyu xuexi," 112.

88 Reported both in ivi, 111–12.

89 On Yan Fu, refer also to Max Ko-wu Huang, "The War of Neologisms: The Competition between the Newly Translated Terms Invented by Yan Fu and by the Japanese in the Late Qing," in *China and Its Others: Knowledge Transfer through Translation, 1829–2010*, ed. James St. André and Peng Hsiao-yen (Amsterdam, New York: Rodopi, 2012). On the relation between Yan Fu, science, and translation, consult Shen Guowei, *Yan Fu yu kexue* (Nanjing: Fenghuang chubanshe, 2017) and Wang Shi et al., *Lun Yan Fu yu Yan yi mingzhu* (Beijing: Shangwu yinshuguan, 1982).

90 On the dating of the edition of the ECSD, consult the bibliographical reference of footnote no. 1 in Sun Zengde, "Yan Huiqing yu *Ying-Hua da cidian*," *Shizhi xuekan*, no. 3 (2015): 50.

an important role in the formation of Chinese scientific and technical lexicon, and in later textual production: their authoritativeness permeated the linguistic and scientific circles of late Qing China. The *TVM* was conceived with a rigorous lexicological awareness and an important objective in mind: to contribute to the standardisation of scientific and technical nomenclature. As a result, it substantially helped in the circulation of scientific knowledge, and from this perspective, its role was greatly amplified by the works considered.

CHAPTER 8

The Translator's Vade-mecum: Later Aftermath

1 Introduction

The importance of a lexicological work cannot be judged solely by the influence it has immediately after its publication. The traces of the TVM are still visible in later texts that, to a different degree, adopted the nomenclature of its glossaries, contributing to its diffusion. The importance that the glossaries Fryer and his colleagues edited had on modern Chinese scientific and technical lexicon, and in the expression of concepts and ideas behind it, was directly or indirectly acknowledged by later relevant textual production.

The complicated connection between the TVM's lexicological features, the evolution of its manuscripts, and the transmission of texts published at a later stage reflects the formation of a community of translators of scientific and technical texts and the circulation of relevant knowledge in late Qing China. The interactions at work in the lexicographical production of late Qing dynasty and early Republican China are an extension of the TVM's influence as analysed in the previous chapter. Its role began to fade over time, but even if Chinese scientific and technical nomenclature gradually developed autonomously, the TVM was still recognised as a milestone work by later authors, leaving numerous traces of its role in the first half of the twentieth century as well.

In chapter 8, therefore, the research is extended to the first decades of the twentieth century. A selection of texts is considered, showing how the TVM's influence gradually decreased decades after its publication.[1] The criteria adopted for their choice are the similarity of contents with the TVM and their accessibility; texts composed in these decades are so numerous that it was necessary to make a selection based on their availability. Nevertheless, its authority would also be recognised in works that did not directly use it as a reference. The first section is devoted to the field in which the influence of the nomenclature Fryer and his colleagues adopted was more evident: chemistry.[2]

1 Some of the texts analysed are also listed in Xia, "Wan Qing keji shuyu de fanyi," 121–22.

2 Refer also to Wright, *Translating Science*, 178–79, and Zeng Zhaolun, "Ershi nian lai Zhongguo huaxue zhi jinzhan," *Kexue* 19, no. 10 (1935).

 | DOI:10.1163/9789004443211_010

2 Chemistry: Texts by the Guoli Bianyiguan and Other Committees

The first text considered, the *Huaxue yuhui* 化學語彙 (Chemistry glossary), represents an intermediate stage in the passage from the TVM's translation choices to those employed by other organisations and committees that, at a later phase, would deal with the standardisation of scientific and technical nomenclature, such as the Guoli bianyiguan.[3] In the *HXYH*, those lexicological innovations that will be explained in detail in texts published at a later stage, such as the *Huaxue mingming yuanze* 化學命名原則 (Principles of chemical nomenclature) in its various editions, are already present.[4] Nevertheless, some terms are patently influenced by the TVM and correspond to the solutions adopted in the latter:

TABLE 37 *HXYH* and the *TVM*

英 (English)	中 (Chinese)	日 (Japanese)	*TVM*[a]
Aldehyde, 1	阿勒弟海特	アルデヒト	阿勒弟海特
Amalgam, 2	汞膏	アマルガム	汞膏
Ammonia, 2	阿摩尼亞	アムモニア	阿摩尼阿 即淡輕$_{\text{三}}$
Atom, 3	原子		質點
Cupric oxide, 8	養化第二銅	酸化第二銅	銅養
Cuprous oxide, 8	養化第一銅	酸化第一銅	銅養
Gallic acid, 12	沒食子酸		加里酸 沒石子酸
Nitroglycerine, 18	尼脫路各里司林	ニトロ =グリセリン	淡養$_{\text{五}}$各里司里尼
Oil of bitter almonds, 18	苦杏仁油	苦扁桃油	苦杏仁油
Toluidine, 26	多路以弟恩	トルイヂン	朶路以弟尼
Valerianate, 26	纈草酸鹽		發里里阿尼酸

a In the tables included in this chapter, the glossaries or the pages in which entries are attested in the TVM are not specified, unless considered necessary. Chinese translations are drawn from those in the TVM corresponding to the original lemma of the texts taken into account in every single table. In case this was not possible, they are quoted from the lemma in the TVM considered the closest in meaning in every circumstance.

3 Xuebu shendingke, *Huaxue yuhui* (Shanghai: Shangwu yinshuguan, 1908).

4 Guoli bianyiguan, *Huaxue mingming yuanze*.

In the *HXYH*, elements already present in the *TVM* can be found, while others are more similar to texts published at a later stage. As an example, in the first entry in table 37, the phonemic loan for "Aldehyde," *aledihaite* 阿勒弟海特, is precisely the same used on page 2 of both the *XYDCYP* and the *HXCL* to translate the eponymous lemmata.

As particularly inferred in the case of *amonia* 阿摩尼阿, a neologism attested for the first time in the *TVM*, the substitution of phonemic loans with other translation choices, adamantly supported by Fryer, had not been completely carried out in the *HXYH*.[5] In some circumstances, as evidenced from the lemma "Toluidine," characters used for the phonemic loans had already undergone modifications, changing from *duoluyidini* 朶路以弟尼 to *duoluyidi'en* 多路以弟恩. In other entries, for instance "Atom," translation choices adopted in the *TVM* have been substituted with others closer to contemporary scientific and technical Chinese nomenclature.[6] A relevant example is "Valerianate," a salt of valeric acid, for which Fryer and his colleagues still employed a phonemic loan. The solution within the *HXYH*, *xiecao suanyan* 纈草酸鹽, composed of *xiecao*, indicating Valeriana officinalis, from which the acid is extracted, and *suanyan*, generally referring to the salt of an acid, is closer to the contemporary equivalent, *wu suan* 戊酸, than the phonemic loan *falili'ani suan* 發里里阿尼酸.

On the topic of phonemic loans and the connection between the *TVM* and the *HXYH* nomenclature, it is important to note that a reference to the latter's translation choices, particularly pertaining to the excessive use of phonemic loans, can be read in the *Huaxue mingming yuanze* and its revised edition: "其在日本，殫數十化學專家之力，所發表之化學語彙，亦僅無機化學而止，至於有機化學名詞，概從音譯。茲事繁重，於焉可知" ([…] published in Japan, thanks to the strenuous effort of dozens of experts of chemistry, a glossary of chemical terms; it was, though, limited to inorganic chemistry, while terms in organic chemistry were all based on transliteration. It is simple to realise that such a task was exceedingly difficult).[7] Even though the *HXYH* already substituted numerous phonemic loans adopted in the *TVM*, it

5 *Amonia* 阿摩尼阿 is attested for the first time as a single lemma as the translation of "Ammonia" on page 3 of the *HXCL*; *amonia* was in reality already attested as part of other translations within the *JS*. According to the *Zhongguo jiben guji ku*, the term was later recovered in Chen Longchang, *Zhong-Xi binglüe zhizhang*, in *Kuangxue xinyao xinbian* 礦學心要新編 [Essentials of mineralogy: a new collection], by Song Gengping 宋賡平 (n.d.), and in various passages of the *Shiwu tongkao*.

6 Refer to the table in Wright, *Translating Science*, 354.

7 Respectively Guoli bianyiguan, *Huaxue mingming yuanze*, preface, iii, and Idem, *Huaxue mingming yuanze (zengding ben)* (Nanjing: Zhengzhong shuju, 1947).

was retrospectively blamed for their excessive adoption in the field of organic chemistry.

Finally, the two lemmata "Cupric oxide" and "Cuprous oxide," both translated by Fryer and his Chinese colleagues as *tongyang* 銅養, are emblematic of how some terms of the *TVM* were rejected for obvious inconsistencies connected to chemical knowledge.[8] In the *HXYH*, the two entries have been rendered as *yanghua di er tong* 養化第二銅 and *yanghua di yi tong* 養化第一銅, more similar to those adopted in contemporary Chinese scientific nomenclature, *yanghua tong* 氧化銅 and *yanghua ya tong* 氧化亞銅, rather than *tongyang*. Nevertheless, the two solutions annotated in the *HXYH* still reveal traces of the *TVM*'s influence in the adoption of the character *yang* 養 in place of *yang* 氧.[9]

The next work we will consider is *Chemical Terminology: Serial No. II*, published in 1922.[10] In the text, scientific and technical terms are listed, mostly connected to the field of chemistry. The English term is provided first, followed by German and Japanese renderings, finally by Chinese translations approved by the Committee.

With the exception of the entries listed in table 38, corresponding between *Chemical Terminology: Serial No. II* and the *TVM*, extremely exiguous in number, the Committee rejected the nomenclature devised by Fryer and his colleagues. The scientific and technical lexicon of chemistry was already following another path and indeed, in the majority of cases, the original entries in English do not correspond in the two texts.[11]

8 *Tongyang* 銅養 was attested in the *XYDCYP* on page 17 as the translation of "Copper oxide" and "Copper protoxide." In the *HXCL*, *tongyang* was listed on page 11 to translate as "Copper oxide," and on page 12 as "Cupric oxide" and "Cuprous oxide." *Tongyang* was in reality originally annotated in the third column of the *JS*, "Terms in general use and chief elements," as part of translations, in most cases, of compounds indicating the principal element within a certain mineral.

9 As confirmed, particularly pertaining to chemical elements, in Wright, *Translating Science*, 350.

10 *Chemical Terminology: Serial No. II—Chemical Terms: Approved by the Ministry of Education—Huaxue mingci (er)—Huaxue shuyu—Jiaoyubu shending* (General Committee on Scientific Terminology, 1922). The first part of the text is *Chemical Terminology: Serial No. I—Elements: Approved by the Ministry of Education—Huaxue mingci (yi)—Yuanzhi—Jiaoyubu shending* (General Committee on Scientific Terminology, 1922).

11 The reader can also consult Gregory, Thomas G., and Gao Xian, *Huaxue yaopin cidian* (Shanghai: Xinya shudian, 1946).

TABLE 38 *Chemical Terminology: Serial No. II* and the TVM

英 (English)	德 (German)	日本 (Japanese)	決定 (Final decision)	*TVM*
Air pressure, 1	Luftdruck	空氣壓	空氣壓力	空氣壓力
Alkali, 1	Alkali		鹼	鹻類質[a]
Combination, 3	Vereinigung od. Verbindung	化合	化合	Combined carbon 化合之炭質
Electroplating, 5	Galvanostegie	電鍍	電鍍	Electro-planting 電鍍銀
Element, 5	Element	元素	原質，原素	原質
Energy, 6	Energie		能力	Actual energy 實能力
Heating, 7	Erhitzung	加熱	加熱	Heating power 加熱之力

a For the entries in table 38, only one of the lemmata in which the term is listed in the TVM is provided, with the exception of the second and the last, respectively *jianlei zhi* 鹻類質 and *jiare zhi li* 加熱之力, which only appear in one lemma within different TVM's glossaries. Moreover, the original term within the TVM is listed in the last column only if it does not correspond to the one in *Chemical Terminology: Serial No. II*.

Such a tendency is confirmed in table 39 below. In the *Fifth Report of General Terms in Organic Chemistry* the translation choices included in the TVM were also not used.[12] The text—publishing year unknown—had already brought Chinese chemical nomenclature towards modernity: the presence of phonemic loans is extremely low within the *FRGTOC*, and numerous terms rendered by Fryer and his colleagues through this translation choice were already replaced.

The path indicated by the General Committee on Scientific Terminology was in reality the same pointed out in the *TVM*. Nevertheless, due to time limits, inadequate knowledge, and other objective problems, Fryer and his colleagues were not able to carry through with it. As evidenced in the example "Aloin," all phonemic loans within the *TVM* have been replaced in the *FRGTOC* with solutions that could reflect the etymology of the original lemma. Apparently,

12 General Committee on Scientific Terminology, *Fifth Report of General Terms in Organic Chemistry—Huaxue mingci shenchazu di wu ci shenchaben—Youji huaxue putong mingci* (publication place and year unknown).

TABLE 39 *Fifth Report of General Terms in Organic Chemistry* and the TVM

英文名 (English term)	化學式 (Chemical formula)	決定 (Final decision)	TVM
Acetic aldehyde, 1	$CH_3 \cdot CHO$	醋醛	醋酸阿勒弟海特
Aesculin, 1	$C_{15}H_{16}O_9$	七葉樹素	愛司古里尼
Aloin, 1	$C_{16}H_{16}O_7$ 或 $C_{16}H_{18}O_7$	蘆薈素	啞囉以尼
Arabin, 2		阿剌伯膠素	阿拉伯尼
Benzoic acid, 2	$C_6H_5 \cdot CO_2H$	安息香酸	徧蘇以克酸 安息酸[a]
Caffeic acid, 3	$C_6H_3(OH)_2 \cdot CH{:}CH \cdot CO_2H$	咖啡酸	咖啡以可酸 加非以克酸
Chloroform, 3	$CHCI_3$	哥羅仿	格羅路福耳密 哥囉昉[b]
Formic acid, 5	$H \cdot CHO_2$	蟻酸	福耳密克酸 即蟻酸
Gallic acid, 5	$C_6H_2(OH)_3 \cdot CO_2H + H_2O$	沒食子酸	沒石子酸
Malic acid, 6	$CO_2H \cdot CH_2 \cdot CH(OH) \cdot CO_2H$	蘋果酸	蘋果酸
Marsh gas, 7	CH_4	沼氣	炭$_{二}$輕$_{四}$ 炭$_{二}$輕$_{四}$氣[c]

a In this circumstance, *biansuyike suan* 徧蘇以克酸 and *anxi suan* 安息酸 are annotated for the entry "Acidum benzoicum" on page 1 of the XYDCYP, whereas on page 9 "Benzoic acid" is translated as *biansuyike suan* 徧蘇以克酸. In the HXCL, "Benzoic acid" is instead rendered as *biansuyike* 徧蘇以克. For the subsequent entry within table 39, to wit "Caffeic acid," the two translations provided, *kafeiyike suan* 咖啡以可酸 and *jiafeiyike suan* 加非以克酸, are listed respectively on page 11 of the XYDCYP and on page 7 of the HXCL.

b The first translation, *geluolufuermi* 格羅路福耳密, is attested on page 10 of the HXCL; the second, *geluofang* 哥囉昉, is part of translations of numerous lemmata within the XYDCYP.

c The two translations, *tan$_{er}$qing$_{si}$* 炭$_{二}$輕$_{四}$ and *tan$_{er}$qing$_{si}$ qi* 炭$_{二}$輕$_{四}$氣, are listed respectively on page 21 of the HXCL and on page 26 of the XYDCYP.

Fryer and his colleagues simply did not grasp the etymological proximity of some related terms, as is the case of "Convolvulaceæ"; the latter was translated on page 17 of the XYDCYP as *kangfuwulaxiyi* 康弗勿拉西依 and *xuanhua ke* 旋花科, but they only provided the phonemic loan *kangfuwulini* 康弗勿里呢 for "Convolvulin," which is translated in the FRGTOC as *xuanhua su* 旋花素. Analogously, translating "Aloin" as *luhui su* 蘆薈素 rather than the phonemic loan *yaluoyini* 啞囉以尼 makes it easier for the reader to infer that this compound is extracted from aloe. It should be noted, though, that Fryer and his colleagues coherently translated lemmata in the TVM related to aloe by precisely employing the phonemic loan *yaluo* 啞囉.[13]

In some circumstances though, as with the case of "Arabin," in the *FRGTOC* phonemic loans have also been used, adopting characters different from those listed in the TVM. Apparently, the hybrid *alabo jiaosu* 阿剌伯膠素 provides more precise indications pertaining to the referent than the phonemic loan *alaboni* 阿拉伯尼. Nevertheless, the arabin, *alaboni*, is simply the main component of gum arabic, *alabo jiaosu*, and does not correspond to it.

In other circumstances, as with "Benzoic acid," the solutions put forward in the TVM were adequate and extremely close to contemporary Chinese renderings. *Anxi suan* 安息酸, one of the translations for "Benzoic acid," diverges from the solution adopted by the General Committee on Scientific Terminology, *anxixiang suan* 安息香酸, only for the absence of the character *xiang* 香. In the case of "Marsh gas," it would have been sufficient to create a calque of the original term, *zhao qi* 沼氣. Fryer and his colleagues preferred to supply the chemical formulas *tan*$_{er}$*qing*$_{si}$ 炭$_{二}$輕$_{四}$ and *tan*$_{er}$*qing*$_{si}$ *qi* 炭$_{二}$輕$_{四}$氣, since in numerous chemistry texts published in the nineteenth century, the marsh gas formula was considered to be C_2H_4, the same reflected in the two translations provided.[14] In other lemmata, such as "Caffeic acid," in the TVM redundant characters were added to reproduce the original term as accurately as possible, such as *kafeiyike suan* 咖啡以可酸 and *jiafeiyike suan* 加非以克酸. The terms thus created are formed of an excessive number of characters, an option Fryer himself was opposed to. In other instances, such as "Chloroform," the choice to insert the radical *kou* in the characters adopted for phonemic loans, led to the creation of new characters or the adoption of extremely rare ones, as in *geluofang* 哥囉昉. In this case as well, however, such a term can already be found in the Japanese medical journal *Iji shinbun* 醫事新聞 (Medical journal), more precisely on page 30 of issue 97, published in 1883, four years before the XYDCYP; it was later adopted in juan 31 of the *Shiwu tongkao*, dedicated to *yixue* 醫學 (medicine). This adds another link to the interactions between Japanese sources and lexical innovation analysed in this work.[15]

---

13 The phonemic loan *yaluo* 啞囉 always translates as the concept of "aloe" in the numerous cases in which it is attested as part of renderings within the TVM. In the only two instances in which *yaluo* 啞囉 appears as an independent lemma, as the translation of "Aloes," on page 3 of the XYDCYP and on page 2 of the HXCL, the phonemic loan *yaluo* 啞囉 is supplied first, while *luhui* 蘆薈 follows as an alternative rendering.

14 Refer, for example, to Bloxam, *Chemistry Inorganic and Organic*, 62, 88, and passim.

15 The term was later also employed in juan 5, issue number 4, p. 122, of *Beiping tushuguan guankan* 北平圖書館館刊 [Journal of Peiping National Library]; the periodical was published between 1928 and 1937.

There are also circumstances in which an incorrect homophone character was employed in the *TVM*, as in the rendering of "Gallic acid"; *meishizi suan* 沒食子酸, the solution within the *FRGTOC*, diverges from the translation Fryer and his colleagues supplied, *meishizi suan* 沒石子酸, only on the second character.[16] Finally, at times, terms in the *TVM* and the *FRGTOC* completely coincide, as in the case of "Formic acid," rendered in both as *yi suan* 蟻酸. Nevertheless, the distance between the translations provided in the *TVM* and by the General Committee on Scientific Terminology is evident in the entire *FRGTOC*, published numerous years after the *TVM*, all the more so particularly for the field of organic chemistry.[17]

Such differentiation is even more remarkable in *Youji huaxue putong mingci cao'an* 有機化學普通名詞草案 (A draft of common names for organic chemistry).[18] Within the text, in which the principles followed for the creation of a Chinese nomenclature for organic chemistry are meticulously reported, a typical entry is as follows:

TABLE 40 *Youji huaxue putong mingci cao'an*

化學式 (Chemical formula)	英文名 (English term)	擬定 (Provisional decision)	備考 (Notes)
$C_3H_5(OH)$, 1	Alcohol Allyl	葱醇	由Allium葱屬

In the *Youji huaxue putong mingci cao'an*, as a whole, translations are provided to refer to the etymology of the referent of the original term as much as possible, as also explained in the notes of the individual entries in the text: "Alcohol Allyl" is rendered as *cong chun* 葱醇, since it is mostly found in the family of Amaryllidaceae. For the lemma above, the translation supplied in the *TVM*, on page 2 of the *HXCL*, was *alailike chun* 阿來里克醕 instead. It is evident from this example how, in those circumstances in which Fryer and his colleagues followed the path of phonemic loans or, as in this case, of hybrids, both for time limit or scarce scientific and technical knowledge, they were often far from ideal translation solutions.

16 The term was already attested in the *Puji fang*.
17 As pointed out in Wang Yangzong, "A New Inquiry," 271–83.
18 Year and place of publication unknown; the text consulted is stored in the Shanghai Library.

It is obvious, in any case, that the temporal distance between the publication of the two texts should not be ignored. For example, in the *Sixth and Seventh Reports of the General Committee on Scientific Terminology—Chemical Terms* (*Systematic Names of Organic Chemistry*), with the Chinese title *Huaxue mingci shenchazu di liu qi ci shenchaben—Youji huaxue xitong mingci*—化學名詞審查組第六七次審查本—有機化學系統名詞, by the General Committee on Scientific Terminology—*Kexue mingci shenchahui* 科學名詞審查會, questions pertaining to inorganic chemistry terminology are presented in a more articulate and extended manner, touching on concepts and topics that had still not been dealt with in the *TVM*.

The next text considered, the *Huaxue mingming yuanze*, helps the reader confirm that the path followed by chemical terminology was destined to be, as a whole, different from that indicated by the *TVM*.

At times, translations annotated in the *TVM* were less adequate than those adopted within other coeval texts, even for example the "Terms used in Dr. Macgowan's translation" of the *JS*. As can be inferred from the first item in table 41, *yansuan qi* 鹽酸氣, the rendering of "Acid muriatic" in the *Jinshi shibie* by Hua Hengfang and Macgowan is certainly closer to the solution adopted by the Guoli bianyiguan, *cu yansuan* 粗鹽酸, than *qinglü qi* 輕綠氣, provided in the third column of the *JS*, "Terms in general use and chief elements," corresponding to the solution Fryer and his Chinese colleagues endorsed.

TABLE 41 *Huaxue mingming yuanze* and the *TVM*

英文 (English)	中文 (Chinese)	*TVM*
Acid, muriatic (8),[a] 4	粗鹽酸	鹽酸氣[b] 輕綠氣
Gum, tragacan (5),[c] 74	托辣甘樹膠	脫拉茄看得樹膠, 特拉茄看得膠
Turmeric (4),[d] 154	鬱金, 薑黄	薑黄, 鬱金;[e] 托末里客即薑黄

a The number in brackets, as reported in the legend of the text, indicates that the entry belongs to the sector of acids and bases.

b In this case, *yansuan qi* 鹽酸氣 is listed in the second column of the *JS*, to wit "Terms used in Dr. Macgowan's translation."

c The number in brackets indicates the entry belongs to the sector of tanning.

d The number in brackets indicates the entry belongs to the sector of inks and paints.

e The first two translations are listed on page 64 of the *XYDCYP*, the following two on page 36 of the *HXCL*.

Moreover, in the second line, it is possible to note that the Guoli bianyiguan, at times, also resorted to phonemic loans and hybrids. Nevertheless, characters adopted were frequently different from those of the TVM. This can be appreciated in the lemma "Gum, tragacan," attested as "Gum tragacanth" on page 27 of the XYDCYP and rendered with the hybrid *tuolaqiekande shujiao* 脫拉茄看得樹膠, while on page 17 of the HXCL it was translated in the namesake entry as *telaqiekande jiao* 特拉茄看得膠; the hybrid chosen by the Guoli bianyiguan is instead *tuolagan shujiao* 托辣甘樹膠.[19]

In other circumstances, terms annotated in the TVM and the *Huaxue mingming yuanze* totally coincide, as in the third item of table 41, even in both the solutions provided. From these few examples, however, it would be difficult to establish with certainty the hypothesis that the *Huaxue mingming yuanze* has drawn its nomenclature from the TVM.[20]

3 Beyond Chemistry: Other Branches of Science

Extending the analysis to other branches of knowledge, consider the *Yixue mingci huibian* 醫學名詞彙編 (Collection of medical terms).[21] The text is an extremely specialised medical glossary; apart from specific entries, traces of the employment of the TVM in the text are not found.

19 Nevertheless, it is important to note that in the TVM the original term is annotated in a slightly different way, therefore accounting for, at least partially, discrepancies in the phonetic portion of the hybrid.

20 The same can also be considered valid for numerous other texts consulted. Among these, the following can be cited: General Committee on Medical Terminology, *Yixue mingci shenchahui: Second Report of Chemical Terms—Huaxue mingci shenchazu di er ci shenchaben—Huaxue shuyu* (publication place and year unknown), which corresponds almost entirely to the aforementioned *Chemical Terminology: Serial No. II*, with the only exception being the entry "Reagent" on page 12, not listed in the first, but annotated in the second; Wu Qiu, trans., *Zhiwu shenglixue—Nongxue congshu*, 3rd ed. (Shanghai: Xinxue shehui, 1913); Idem, trans., *Shengwu shenglixue—Gaoji nongxue xiaoyong*, 6th ed. (Shanghai: Xinxue shehui, 1926); and Huaxue mingci shencha weiyuanhui gongzuo xiaozu, *Huaxue mingci cao'an: A-L* (Zhongguo huaxuehui, publishing year unknown), which only represents a draft, with all the English lemmata that are handwritten in the final section of the text. Even though it is included in the Shanghai Library catalogue, the second part of the latter, namely Idem, *Huaxue mingci cao'an: M-Z* (Zhongguo huaxuehui), could not be located.

21 Kexue mingci shenchahui, ed., *Yixue mingci huibian: La-Ying-De-Han duizhao—A Latin-English-German-Chinese Medical Terminology* (Shanghai: Kexue mingci shenchahui, 1931; republished in 1940).

TABLE 42 *Yixue mingci huibian* and the TVM

古文名 (Old term)	英文名 (English term)	德文通用名 (Common German term)	日譯	參考名 (Term for reference)	决定名 (Term agreed upon)	*TVM*
Aether aceticus, 13	Acetic ether	Essig-aether	醋依的兒		醋醚, 醋酸以脫	醋酸以脫
Kreosotum, 251	Creosote	Kreosot			木焦油	格里亞蘇特
224	Hydrogen peroxide	Hydrogenbioxyd; Wasserstoff-hyperoxyd	過酸化水素液		過氯化氫	輕養二
Mel depuratum, 284	Purified honey	Gereinigter Honig	精製蜂蜜		精製蜂蜜	提淨蜜糖

The same also applies to the field of botany and zoology,[22] as evidenced when consulting the *Dong zhiwu mingci huibian* 動植物名詞彙編 (Collection of

22 As an example of dictionaries of toponyms and anthroponyms, He Songling, *Biaozhun hanyi waiguo renming diming biao: fu hanwen suoyin, xifang yiming zongfen ge biao* (Shanghai: Shangwu yinshuguan, 1924; republished in 1925), can be mentioned. The text resumes numerous original terms, in most cases from English, already listed in the RMDM, both for the section of toponyms and anthroponyms. Moreover, numerous entries included in the text, which is extremely accurate and detailed, also coincide with those annotated in volume ten of the manuscripts stored in the EAL. Nevertheless, characters used in He, *Biaozhun hanyi waiguo renming diming biao*, are regularly different from those adopted by Fryer and his colleagues; the use of the TVM can therefore be excluded in this case as well. Same results are produced through the analysis of other texts, covering different scientific branches, from which, though, examples will not be listed in this work. More specifically, the texts consulted are: for physics, Guoli bianyiguan, *Wulixue mingci* (Shanghai: Shangwu yinshuguan, 1934); for astronomy, Idem, *Tianwenxue mingci* (Shanghai: Shangwu yinshuguan, 1934); for bacteriology and immunology, Idem, *Xijunxue mianyixue mingci* (Shanghai: Shangwu yinshuguan, 1937); for electrotechnics, Idem, *Dianji gongcheng mingci—Dianhua bu* (Shanghai: Zhengzhong shuju, 1946) and Idem, *Dianji gongcheng mingci—Dianli bu* (Shanghai: Zhengzhong shuju, 1947); for mineralogy, among dictionaries, Idem, *Kuangwuxue mingci* (Shanghai: Shangwu yinshuguan, 1936), and among textbooks, Xie Jiarong, ed., *Dizhixue* (Shanghai: Shangwu yinshuguan, 1926) and Yu Wuheng and Du Ruocheng, ed., *Dizhixue* (Shanghai: Shangwu yinshuguan, 1932); for chemical engineering, Guoli bianyiguan, *Huaxue gongcheng mingci* (Shanghai: Zhengzhong shuju, 1946); for mechanical engineering, Zhongguo jixie gongcheng xuehui,

TABLE 43 *Dong zhiwu mingci huibian* and LBT

拉丁名 (Latin term)	英文名(德文名附) (English term with German term)	参考名 (Term for reference)	决定名 (Term agreed upon)	LBT
Abutilon avicennae, Gaertn., 2	Stamp-weed (Sommer-schönmalve)	莔麻, 蔶, 青麻, 檾麻	莔麻 (錦葵科)	桐麻, 青麻, 莔麻, 檾麻, 莔麻
Abutilon indicum, G. Don., 2	Indian mallow	冬葵子	冬葵子 (錦葵科)	冬葵子
Aleurites cordata, Steud., 11	Tung-oil Plant; Chinesischer Ölfirnisbaum (G)	桐子樹, 油桐, 罌子桐	油桐 (大戟科)	桐子樹, 罌子桐

zoological and botanical terms).[23] In this case as well, terms drawn from the *TVM* are not present. Nevertheless, when comparing the *Dong zhiwu mingci huibian* with manuscript number eight of the EAL, the LBT, the coincidence of numerous translations can be noticed.

Consider the first example: in the *Dong zhiwu mingci huibian*, the entry "Abutilon avicennae" is annotated on page 2. Within the texts consulted by the General Committee on Scientific Terminology (Kexue mingci shenchahui) for the entry, reported in the third column, four solutions are provided: *mengma* 莔麻, *qing* 蔶, *qingma* 青麻, and *qingma* 檾麻, the final decision of the Committee being *mengma* 莔麻. In the LBT, in the namesake entry, five translations can be read: *tongma* 桐麻, coloured red, "Shai & Sz chuen list"; *qingma* 青麻, coloured indigo and green, namely "Dr. Henry's list"; *mengma* 莔麻, identified by purple, to wit the "Japanese list"; *qingma* 檾麻 and *mengma* 莔麻, once more marked indigo and green. With the exception of *tongma* 桐麻, the remaining four translations are the same in both the *Dong zhiwu mingci huibian* and the LBT.

ed., *Ying-Han duizhao jixie gongcheng mingci* (Shanghai: Shangwu yinshuguan, 1949); for hydraulic engineering, Quanguo jingji weiyuanhui shuili chu, ed., *Shuili gongcheng mingci cao'an* (Nanjing: Quanguo jingji weiyuanhui, 1935).

23 Kexue mingci shenchahui, ed., *Dong zhiwu mingci huibian: kuangwu mingci fu* (Shanghai: Kexue mingci shenchahui, 1935).

There are different reasons why the terms included in the *TVM* were not adopted in the texts considered here. The first pertains to a lexicological factor already mentioned. Phonemic loans, such as the ones listed in table 42 for translating "Creosote," were composed of too many characters and could not be incorporated in the lexical texture of Chinese. The second factor is related to the first and strictly pertains to chemical terminology—it would be impossible to express all substances, as Fryer and his colleagues frequently did, with terms reproducing their chemical formulas. It would theoretically be feasible to represent "hydrogen peroxide" with characters based on its relatively easy formula. However, when extending the reasoning to more complicated substances, the principle would simply not apply. Finally, as exemplified in table 43, for many lemmata the synonymous terms provided were too numerous, frequently implying a semantic opacity. As seen in this work, such confusion was at times generated by ambiguities inherent to the scientific and technical knowledge of the time that Fryer and his colleagues compiled the *TVM*, and also by the nature of the compilation of the *TVM* itself, for which heterogeneous sources were consulted.

To sum up, the reason for the rejection of the *TVM*'s terminology, not necessarily only by the texts considered here, lies in the impossibility of reconciling two peculiarities of a modern language: "productivity" and "transmission capability."[24] Conversely, those terms that survived and were adopted into the scientific and technical lexicon are exactly those conforming to these characteristics.

Even so, all the examples provided confirm the validity of the work carried out in the *TVM* by Fryer and his colleagues, particularly the systematisation of a method that could be applied even in those cases where terms, whether drawn from other sources or coined from scratch, would prove to be inadequate and replaced by others. Analysing some of the aforementioned texts, it is easily perceivable that the discrepancies with the *TVM*'s renderings frequently lie in the choice of single characters, rather than in translation choices themselves. For example, which characters should be used as right-hand components of words, in the field of mineralogy, either *shi* 石 (stone) or *yan* 岩 (rock) and *ni* 泥 (mud) or *tu* 土 (earth), or which characters should be adopted in phonemic loans.[25]

24 Shen, "Chinese Language and the Modern," 101–102.

25 Refer to the term "Loam," which in Xie, *Dizhixue*, 105, is still rendered with a phonemic loan, *lumu* 壚坶, while the one annotated by Fryer and his colleagues on page 21 of the *HXCL* corresponds to *luomo* 羅末.

It should be restated here that the significance of the *TVM*'s contribution lies also in the stabilisation of the use of some characters as right-hand components of words. For instance, the term "Alcohols," rendered on page 2 of the *HXCL* as *chunlei* 醕類, would still be translated as *chunlei* 醇類 in later texts, with the first character, *chun* 醇, a graphical variant of the one adopted in the *HXCL*.[26] The same also holds true for texts including much more advanced content in comparison with those of the *TVM*, particularly in the field of organic chemistry; some terms still overlapped with the *TVM*'s nomenclature. On page 2 of the *HXCL*, for example, "Alkaloid organic" was rendered as *shengwu sijian lei* 生物似鹻類, and in later works pertaining to organic chemistry, "alkaloids" will be similarly translated as *shengwu jian lei* 生物鹼類.[27]

Therefore, even though the nomenclature of the *Dong zhiwu mingci huibian*, as well as that of other texts considered, does not imply the Committee's adoption of the *TVM*'s terminology, table 43 demonstrates that the entangled intersection between lexical innovation, a reference to Chinese traditional texts and works translated by foreign missionaries and scholars with the help of Chinese colleagues, is much more intricate than depicted until now. Consequently, the examples provided further certify the importance of examining both the *TVM*, its manuscripts, and other relevant works, to draw a better picture of the interactions between the lexicographical output of late Qing China and texts printed at a later stage.

At the same time, table 43 indicates research paths that can be further explored. As demonstrated with *Dong zhiwu mingci huibian*, the sources consulted by the General Committee on Scientific Terminology apparently are the same ones Fryer referred to while compiling the LBT; some of the solutions adopted within these sources have not merged in the *TVM*, as was the case for others. The question is further complicated when comparing the nomenclature adopted in the *Dong zhiwu mingci huibian* with that annotated in other EAL manuscripts. For example, the employment of the BTWDCL within the *Dong zhiwu mingci huibian* can be excluded. In other circumstances, terms listed in *Dong zhiwu mingci huibian* are at times coincident and at times divergent with those of the manuscripts, as in the case of the "Shai & Sz chuen list."

In conclusion, the interactions considered here, also applying to texts different from the ones presented in this section, are multifaceted and are not a mere transmission of lexical items from one translator to the other, from one glossary to the following one. They imply exchanges at the level of ideas, and

26 Zheng, *Huaxue mingming*, 23.
27 Guoli bianyiguan, *Huaxue mingming yuanze*, 43.

therefore concepts, with the relevant issues pertaining to textual production and circulation of knowledge that are worth analysing in detail.

4 Lexicological Legacy: Acknowledgement and Appraisal

Traces of the work carried out in the TVM are not limited to the works illustrated in chapter 7. The achievements by Fryer and his Chinese colleagues are also acknowledged in texts published decades after the TVM, even in those cases where their solutions were rejected, due to intrinsic reasons lying within adopted translation choices, or pertaining to the evolution experienced by the scientific fields covered in the TVM.

Among the different examples, the *Kuangwuxue mingci* 礦物學名詞 (Terms in mineralogy) can be mentioned first.[28] Important indications are annotated in the first page of the preface: "我國之有礦物學譯名專著，自清光緒九年江南製造局刊行華蘅芳迻譯之金石表一書始。書中譯名凡一千七百則，除音譯部份外，今尚多所承用" (Treatises including translated terms for mineralogy were composed in China starting from the ninth year of the reign of Guangxu, during the Qing dynasty, when the Jiangnan Arsenal published a glossary of minerals and stones translated by Hua Hengfang. In the book, a total of one thousand and seven hundred translations was provided; apart from transliterations, the great majority is still employed nowadays). As clearly expressed in the passage, the publication of texts in mineralogy started in China with the publication, by the Jiangnan Arsenal, of a glossary of technical terms attributed to Hua Hengfang. Even though the preface of the *Kuangwuxue mingci* indirectly refers to the *Jinshi shibie*, the translation of Dana's *Manual of Mineralogy* by Hua Hengfang and Macgowan, it is in reality indicating the JS. The ninth year of the reign of Guangxu corresponds to 1883, the year the JS was published, in which terms from the *Jinshi shibie*, printed in 1871, were also included.

In the preface quoted above, it is affirmed that terms adopted within the JS were still in use at the time of the compilation of the *Kuangwuxue mingci*, with the exception of phonemic loans. Even though the nomenclature devised by the Guoli bianyiguan in the *Kuangwuxue mingci* is different from that of the TVM, the pioneering achievements of the latter are clearly, though, indirectly, acknowledged.

Analogous references to the TVM can also be found in *Huaxue gongcheng mingci* 化學工程名詞 (Terms in chemical engineering), even if the glossary

28 Guoli bianyiguan, *Kuangwuxue mingci*.

does not employ the *TVM*'s nomenclature.[29] In the *Huaxue gongcheng mingci* preface, the role of Fryer's work is once more appreciated: his name is mentioned, in addition to one of the glossaries edited with his colleagues, namely the *HXCL*. Moreover, in the preface to the text, the unsuccessful resolution of the problems related to the unification and standardisation of scientific and technical nomenclature is also stressed: "及遜清咸同年間，江南製造局出版西人傅蘭雅所譯之化學工藝、化學器、化學材料名目表諸書出，新式化學工程之著作，始見其端倪，然名詞術語之用，紛歧雜沓，直至近年尚未統一也" (In the years between the end of the Qing dynasty and the Xianfeng and Tongzhi reigns, the Jiangnan Arsenal published numerous texts rendered by the Western [translator] John Fryer, including some on chemical technology, chemical instruments, and a glossary of names of chemical substances. The new style of the treatises on chemical engineering, therefore, was seen for the first time […]; nevertheless, there were unnumbered complications in the use of technical terminology, [so much that] until recently it still has not been unified).[30]

More proof that the editors of the Guoli bianyiguan were familiar with the *TVM*, and consulted it for the compilation of its texts, comes from the preface to the *Jixie gongcheng mingci* 機械工程名詞 (Terms in mechanical engineering).[31] In this case as well there is a specific reference to one of the other glossaries edited by Fryer and his colleagues, in this circumstance, the *QJ*, as well as to the texts of the Ming dynasty *Yuanxi qiqi tushuo* 遠西奇器圖說 (Diagrams and explanations of the wonderful machines of the Far West), by the German astronomer Johann Schreck (Deng Yuhan 鄧玉函, 1576–1630) and the Chinese scholar Wang Zheng 王徵 (1571–1644), and *Huogong qieyao* 火攻挈要 (Essentials of gunnery), by the German Jesuit missionary Johann Adam Schall von Bell (Tang Ruowang 湯若望, 1591–1666) and the Chinese scholar Jiao Xu 焦勛 (n.d.): "機械工程名詞之迻譯遠自明王徵之遠西奇器圖說，焦勛之火攻挈要，清季江南製造局之汽機中西名目表等，卽已粗具基礎矣" (The translation of terms in mechanical engineering derives as far as the *Yuanxi qiqi tushuo* of Wang Zheng and the *Huogong qieyao* of Jiao Xu, of the Ming dynasty, and the *Vocabulary of Terms Relating to the Steam Engine* of the Jiangnan Arsenal, during the Qing dynasty, which all laid its foundation).

Even though the *Jixie gongcheng mingci*, rather than quoting the *TVM*, seems to improve its translations, the importance the *TVM* had in the history of the

29 Guoli bianyiguan, *Huaxue gongcheng mingci*.

30 Ivi, preface, 1.

31 Guoli bianyiguan, *Jixie gongcheng mingci* (*putong bu*) (Chongqing: Zhengzhong shuju, 1946).

compilation of bilingual glossaries, as well as its pioneering value, has been certified once more, even decades after its publication.[32] Furthermore, in the prefaces of numerous texts consulted, even those composed in the forties of the twentieth century, it is often noted that, in translating scientific and technical terms, it is preferable to avoid phonemic loans, resorting to them only when necessary. This again confirms the far-sightedness of the choice by Fryer and his colleagues, restated in various occasions, and frequently against the preference of other translators.

The direct acknowledgements of the work carried out by Fryer could not be seen in those works, and by those authors who employed the scientific and technical nomenclature of the *TVM* more extensively, Yan Huiqing included. This is due to two reasons: the first is ego, which led numerous authors to not quote sources consulted for the compilation of their own lexicological works.[33] The difficulty in locating direct sources of the lexicographical production by Jesuit missionaries and Protestant or secular translators is a problem faced by all scholars conducting research in this field. The second reason is the necessary temporal distance that such recognition requires; as stated by Fryer himself, the suitability of different competing nomenclatures would be certified only when put to the test of time. This principle also applies to the lexical innovations arising every day in all languages, and in different sectors of the lexicon, not only those of Chinese scientific and technical nomenclature presented. Such a test was successfully passed by the *TVM*.

32 It should be noted that all lemmata listed in the first page of the CT for which Fryer has not supplied a Chinese rendering, are all invariably included within Guoli bianyiguan, *Jixie gongcheng mingci* (*putong bu*).

33 This hesitancy also resulted in cases of plagiarism, such as the well-known one by the French sinologist Chrétien Louis Joseph de Guignes (Xiao Dejin 小德金, 1759–1845), who adopted for his dictionary, without credit, the manuscript composed by the Franciscan missionary Basilio Brollo (Ye Zongxian 葉宗賢, also known as Ye Zunxiao 葉尊孝, 1648–1704).

Conclusions

The work Fryer and his Chinese colleagues accomplished in the compilation of the *TVM* is a milestone and an important research instrument from different perspectives: translation studies, lexicology, cultural exchanges between the West and China, China's modernisation, the history of science, and more generally, linguistics and history. By adopting existing nomenclature, creating neologisms whenever deemed necessary, and substituting them in case more suitable terms could be found, the translations carried out at the Jiangnan Arsenal and crowned with the *TVM* all contributed to enriching the Chinese lexicon in a crucial phase of its development. The *TVM* later influenced textual, and more specifically, lexicographical production for a considerable period of time.

Even so, the *TVM*'s journey through such an important phase in the formation of modern Chinese scientific and technical lexicon was hindered by different reasons, and finally came to an end. Strictly pertaining to the linguistic and lexicological obstacles, they can be summed up in three factors: the presence of too many phonemic loans, the impossibility of expressing all substances through terms reproducing their chemical formulas or scientific peculiarities, and finally, a semantic opacity pertaining to the terms provided in the *TVM*, frequently generated by the scientific and technical knowledge of the time. This all led to a portion of the *TVM*'s nomenclature to be deemed unproductive, and therefore unable to adequately convey certain notions.

Pertaining in particular to the overuse of phonemic loans, it is important to note that this flaw did not affect another fundamental missionary translation project in China's history—the one carried out by the Jesuits in the seventeenth and eighteenth centuries. Even the most expert Jesuit translators had to face the scarcity of adequate terminology for religious texts they intended to render in Chinese, as well as delicate issues regarding the translation of religious concepts. Nevertheless, the adoption of phonemic loans in this specific sector was rejected by Chinese readers, who in the case of texts compiled by Jesuits were for the most part high representatives of the Ming court.[1] Jesuits, therefore, preferred to resort to terms existing in Chinese classical texts that could cover the semantic field of the referent to be translated, conferring a slightly shifted meaning. Consequently, phonemic loans were avoided as much as possible and confined to toponyms and other less sensitive concepts.[2]

1 Kurtz, *The Discovery of Chinese Logic*, 27–28.

2 *Ibidem*, and Elman, *On Their Own Terms*, xxvii.

 | DOI:10.1163/9789004443211_011

This was something Fryer himself had also frequently advocated for scientific and technical nomenclature; at times, though, the adoption of phonemic loans was unavoidable. However, since numerous terms he and his colleagues at the Jiangnan Arsenal rendered in Chinese belonged to the scientific and technical lexicon, the reasons for the rejection of phonemic loans were purely lexicological, and no religious implications were at work. This surely helped many others to be more easily accepted in late Qing textual production as presented in this work.

However, as seen in the case of the Publication Committee of the Educational Association of China, purely linguistic considerations were not only at work in the struggle for the survival of the fittest terms: power, ideological, and personal reasons also played an important role. As particularly exemplified by Mateer and Hayes, personal hostilities were disguised as lexicological contrapositions; their criticism of the nomenclature for chemical elements Fryer and Xu Shou devised proved in the end to be groundless. In the context of power and personal interactions, it should be noted that one episode particularly contributed to Fryer's alienation from the community of missionaries, and also to the loss of his initial status as a missionary. Rumours of an affair between his fiancée Anna Roleston and the captain of the ship she was aboard heading to China spread in the religious community of foreigners living in China. This resulted in Fryer's isolation among the missionary milieu and influenced his whole life,[3] as confirmed by Fryer himself in a letter to Edkins written on July 28, 1868: "The Church Mission people never write to me now. Most likely they have forgotten me."[4] Fryer's other life choices also contributed to the end of the *TVM*'s journey; his departure from China in 1896 made the creation of a second volume of the *TVM* and the revision of existing nomenclature almost impossible.

The last relevant factor for the rejection of the terms listed in the *TVM* was China's defeat in the First Sino-Japanese War. After this turning point, China's demand for scientific and technical texts skyrocketed; a portion of the *TVM*'s nomenclature was adopted, other terms were rejected. Another factor emerged in Chinese scholars' demand for ideas from the West: Japanese language and relevant texts would contribute the most to completing the evolution of Chinese scientific and technical nomenclature.

It is exactly from the last factor that new research perspectives are highlighted; by examining the lexical production of Western scholars and missionaries in late Qing China, important information about the textual and

3 As indicated in Wright, *Translating Science*, 107–109.

4 Dagenais, *The John Fryer Papers*, I, 374.

lexicological interactions of different sources can be derived. The *TVM* is a mirror reflecting the lexicological status quo of the scientific and technical lexicon in China at the end of the nineteenth century; however, it is worth researching not only for its lexicological achievements. The *TVM* shows, in an indirect and condensed form, both contemporary and earlier texts that foreign translators and their Chinese colleagues could use as a reference for their translations. In the case of the *TVM*, such texts included Japanese sources; these at times represented translations of original European works realised in the framework of the *Rangaku*, at times quoting other Chinese autochthonous texts.

Fryer consulted all these texts because he considered them necessary for the compilation of bilingual English-Chinese scientific and technical glossaries, with the addition of other lexicographical sources compiled by contemporary Western missionaries and scholars. Such a method reflects one of the peculiarities of modern Chinese lexicon, highlighted in secondary literature: it is the result of interactions between linguistic, social, political, and economic factors rooted before the May Fourth Movement.[5] The study of the *TVM* shows the broader scholarly readership of historians and Asianists the complicated link in the history of missions between religious intentions, lexicographical production, and power dynamics with other translators. At the same time, this research certifies that the work by missionaries and scholars was appreciated and acknowledged by later Chinese translators and literati. The linguistic tradition of Western scholars[6] allowed them to grasp some features of Chinese lexicon and build suitable terminology conveying new ideas, later adopted by Chinese translators who guided China and Chinese lexicon on the path to modernity.

The *TVM*'s importance lies exactly in the fact that it is one of the first steps in this process. As the first systematic collection of scientific and technical glossaries with English-Chinese parallel translation, this underestimated text is a milestone in the lexicological and philological production of late Qing China. Its role, born at the intersection of the transmission of scientific ideas, as well as in the broader frame of cultural exchanges between the West and the East, the history of translation, and the strategies adopted by late Qing foreign missionaries and scholars, can hardly be overrated.

The historical frame considered in this text represents the third major wave of translations, and consequently of language contacts, in China's history. This was a crucial phase of translations in Chinese, which scholars have only

5 Masini, *The Formation of Modern Chinese*, preface, iii–iv.

6 Uchida, *A Study of Cultural Interaction*, 11.

recently started to focus on.[7] Other than its purely lexicological contribution, the *TVM* particularly helped the formation in late Qing China of a systematic awareness of the problems behind translation practice and theory, as well as of a community of Chinese scholars addressing these issues. As a vector of ideas, it helped pave the way for the linguistic evolution the Chinese language would face in the twentieth century.

The glossaries included in the *TVM* also played an important role in the history of the Chinese scientific and technical lexicon because they circulated scientific notions for a considerable period of time. The translation work by foreign missionaries and scholars in China within the bigger frame of the *Xixue Dongjian* was from the very beginning nothing but a vector for conveying ideas; these, in turn, could only be expressed by terms. Such terms required certain peculiarities to serve their purpose, to be productive, and finally to be accepted. The intentions and contents of the translations influenced the linguistic attitude of foreign missionaries and scholars involved in this task from the very beginning. This is exemplified by Fryer, whose status was at first that of a religious person, and later turned to more secular objectives.

By analysing nomenclature and terminology, we are not only studying terms and words—the latter are a vehicle through which it is possible to recognise and interpret history and human society. Fryer and the literati of late Qing China tried to express the scientific and technical world around them through the compilation of glossaries. Other than being a personal instrument for organizing available lexicological materials, these glossaries also served as a fundamental reference for other contemporary translators. It can be stated that the outcome of the *TVM* even exceeded its objectives; its influence in the lexicological production and transmission of ideas extended until the first part of the twentieth century and was directly and indirectly acknowledged by other eminent Chinese and foreign scholars.

From a lexicological perspective, starting from Fryer's and Xu Shou's creation of a method for the nomenclature of chemical elements, which included the creation of characters (*zi*), Fryer realised that in order for new terms to be incorporated into the lexicon, they should combine existing characters, thus creating new words (*ci*) and meanings. This method conformed to that of Japanese scholars. Even if, due to the success of the terminology for chemical elements, some translators in China continued to create new characters

7 Cao Guangshun and Yu Hsiao-jung, "Language Contact and its Influence on the Development of Chinese Syntax," in The *Oxford Handbook of Chinese Linguistics*, ed. William S-Y. Wang and Chaofen Sun (Oxford: Oxford University Press, 2015), 205.

to express new ideas, they were later destined to be replaced by terms used in Japanese.[8]

Fryer was aware of this peculiarity of the Chinese lexicon, as evidenced by the consultation of Japanese sources. The *TVM* is therefore connected from a temporal and lexicological perspective to Protestant authors, earlier Jesuit sources, Japanese translations, and original Western texts. The range of the scientific branches considered in the *TVM* is extremely broad: from the early botanical texts compiled by Jesuits to later sectorial texts compiled by Japanese literati, Fryer and his colleagues served the role of a cultural and lexicological bridge for ideas to spread across centuries. Their work helped to connect terms of the Chinese tradition, those employed by Jesuits, and Japanese sources to dictionaries and glossaries, bilingual or monolingual, composed in the late Qing dynasty and the Republican era. This process reached its peak particularly within encyclopaedic and collectaneous works that adopted the *TVM*'s terminology. Their importance in the circulation of scientific and technical knowledge also highlights the role of the *TVM* in the standardisation of Chinese scientific and technical lexicon.

It is true that the position of the *TVM* can be evaluated only in relation to other texts; this is one of the reasons for analysing its lexicological and textual connections with those other works considered in this work. If the lexicological and morphological innovations of the *TVM* had remained confined to the text itself, they would have unlikely spread into Chinese lexicon. The historical circumstances of late Qing China, particularly the First Sino-Japanese War, spurred the demand for new kinds of publications. The lion's share was represented by encyclopaedias and journals—their publication was in turn connected to the thriving lithographic printing technologies. In a virtuous circle, the demand for scientific knowledge from the West was triggered by historical circumstances and this, in turn, contributed to improving printing. Late Qing encyclopaedias were a commercial success, and at the same time, answered the need for new scientific and technical knowledge for a broad readership, using the Chinese language. The new notions required by these publications were available in the form of bilingual glossaries compiled by Fryer and his Chinese colleagues.

Furthermore, the fact that new publications were used as teaching material in the framework of the educational reforms of the first years of twentieth-century China also led other Chinese scholars to value and acknowledge Fryer's work. If personalities such as Yan Huiqing, Dai Liuling, and Yan Fu directly or indirectly used the *TVM*, it is because they recognised its

8 Shen Guowei, "Xifang xin gainian," 121–22.

lexicological value and its usefulness for their educational, political, and translation objectives.

Such a connection between historical factors, lexicological and textual production, and the "transfer of Western knowledge to the East", confirms the validity of the study of materials composed by Western missionaries in late Qing China, highlighted by the theoretical framework of the *shūen kara no apurōchi*, the "peripheral approach." Starting from the analysis of the glossaries, it is clear that the TVM was not limited to the chronological and spatial context of the Jiangnan area. Its lexical innovations played a key part in the formation of Chinese scientific and technical lexicon, and in later textual production, promoting the circulation of Western knowledge in late Qing China.

Despite its rigorous lexicological methods and the TVM's theoretical framework, Fryer personally underlined in different circumstances that neither he, his generation, nor the following one, would create a definite nomenclature for Chinese scientific and technical lexicon.[9] The TVM's importance should rather be attributed to the interpretation and use of some lexicological, and partially also morphological, features of the Chinese language. Such peculiarities would later allow the language to express the reality of the scientific and technical world to the fullest, proving many translators and missionaries wrong, who, with certainty, opposed this possibility. The aim of the TVM was fundamentally none other than to be a lexicological and lexicographical landmark for texts that would be subsequently published. Such a purpose was acknowledged not only in China but also other countries, as with the Japanese ambassador Yanagihara Sakimitsu's 柳原前光 (1850–1894) mission to China, expressly for the purpose of purchasing books translated at the Jiangnan Arsenal.[10] Among the works bought, numerous books in the field of chemistry edited by Fryer and Xu Shou were included; these texts already adopted the nomenclature that would later be systematically gathered in the TVM.[11] Of course, not only

9 Fryer, "Scientific Terminology," 545–46.

10 Mentioned in Xu Zhenya, "Xu Shou fuzi dui Zhongguo jindai huaxue de gongxian," *Daxue huaxue* 15, no. 1 (2000): 59, and David Wright, "Careers in Western Science in Nineteenth-century China: Xu Shou and Xu Jianyin," *Journal of the Royal Asiatic Society*, third series, 5, no. 1 (1995): 67, footnote no. 73.

11 For the circulation of the texts translated at the Jiangnan Arsenal in Japan, refer also to Chen Liwei, "*Jiangnan zhizaoju de chubanwu zai Riben de chuanbo he liyong*," in *Lishi shang de Zhongguo chuban yu Dongya wenhua jiaoliu*, ed. Fudan daxue lishi xi chuban bowuguan (Shanghai: Shanghai baijia chubanshe, 2009). Consult also Shen Guowei, "*Xiandai hanyu zhong de riyu jieci zhi yanjiu—Xushuo*," *Riyu xuexi yu yanjiu*, no. 5 (1988): 15 and 18, where the employment by Japanese translators of English and Chinese dictionaries compiled by foreign missionaries and scholars is referred to, as well as the diffusion in Japan of translations realised under the sponsorship of the Jiangnan Arsenal.

texts pertaining to chemistry circulated extensively in Japan; others in different fields also contributed to the exchange of ideas and terms conveyed in the period from the late Qing dynasty to the beginning of the twentieth century.[12]

It is possible to affirm with certainty that the purpose Fryer and his colleagues established for the *TVM* was achieved, given that such a consistent number of representative texts printed after the *TVM* adopted its nomenclature. This is also shown by the fact that decades later, the works that contributed to the renovation of Chinese scientific and technical terminology recognised the *TVM*'s achievements, even when following a different path. The *TVM*'s aforesaid role of a lexicological and lexicographical reference went at the same pace as the fulfilment of the standardisation of Chinese scientific and technical lexicon; this is exactly the ultimate aim of all kinds of scientific and technical glossaries, which Fryer and his colleagues consecrated a consistent portion of their lives and work to.

Further lexicological, philological, and historical research will consider other outputs of the textual production, within and beyond, the time frame considered in this work. The mosaic of research conducted on the formation of modern Chinese lexicon, and on the cultural exchanges in the late Qing dynasty and Republican era, will also lead to other major outcomes, pertaining both to the circulation of scientific and technical knowledge and its connection with native Chinese literature. This involves different directions, *ante* and *post quem* the time frame considered here. Lexicologists, lexicographers, historians, and scholars studying the diffusion of ideas and terms conveyed between China and the West will, hopefully, hone the new insights that have been presented here.

---

12 An extremely high number of works translated by Fryer and his Chinese colleagues is gathered in various libraries and institutions in Japan. For the most complete list, refer to Yatsumimi Toshifumi, "Shin makki seijin choyaku kagaku kankei chūgokusho oyobi wakokubon shozai mokuroku," *Kagakushi kenkyū* (1995).

APPENDIX

Synoptic Tables and Reproductions—Contents of the *TVM*

TABLE A Structure of the *TVM*—Published glossaries

Glossary	Publication year	Pages	Number of entries
Vocabulary of Mineralogical Terms	1883	36	1,672
Vocabulary of Names of Chemical Substances	1885	37	3,362
Vocabulary of Names of Materia Medica	1887	68	6,073
Vocabulary of Terms Relating to the Steam Engine	1889	58	5,061
Vocabulary of Proper Names	1887	22	3,067

TABLE B Structure of the *TVM*—Manuscripts

Glossary	Compilation year	Pages	Notes
List of geographical names (vol. 1)	Around 1892	97	The manuscript, originally drafted by Lyman Dwight Chapin, was probably given to Fryer.
Syllabary for the transfer of foreign names into Chinese (vol. 2)	Around 1892	28	According to the LGN preface, it was compiled by Chapin. Another version can be found, written with different ink.
Botanical Terms from William's Syllabic Dictionary of the Chinese language (vol. 3)	Before 1892	199 + 8 + 7	Terms from three texts are comprised: Samuel Wells Williams' *SDCL*, the *GDSE* by Walter Henry Medhurst, and Parker's *UYT*.

 | DOI:10.1163/9789004443211_012

TABLE B Structure of the *TVM*—Manuscripts (*cont.*)

Glossary	Compilation year	Pages	Notes
Vocabulary of proper names (vol. 4)	Around 1887	129	Entries are annotated on strips of paper with a colour key: mustard, dark sky blue, light sky blue, purple, olive green, and wine. Strips without any colour are also present; the colour key is not reported. A group of strips do not have any colour; their paper is more yellowed than the others.
Vocabulary of names of Materia Medica (vol. 5)	Before 1887	210	Original terms for the most part are written in purple and, more rarely, black ink. Chinese translations are written in black ink.
Vocabulary of chemical terms (vol. 6)	Before 1885	102	Followed by 96 pages of botanical terms and 17 pages of zoological terms.
Vocabulary of geological terms (vol. 7)	Around 1890	114	Numerous original terms, written on strips of paper glued onto the volume, are not combined with a Chinese translation. The majority of original terms are written with a typewriter, even if Chinese renderings are handwritten. A colour key is present: red, yellow, purple, blue, sepia, brown, light green, and light violet, but sources are not indicated.
List of botanical terms (vol. 8)	Around 1892	172	Sources from which Fryer drew his translations are presented in detail with a colour key.
Vocabulary of terms relating to the steam engine (vol. 9)	Before 1889	158	Entries are written on strips of paper; colour key of entries in blue, sky blue, wine, red, and green.

TABLE B Structure of the *TVM*—Manuscripts (*cont.*)

Glossary	Compilation year	Pages	Notes
Vocabulary of Terms in Naval Architecture (vol. 9)	Around 1889	178	Entries are written on right-hand pages, directly on the volume; English lemmata and Chinese translations are mostly written in black ink.
Vocabulary of proper names (vol. 10)	After 1887	73	Annotated on strips of paper glued onto the volume. Two different kinds of strips: strips of yellowed paper, with the original terms typewritten and their Chinese renderings handwritten or typewritten; whiter strips with both original terms and relevant Chinese renderings handwritten. Colour key is not annotated.

TABLE C Original texts and Chinese translations adopted in the *TVM*[1]

Original text	Chinese translation (when present)	*TVM*
David Ames Wells, *Principles and Applications of Chemistry; For the Use of Academies, High-Schools, and Colleges: Introducing the Latest Results of Scientific Discovery and Research, and Arranged with Special Reference to the Practical Application of Chemistry to the Arts and Employments of Common Life*, 5th ed. (New York: Ivison, Phinney and Co., 1862)	John Fryer and Xu Shou, trans., *Huaxue jianyuan* (Shanghai: Jiangnan zhizaoju, 1871)	*HXCL*

1 The references listed in this table represent a convenient instrument for the reader and include only those explicitly mentioned or inferred as direct sources by Fryer; more detailed analyses can be found in the relevant chapters of this text.

TABLE C Original texts and Chinese translations adopted in the *TVM* (*cont.*)

Original text	Chinese translation (when present)	*TVM*
Charles Loudon Bloxam, *Chemistry Inorganic and Organic. With Experiments and a Comparison of Equivalent and Molecular Formulæ* (London: John Churchill and Sons, 1867)	John Fryer and Xu Shou, trans., *Huaxue jianyuan bubian* (Shanghai: Jiangnan zhizaoju, 1882)	*HXCL*
Charles Loudon Bloxam, *Chemistry Inorganic and Organic. With Experiments and a Comparison of Equivalent and Molecular Formulæ* (London: John Churchill and Sons, 1867)	John Fryer and Xu Shou, trans., *Huaxue jianyuan xubian* (Shanghai: Jiangnan zhizaoju, 1875)	*HXCL*
James Dwight Dana, *Manual of Mineralogy, Including Observations on Mines, Rocks, Reduction of Ores, and the Application of the Science to the Arts* (edition adopted: unknown)	Daniel Jerome Macgowan and Hua Hengfang, trans., *Jinshi shibie* (Shanghai: Jiangnan zhizaoju, 1871)	*JS*
John Forbes Royle and Frederick William Headland, *A Manual of Materia Medica and Therapeutics; Including the Preparations of the British Pharmacopœia, and Many Other Approved Medicines*, 5th ed. (London: John Churchill & Sons, 1867)	John Fryer and Zhao Yuanyi, trans., *Xiyao dacheng* (Shanghai: Jiangnan zhizaoju, 1879)	*XYDCYP*, *RMDM*
Thomas John Main and Thomas Brown, *The Marine Steam-Engine. Designed Chiefly for the Use of the Officers of Her Majesty's Navy*, 3rd ed. (London: Longman Brown, Green and Longmans, 1855)	Alexander Wylie and Xu Shou, trans., *Qiji faren* (Shanghai: Jiangnan zhizaoju, 1871)	*QJ*
John Bourne, *A Catechism of the Steam Engine in its Various Applications to Mines, Mills, Steam Navigation, Railways, and Agriculture; With Practical Instructions for the Manufacture and Management of Engines of Every Class*, 11th ed. (London: Longman, Green, Longman, Roberts, & Green, 1865)	Fryer John and Xu Jianyin, trans., *Qiji biyi* (Shanghai: Jiangnan zhizaoju, 1872)	*QJ*

TABLE C Original texts and Chinese translations adopted in the *TVM* (*cont.*)

Original text	**Chinese translation (when present)**	*TVM*
John Bourne, *A Catechism of the Steam Engine in its Various Applications to Mines, Mills, Steam Navigation, Railways, and Agriculture; With Practical Instructions for the Manufacture and Management of Engines of Every Class*, 11th ed. (London: Longman, Green, Longman, Roberts, & Green, 1865)	Fryer John and Xu Jianyin, trans., *Qiji biyi fujuan* (Shanghai: Jiangnan zhizaoju, 1872?)	
Nicholas Procter Burgh, *Pocket-book of Practical Rules for the Proportions of Modern Engines and Boilers for Land and Marine Purposes* (London: E. and F. N. Spon, 1864)	John Fryer and Xu Jianyin, trans., *Qiji xinzhi* (Shanghai: Jiangnan zhizaoju, 1872)	*QJ*
Richard Sennett, *The Marine Steam Engine. A Treatise for the Use of Engineering Students and Officers of the Royal Navy* (London: Hebert, Cheapside; Portsea: Woodward, Common Hard, 1852)	John Fryer and Hua Beiyu, trans., *Bingchuan qiji* (Shanghai: Jiangnan zhizaoju, 1894)	*QJ*
Heinrich Paasch, *From Keel to Truck—A Marine Dictionary in English, French and German Amply Illustrated by Explanatory Diagrams of the Most Important Details—For the Use of Ship-Owners, Builders, Brokers, Insurance-Societies, Average-Staters, Barristers, Solicitors, Translators, Surveyors, Masters, Mates, Engineers, Naval-schools, Nautical Journalists, etc.* (Antwerp: Ratinckx Frères, 1885)		*QJ*, CT
Justus Doolittle, *Ying-Hua cuilin yunfu—A Vocabulary and Hand-Book of the Chinese Language, Romanized in the Mandarin Dialect* (Fuzhou: Rozario, Marcal, and Company, 1872–1873)		*QJ*, LBT
Samuel Wells Williams, *A Syllabic Dictionary of the Chinese Language; Arranged According to the Wu-Fang Yuen Yin, With the Pronunciation of the Characters as Heard in Peking, Canton, Amoy and Shanghai* (Shanghai: American Presbyterian Mission Press, 1874)		LBT, BTWDCL

TABLE C Original texts and Chinese translations adopted in the *TVM* (*cont.*)

Original text	Chinese translation (when present)	*TVM*
Edward Harper Parker, *Up the Yang-tse* (Hong Kong: China Mail Office, 1891)		LBT, BTWDCL
Walter Henry Medhurst, *General Description of Shanghae and its Environs, Extracted from Native Authorities* (Shanghai: Mission Press, 1850)		LBT, BTWDCL
Emil Bretschneider, *Early European Researches into the Flora of China* (Shanghai: American Presbyterian Mission Press, 1881)		LBT
Elijah Coleman Bridgman, *A Chinese Chrestomathy in the Canton Dialect* (Macao: Samuel Wells Williams, 1841)		LBT
Frederick Porter Smith, *Contributions Towards the Materia Medica and Natural History of China. For the Use of Medical Missionaries & Native Medical Students* (Shanghai: American Presbyterian Mission Press; London: Trübner, 1871)		LBT
Catalogue of the Chinese Imperial Maritime Customs Collection, at the United States International Exhibition, Philadelphia, 1876. Published by Order of the Inspector General of Chinese Maritime Customs (Shanghai: Statistical Department of the Inspectorate General of Customs, 1876)		LBT
Augustine Henry, *Notes on Economic Botany of China* (Shanghai: Presbyterian Mission Press, 1893)		LBT
Udagawa Yōan, *Shokugaku keigen* (Edo: Suharaya Ihachi, 1833?)		LBT
Itō Keisuke, *Taisei honzō meiso* (Owari: Kajō Shooku zōhan, 1829?)		LBT
Iinuma Yokusai, *Shintei Sōmoku zusetsu sōbu nijū kan* (Ogaki: Hirabayashisō, 1874)		LBT

TABLE 44 Neologisms in the *TVM*, *Jinxiandai ciyuan*, and the *Zhongguo jiben guji ku* database
Synoptic table including neologisms attested for the first time in the *TVM*, including: 1) the Chinese term; 2) the original term, the occurrence, and the glossary in the *TVM*; 3) page, definition, and attestation in *Jinxiandai ciyuan*; 4) number of attestations and the earliest attestation for the term in the *Zhongguo jiben guji ku* database

Chinese term	Attestation in the *TVM*[2]	Page, definition, and attestation in the *Jinxiandai ciyuan*	*Zhongguo jiben guji ku*; number of attestations, and the earliest one[3]
泥土	Terra japonica 日本泥土, 61: *XYDCYP*	549; 黏土; 1930, *Zhonghua baike cidian* 中華百科辭典 (Chinese encyclopaedic dictionary)[4]	1999; Han (2); 0
鋅板	Amalgamated zinc plates 飽足汞之鋅板, 2: *HXCL*	824; 用锌制成的印刷版; 1913 *Xin zidian—Suoyin xin zidian shuoming* 新字典—縮印新字典說明 (New dictionary—Explanation of the New dictionary in reduced edition)	150; Qing (150); 0

2 For the purpose of this and the next table, in the second column only the lemmata in which the Chinese term is listed in the *TVM* as an independent term are annotated, therefore not specifying entries in which it is part of other translations; if the Chinese term is not present as an independent lemma, such entries are in turn annotated. Furthermore, the author has not indicated, unless considered useful for the objective of this work, whether in the *TVM* the Chinese term is a translation alternative to another or whether it was followed by other renderings. Graphical variants of characters are also pointed only if considered indispensable.

3 In the fourth column of this and the following table, the total number of attestations in the *Zhongguo jiben guji ku* is first provided, then the dynasty or period of the first attestations and, in case terms are also provided in the second column of the table as listed in the *TVM*, the total number of their attestations and the relevant dynasty or period. Due to the extremely high number of earlier or later attestations for some of the lemmata, in this and the following table there might be cases in which the combinations of characters is fortuitous and unrelated to the item listed; this has been signalled only when the risk of ambiguity was higher. To analyse the exact meaning of each term in the context of the original texts would require dozens of volumes and exceed the purpose of this work. Figures are nevertheless provided since, notwithstanding an unavoidable percentage of errors, they help to prove the hypotheses on the role of the *TVM* presented in this work, better explaining its lexicographical connections and indicating further research perspectives.

4 Shu Xincheng, ed., *Zhonghua baike cidian* (Shanghai: Zhonghua shuju, 1930).

TABLE 44 Neologisms in the *TVM*, *Jinxiandai ciyuan*, and the *Zhongguo jiben guji ku* database (*cont.*)

Chinese term	Attestation in the *TVM*	Page, definition, and attestation in the *Jinxiandai ciyuan*	*Zhongguo jiben guji ku*; number of attestations, and the earliest one
硫酸	Sulphuric acid, 32: *JS*	494; 一种无机化合物; 1889, *Youli Riben tujing yuji* 遊歷日本圖經餘記 (Additional records of a chronicle of Japan with maps)[5]	26; Qing (26)
功率	Duty of engine 汽機功率, 17: *QJ*	274; 指物体在单位时间内所做的功; 1934, *Ziran kexue cidian* 自然科學辭典 (A dictionary of natural science)	400; Zhou (2); 1; Qing
發動	Driving coupling 發動接盤, 17: *QJ*	198; 使机器运转; 1917, *Qingbai leichao—Wupin lei—Qiji* 清稗類鈔—物品類—汽機 (A Qing history of trivial matters—Objects—Steam engine)	3951; Spring and Autumn—Warring States (7); 0
發動機	Driving gear 發動機器齒輪, Driving shaft 發動機器軸, Driving springs 發動機器簧, 17; Gear 發動器, 24: *QJ*	198; 能够把一种形式的能转化为另一种更有用的能的机器; 1902, *Hua-Ying yinyun cidian jicheng* 華英音韻辭典集成 (A Chinese-English pronunciation dictionary)	13; Qing (12); 0; 0; 0; 5 (Qing)
靛藍	Indigo blue, 30: *XYDCYP*. Indigo blue, 18; Saxony blue, 30:[6] *HXCL*	169; 有机染料，深蓝色; 1890, "Gezhi shiqi 格致釋器 (The investigation of things and extension of knowledge—Explanations of objects)," *Gezhi huibian*, vol. 4	21; Ming (1)
釘錘	Hammer riveting 帽釘錘, 26: *QJ*	171; 钉钉子用的锤子; 1934, *Ziran keben* 自然課本 (A textbook of nature), vol. 3	4; Song (1); 0

5 Fu Yunlong, *Youli Riben tujing yuji* (1889).

6 In the MCST database the first attestation, listed as *dianlan se* 靛藍色, is traced back to Young John Allen and Zheng Changyan, trans., *Gezhi qimeng gewuxue* (Shanghai: Jiangnan zhizaoju, 1880).

TABLE 44 Neologisms in the *TVM*, *Jinxiandai ciyuan*, and the *Zhongguo jiben guji ku* database (*cont.*)

Chinese term	Attestation in the *TVM*	Page, definition, and attestation in the *Jinxiandai ciyuan*	*Zhongguo jiben guji ku*; number of attestations, and the earliest one
毒性	Antiaris toxicaria 毒性安替阿里司, 5; Nightshade, deadly 毒性啤啦吋㗱, 41; Physostigma venenosum 毒性非蘇司替格瑪, 46; Tephrosia toxicaria 毒性替勿路西亞, 61: *XYDCYP*	180; 毒物危害人或其他生物的特性; 1892, "Ren yu weishengwu zhengzhan lun 人與微生物爭戰論 (The theory of the war between humans and microorganisms)," *Gezhi huibian*, vol. 6	169; Northern and Southern (4); 0; 0; 0; 0
軟膏	Cataplasma carbonis 炭軟膏 Cataplasma conii 哥尼由末軟膏, Cataplasma fermenti 麥酒酵軟膏, Cataplasma lini 胡麻子軟膏, Cataplasma sinapis 芥末軟膏, Cataplasma sodæ chloratæ 含綠氣之鈉養軟膏, 13; Charcoal poultice 炭軟膏, 14; Conii cataplasama 哥尼由末軟膏, 17; Soda chlorinated poultice 含綠氣之鈉養軟膏, Sodæ chlorate cataplasma 含綠氣之鈉養軟膏, 56: *XYDCYP*	636; 一种外用药物; 1909, *Like tongzheng—Dongwu pian—Haishou* 理科通證—動物篇—海獸 (Unified demonstration of sciences—Animals—Maritime beasts)	11; Ming (5); 0; 0; 0; 0; 0; 0; 0; 0; 0; 0
石墨	Black lead, 6; Plumbago, or Graphite, 26: *JS*	672; 一种矿物; 1917, *Qingbai leichao—Nongshang lei—Shangpin* 農商類—商品 (A Qing history of trivial matters—Agriculture and commerce—Commodities)	2531; Han (4)

TABLE 44 Neologisms in the *TVM*, *Jinxiandai ciyuan*, and the *Zhongguo jiben guji ku* database (*cont.*)

Chinese term	Attestation in the *TVM*	Page, definition, and attestation in the *Jinxiandai ciyuan*	*Zhongguo jiben guji ku*; number of attestations, and the earliest one
電表	Electric indicator 顯電表, 18: *QJ*	158; 电器仪表, 用来测量电压、电流、 电功率等; 1901, *Gewu zhixue—Dianxue* 格物質學—電學 (Investigation of things and their properties—Electricity)[7]	
電鍍	Electro-planting 電鍍銀, 13: *HXCL*	160; 利用电解原理在某些金属表面镀上一层薄的金属或合金的过程; 1890, "Gezhi shuyuan xin dao gezhi jiaoxi 格致書院新到格致教習 (On the new teachers for the investigation of things and extension of knowledge at the Polytechnic institution)," *Gezhi huibian*, vol. 4	21; Qing (21); 0
哥伦比亚	Columbia 哥侖比亞, Columbian 哥倫比亞者, 16: *RMDM*	257; 南美洲西北部的一个国家; 1915, *Feilübin saihui ji* 斐律濱賽會記 (Records of the religious processions of the Philippines)	20; Qing (20); 1 (Qing); 0
乳頭	Achras mammosa 乳頭形阿格拉司, 1; Lucuma mammosa 乳頭形羅故瑪, 37: *XYDCYP*	635; 像乳头的东西; 1907, *Shengli weisheng xin jiaokeshu* 生理衛生新教科書 (A new textbook on physical hygiene)[8]	618; Jin (5); 0; 0
果汁	Colocynth pulp 哥囉噺果汁, Colocynthidis pulpa 哥囉噺果汁, 17; Succus rhamni 拉磨尼果汁, 59: *XYDCYP*	301; 用鲜果的汁做成的饮料; 1915, *Hao Ci er kexue* (*chuji*) 豪慈兒科學(初集) (Paediatrics by Holt: first collection)[9]	54; Jin (1); 0; 0; 0

7 Alvin Pierson Parker and Xie Honglai, trans., *Gewu zhixue* (Shanghai: Meihua shuguan, 1898).

8 Sun Zuo, *Shengli weisheng xin jiaokeshu* (Shanghai: Shangwu yinshuguan, 1907).

9 Mary Hannah Fulton and Zhou Zhongyi, trans., *Hao Ci er kexue* (*chuji*) (Shanghai: Zhongguo Boyihui, 1915).

TABLE 44 Neologisms in the TVM, *Jinxiandai ciyuan*, and the *Zhongguo jiben guji ku* database (*cont.*)

Chinese term	Attestation in the TVM	Page, definition, and attestation in the *Jinxiandai ciyuan*	*Zhongguo jiben guji ku*; number of attestations, and the earliest one
打胎	Ergotætia arbortifaciens 打胎耳臥達以希阿, 22: *XYDCYP*	120; 人工流产; 1909, *Tuhua ribao* 图画日报 (Illustrated daily)[10]	45; Ming (7); 0
石蕊	Litmus, 36: *XYDCYP*. Litmus 石蕋, 21: *HXCL*	673; 用石蕊制成的蓝色粉末, 溶于水, 在化学上用作指示剂: 1886–1890, *Wanguo yaofang* 萬國藥方 (Prescriptions from the ten thousand nations), juan 1[11]	95; Tang (2)[12]
百合科	Lilaceæ, 35: *XYDCYP*	17; 单子叶植物纲百合亚纲中的一科; 1906, *Bowuxue jiaokeshu* 博物學教科書 (A textbook of natural history)	
花梗	Myrospermum pedicellatum 小花梗米羅司卑麻末, 41: *XYDCYP*	333–334; 花下面生出的小枝; 1906, *Bowuxue jiaokeshu*	43; Song (2); 0
減速	Ease, 17: *QJ*	375; 降低速度; 1936, *Kexue huabao* 科學畫報 (Scientific pictorial), vol. 3, no. 16[13]	69; Tang (2)
焊料	Chryscolla 銲料石, 10; Silver solder 銀銲料; 31: *HXCL*	313; 焊接时用来填充工件接合处的材料; 1955, *Wuxiandian* 无线电 (Radio), no. 12	1; Qing; 0; 0

10 Huanqiushe bianjibu, *Tuhua ribao* (Shanghai: Shanghai guji chubanshe, 1999).

11 Stephen Alexander Hunter, trans., *Wanguo yaofang* (Shanghai: Meihua shuguan, 1886).

12 *Rui* 蕋 is a variant of *rui* 蕊, therefore results shown for *shi rui* 石蕋 are the same.

13 In the MCST database an earlier attestation, with the meaning "accelerative," is backdated to 1860, in *Shuli gezhi* 數理格致 [Principles of number: investigation of things and extension of knowledge], a translation by Alexander Wylie and Li Shanlan of Newton's *Philosophiæ Naturalis Principia Mathematica*. For a history of the *Shuli gezhi*, refer to Han Qi, "*Shuli gezhi* de faxian—Jian lun 18 shiji Niudun xiangguan zhuzuo zai Zhongguo de chuanbo," *Zhongguo keji shiliao—China Historical Materials of Science and Technology* 19, no. 2 (1998).

TABLE 44 Neologisms in the *TVM*, *Jinxiandai ciyuan*, and the *Zhongguo jiben guji ku* database (*cont.*)

Chinese term	Attestation in the *TVM*	Page, definition, and attestation in the *Jinxiandai ciyuan*	*Zhongguo jiben guji ku*; number of attestations, and the earliest one
焊錫	Solder 銲錫, 47: *QJ*	313; 一种锡铅合金; 1955, *Wuxiandian*, no. 12[14]	5; Ming (2)[15]
生石灰	Quick lime, 29: *HXCL*. Quicklime, 28: *JS*	665; 一种无机化合物; 1913, *Gongye changshi* 工業常識 (Elementary knowledge of industry)[16]	22; Yuan (1);
果糖	Sugar fruit, 33: *HXCL*	301; 一种有机化合物; 1913, *Gongye changshi*	34; Song (3);
水表	Water gauge, 56: *QJ*	697; 记录自来水用水量的仪表; 1929, *Richang huaxue shenghuo* 日常化學生活 (Daily chemical life)	158; Han (1)
推力	Indicated thrust of propeller 螺輪所顯之推力, 28: *QJ*	753; 推进的力量; 1949, *Huojian* 火箭 (Rocket)[17]	99; Song (8); o
漏斗	Funnel, 24: *QJ*	496; 把液体或颗粒、 粉末灌到小口的容器里用的器具，一般是由一个锥形的斗和一个管子构成; 1892, "Fangzhi jiqi tushuo 紡織機器圖說 (An illustrated exposition of the textile machine)," *Gezhi huibian*, vol. 6[18]	103; Three Kingdoms (1)

14 In the MCST database the first attestation, with the same meaning, is traced to Mateer, *Technical Terms*.

15 *Han* 銲 is the traditional variant of *han* 焊, therefore results shown for *han xi* 銲錫 are the same.

16 In the MCST database the first attestation is traced to Parker and Xie, *Gewu zhixue*.

17 In the MCST database the first attestation is traced to *Shuli gezhi*, by Alexander Wylie and Li Shanlan, with the meaning "impulse"; the term, in turn, is annotated with the sense of "repulsion" in John Fryer and Zhou Xun, trans., *Dianxue gangmu* (Shanghai: Jiangnan zhizaoju, 1881) and with the connotation of "thrust," in Yan, *An English and Chinese Standard Dictionary* (1910).

18 In the MCST database the attestation for the term *baitie loudou* 白鐵漏斗 is traced to Lemaire and Giquel, *Dictionnaire de poche français-chinois*, specifically on page 358 of its second section, the *Dictionnaire Technique Français-Chinois*, annotated as " 白铁漏斗" with the meaning "Entonnoir (en fer blanc)."

TABLE 44 Neologisms in the *TVM*, *Jinxiandai ciyuan*, and the *Zhongguo jiben guji ku* database (*cont.*)

Chinese term	Attestation in the *TVM*	Page, definition, and attestation in the *Jinxiandai ciyuan*	*Zhongguo jiben guji ku*; number of attestations, and the earliest one
烙鐵	Soldering iron 嵌錫烙鐵, 47: *QJ*	467; 焊接时熔化焊镴用的工具; 1909, *Tuhua ribao*, no. 139	191; Tang (10); 0
墊圈	Washer, 56: *QJ*	168; 垫在被连接件与螺母之间的零件; "Gezhi shiqi," *Gezhi huibian*, vol. 4	7; Ming (2)
哨子	Steam whistle 汽哨子, 49: *QJ*	654; 用金属或塑料等制成的能吹响的器物; 1944, *Si shi tong tang* 四世同堂 (Four generations under one roof)	54; Yuan (3); 0
鉗子	Tongs, 52: *QJ*	601; 用来夹住或夹断东西的器具; 1947, *Kexue huabao*, vol. 13, no. 2[19]	83; Han (2)
花邊	Moulding, 34: *QJ*	333; 带有花纹的边缘; 1936, *Tu'an jiaocai* 圖案教材 (Teaching materials with patterns)[20]	2292; Tang (42)
秤桿	Weigh levers, 56: *QJ*	94; 杆秤的组成部分，一般用木棍制成，上面镶着计量的秤星; 1901, *Gewu zhixue—Jiqi benyuan* 機器本原 (The principle of machines)[21]	58; Ming (6)
廢水	Pipe waste water 廢水管, 37; Waste tank 廢水箱, 56: *QJ*	219; 工业生产或生活过程中所排出的无用甚至有害的水; 1890, "Juzhai weisheng lun 居宅衛生論 (On domestic hygiene)," *Gezhi huibian*, vol. 4	471; Spring and Autumn—Warring States (1); 0; 0

19 In the MCST database the first attestation is traced to Mateer, *Technical Terms*, with the meaning "blacksmith's tongs."

20 In the MCST database the first attestation is traced to Mateer, *Technical Terms*, with the meaning "fancy border (in printing, etc.)" and "molding."

21 In the MCST database the first attestation is traced to the text Fryer composed in 1889, *Zhongxue xuzhi* 重學須知 [Essentials of mechanics], collected within *Xixue gezhi daquan* 1897, with the meaning "beam of a balance."

TABLE 44 Neologisms in the *TVM*, *Jinxiandai ciyuan*, and the *Zhongguo jiben guji ku* database (*cont.*)

Chinese term	Attestation in the *TVM*	Page, definition, and attestation in the *Jinxiandai ciyuan*	*Zhongguo jiben guji ku*; number of attestations, and the earliest one
外徑	Callipers outside 比外徑規, 9; Cylinder external diameter 汽筩外徑, 14; Diameter of crank 拐軸外徑, 15; Outer diameter of tubes 烟管外徑, 35: *QJ*	761; 断面为圆环状的物件外缘的直径; 1903, *Xinji shiwu huitong* 新輯時務匯通 (The revised comprehensive synthesis of the times), vol. 33	832; Zhou (1); 0; 1 (Qing); 0; 1 (Qing);
不丹	Bootan, 15: *RMDM*	65–66; 喜马拉雅山脉南麓的一个国家; 1902, *Shijie dili zhi* 世界地理志 (Records of the world geography)	178; Qing (?)[22]
聖彼得堡	St. Petersburg, 21: *RMDM*	668; 沙皇时代的俄国首都; 1899, "Dawen Zhong-E jiaoqin zhi gu 答問中俄交親之故 (Questions and answers on the reason of the friendship between China and Russia)," *Qingyi bao* 清議報 (The China Discussion), vol. 4	27; Qing (23)
瓦特	Watt, 13: *RMDM*[23]	759; 功率单位; 1931, *Ying-Han duizhao baike minghui* 英漢對照百科名匯 (Encyclopaedic terminology—English and Chinese parallel text)	109; Qing (99)[24]

22 Even though the first attestations date back to the Jin dynasty (266–420), the combination of the two characters has other unrelated meanings; the first attestation with the connotation considered here presumably dates back to the Qing dynasty.

23 As the term is listed in the *RMDM*, a vocabulary of toponyms and anthroponyms, the reference is probably to James Watt (1736–1819). Considering that the unit of measurement carries a proper name deriving from the surname of the Scottish mathematician, the term can be considered a neologism.

24 Even though one attestation during the Song dynasty and another during the Ming dynasty can be found, the combination of the two characters has other unrelated meanings.

TABLE 45 Neologisms in the *TVM*, MCST database, and the *Zhongguo jiben guji ku* database
Synoptic table including neologisms attested for the first time in the *TVM*, including: 1) the Chinese term; 2) the original term, the occurrence, and the glossary in the *TVM*; 3) year and attestation in the MCST database; 4) number of attestations and the earliest attestation for the term in the *Zhongguo jiben guji ku* database

Chinese term	**Attestation in the *TVM***	**Year and attestation in the MCST database**	***Zhongguo jiben guji ku*; number of attestations and the earliest one**
介部	Mollusca, 39: *XYDCYP*	1910,[25] *ECSD*[26]	212; Three Kingdoms (1)
龍涎香	Ambergris 3: *XYDCYP*. Ambergris, 2: *HXCL*	1912, *ECSD*[27]	695; Song (74)
管形	Acacia fistula 管形阿揩西耶, 1; Cassia fistula 管形加西耶, 13; Gamboge pipe 管形籐黃, 25; Winterania canella 小管形温弟里阿那, 67: *XYDCYP*	1904, *TT*[28]	83; Han (3); 0; 0; 0; 0
花鬚[29]	Synanthereæ[30] 花鬚合成管形族, 60: *XYDCYP*	1903, *Xin erya*	1518; Han (2); 0

25 In the lemma "Mollusca," listed in Yan, *An English and Chinese Standard Dictionary* (1908), 1464, and in Idem, *English and Chinese Standard Dictionary* (1925), 641, *jie bu* 介部 is annotated as the third translation, after the English explanation; *jie bu* is also listed as a translation of the namesake entry on page 39 of the *XYDCYP*. The term was located already in the *BCGM*. Nevertheless, in the latter, the character *bu* 部 only indicates the sections in which the *BCGM* is divided. For a detailed description of the taxonomy of the text, refer to Marcon, *The Knowledge of Nature*, 28–50.

26 Yan, *An English and Chinese Standard Dictionary* (1910).

27 Yan Huiqing, *An English and Chinese Standard Dictionary*, 4th ed. (Shanghai: Shangwu yinshuguan, 1912).

28 Mateer, *Technical Terms*.

29 Refer also to Georges Métailié, "The Formation of Botanical Terminology: A Model or a Case Study?" in *New Terms for New Ideas: Western Knowledge and Lexical Change in Late Imperial China*, ed. Michael Lackner, Iwo Amelung, and Joachim Kurtz (Leiden, Boston, Köln: Brill, 2001), 330 and 332, and Idem, "Terminologie et approche interculturelle: terminologies scientifiques du chinois," *Langages*, no. 157 (2005): 122.

30 The term is an obsolete appellative for the family of Compositae or Asteraceae.

TABLE 45 Neologisms in the *TVM*, MCST database, and the *Zhongguo jiben guji ku* database (*cont.*)

Chinese term	Attestation in the *TVM*	Year and attestation in the MCST database	*Zhongguo jiben guji ku*; number of attestations and the earliest one
銅養硫養$_{三}$	Copper sulphate, Copperas blue, 18; Cupri sulphas, 19; Tutia, blue, 65; Vitriol blue, 66: *XYDCYP*. Blue copperas, Blue stone, Blue vitriol, 6; Copper sulphate, Copperas blue 11: *HXCL*. Copper sulphate, 10; Vitriol, blue, 35: *JS*[31]	1898, "Dianxue wenda," *Gezhi yiwen huibao*, no. 1 "電學問答 (Questions and answers on electricity)," *Gezhi yiwen huibao* 格致益聞匯報 (I-Wen-Lou et Revue Scientifique)	
鉛養醋酸	Acetate lead, 1; Lead acetate, Lead sugar of, 34; Plumbi acetas, 48; Sugar of lead, 59: *XYDCYP*. Lead acetate, 20: *HXCL*	1912, *ECSD*	29; Qing (29)
淡輕$_{四}$	Ammonium, 3: *HXCL*	1898, "Miehuo xinfa 滅火新法 (A new method for extinguishing fires)," *Gezhi yiwen huibao*, no. 1	1 (Qing)
淡輕$_{三}$	Ammonia, 3: *XYDCYP*. Ammonia, 3: *HXCL*	1912, *ECSD*	5 (Qing)
鉀養草酸	Oxalate of potash, 44; Potash oxalate, 48: *XYDCYP*	1902, "Xu zhaoxiang xuzhi 續照相須知 (A sequel to the essentials of photography)," *Huibao* 匯報 (Revue pour Tous), no. 212	5 (Qing)

31 For all terms annotated in table 45 as listed in the *JS*, the lemmata in which the Chinese term only appears to indicate the principal element within a certain mineral in the third column—the "Terms in general use and chief elements"—have not been included, unless otherwise specified.

TABLE 45 Neologisms in the *TVM*, MCST database, and the *Zhongguo jiben guji ku* database (*cont.*)

Chinese term	Attestation in the *TVM*	Year and attestation in the MCST database	*Zhongguo jiben guji ku*; number of attestations and the earliest one
鉀養硫養$_{\text{三}}$	Potash sulphate, 48; Potassæ sulphas, 49: *XYDCYP*. Potash sulphate, 28; Sal polychrest, 30: *HXCL*	1898, "Dianxue wenda," *Gezhi yiwen huibao*, no. 1	
打打	Tartar, 34: *HXCL*	1911, *PBXDC*[32]	236; Song (5)[33]
鉀養二果酸	Cream of Tartar, 18; Potash acid tartate, Potash tartrate acid, Potassæ bitartras, 48; Potassæ tartras acida, 49: *XYDCYP*. Cream of tartar, 12; Potash bitartrate, 28; Tartar, 34: *HXCL*	1910, ECSD	24 (Qing)
鉀養二炭養$_{\text{二}}$	Potash bicarbonate, Potassæ bicarbonas, 48: *XYDCYP*. Potash bichromate, 28; Saleratus, 30: *HXCL*	1910, ECSD	
鋅養硫養$_{\text{三}}$	Sal vitrioli, 53; Tutia suffed, 65; Vitriol white, 67; Zinc sulphate, Zinci sulphas, 68: *XYDCYP*. White vitriol, Zinc sulphate, 37: *HXCL*. Vitriol white, 35; White vitriol. 36; Zine [sic] sulphate, 37: *JS*	1898, "Dianxue wenda," *Gezhi yiwen huibao*, no. 1	2 (Qing)
鋅養矽養$_{\text{二}}$	Willemite, 36: *JS*[34]	1910, *ECSD*	

32 Huang, *Putong baike xin da cidian*.

33 Even though one attestation during the Northern and Southern dynasties and another during the Tang dynasty (618–907) can be found, the combination of the two characters has other unrelated meanings.

34 In this circumstance, the term is listed within the third column of the *JS*, referring to the "Terms in general use and chief elements," after the phonemic loan *Weilemide* 韋勒迷得; it indicates the principal elements of the willemite.

TABLE 45 Neologisms in the *TVM*, MCST database, and the *Zhongguo jiben guji ku* database (*cont.*)

Chinese term	Attestation in the *TVM*	Year and attestation in the MCST database	*Zhongguo jiben guji ku*; number of attestations and the earliest one
合强水	Acidum nitrohydrochloricum, 2: *XYDCYP*. Acqua regia, 4; Nitromuriatic acid, 24: *HXCL*	1912, *ECSD*	45 (Qing)
矽養二	Acidum silicicum, 2; Silica, 55: *XYDCYP*. Silica, Silicic acid, 31: *HXCL*. Silica, 31: *JS*	1910, *ECSD*	
果酸	Acidum tartaricum, 2; Tartaric acid, 61: *XYDCYP*. Tartaric acid, 34: *HXCL*	1911, *PBXDC*	121; Tang (2)
草烏頭	Aconite, 2: *XYDCYP*	1912, *ECSD*	725; Han (1)
菖蒲	Acorus calamus, 2; Sweet flag, 60: *XYDCYP*	1912, *ECSD*	9780; Spring and Autumn—Warring States (1)
有刺的	Aralia spinosa 有刺阿拉里亞, 6; Catha spinosa 有刺楷他, 13; Glycyrrzhiza echinata 有刺各里色里薩, 27; Prunus spinosa 有刺普羅奴斯, 49; Thorn apple, 有刺蘋果 62; Toddalia aculatea 有刺托大里亞類, 64: *XYDCYP*	1912, *ECSD*	1 (Qing); 0; 0; 0; 0; 0; 0
淡養五	Nitric acid, 41: *XYDCYP*. Aqua fortis, 4; Nitric acid, Nitric anhydride, 24: *HXCL*. Nitric acid, 24: *JS*	1912, *ECSD*	3 (Qing)

TABLE 45 Neologisms in the *TVM*, MCST database, and the *Zhongguo jiben guji ku* database (*cont.*)

Chinese term	Attestation in the *TVM*	Year and attestation in the MCST database	*Zhongguo jiben guji ku*; number of attestations and the earliest one
種子	Ajava seeds 阿周伐種子, 2; Colchici semina 哥勒枝㗩種子, Colchicum seeds 哥勒枝㗩種子, 16; Menisperaceæ 新月形種子科, 38; Rhus semialata 寬種子如司, 52; Staphisagraiæ semina 司他非薩搿里種子, 58; Stavesacre seeds 司他非薩搿里種子, Stramonii semina 醉仙桃種子, Stramonium seeds 醉仙桃種子, 59; Theobromæ semina 替亞布路瑪種子, 62: *XYDCYP*	1903, *Xin erya*	8394; Han (11); 0; 0; 0; 0; 0; 0; 0
胚乳	Albumen, 2: *XYDCYP*	1903, *Xin erya*	1 (Ming)
海帶	Algae 海帶科, 2; Laminaria 海帶昆布, 33: *XYDCYP*	1912, *ECSD*	464; Spring and Autumn—Warring States (1); 0; 25 (Jin金, 2)
葡萄酒	Vinum, 66; Wines, 67: *XYDCYP*	1911, *PBXDC*	846; Han (2)
血藥	Blood medicines, 9: *XYDCYP*	1912, *ECSD*	982; Jin (2)
藥品	Materia medica, 38: *XYDCYP*	1903, "Guangshuo 光說 (On light)"[35]	816; Northern and Southern (7)
鋁二養三	Alumina, 2: *HXCL*. Alumina, 2: *JS*	1912, *ECSD*	
蒙藥	Anæsthesiants, 4: *XYDCYP*	1912, *ECSD*	72; Spring and Autumn—Warring States (1)
迷蒙藥	Choroform, 14; Chloroformum, 15: *XYDCYP*	1913, *Chinese New Terms and Expressions*[36]	2 (Qing)

35 Kimura Shunkichi, "Guangshuo," in *Xinxue da congshu*, trans. Gong Chai and Xu Bin (Shanghai: Jishan qiaoji shuju, 1903).

36 Morgan Evan, *Chinese New Terms and Expressions, with English Translations, Introduction and Notes* (Shanghai: Kelly & Walsh, 1913). The text was later republished in Idem, *Chinese New Terms: Revised and Enlarged with English Translations, Classifications, Introduction and Index* (Shanghai: Kelly & Walsh, 1932).

TABLE 45 Neologisms in the *TVM*, MCST database, and the *Zhongguo jiben guji ku* database (*cont.*)

Chinese term	Attestation in the *TVM*	Year and attestation in the MCST database	*Zhongguo jiben guji ku*; number of attestations and the earliest one
動物炭	Animal charcoal, 5; Carbo animalis, 12; Charcoal, animal, 14: *XYDCYP*. Animal charcoal, 3, Charcoal animal, 9: *HXCL*	1916, *ECD*	17 (Qing)
子殼	Animirta cocculus 小子殼阿尼沒他, 5; Capsule papaverin 子殼怕巴甫里尼, 12; Caryopsides 麥類子殼, 13; Phytelephas macrocarpa 大子殼非低里法司, 46: *XYDCYP*	1910, *ECSD*	375; Han (4); 0; 0; 0; 0
茴香	Fœniculum vulgare, 25: *XYDCYP*. Fennel, 14: *HXCL*	1912, *ECSD*	3950; Han (1)
止痛藥	Anodynes, 5: *XYDCYP*	1912, *ECSD*	49; Tang (6)
安神藥	Anodynes, 5: *XYDCYP*	1912, *ECSD*	14; Ming (5)
解鹼藥	Antalkalis, 5: *XYDCYP*	1912, *ECSD*	
殺蟲藥	Anthelmintics, 5: *XYDCYP*	1912, *ECSD*	23; Song (1)
解毒藥	Antidotes, 5; Contrajerva, 17; Poisons antidotes to, 48: *XYDCYP*	1910, *ECSD*	195; Jin (1)
肚腹	Antispasmodics 安肚腹腦筋药, 5: *XYDCYP*	1912, *ECSD*	957; Han (3); 0
養淡養$_{五}$	37	1898, "Xin huo san zhi 新獲三質 (On three newly obtained substances)," *Gezhi yiwen huibao*, no. 1	
鄉間	Armoracia rusticana 鄉間阿莫賴西阿, 6: *XYDCYP*	1916, *ECD*	1178; Han (3); 0
芋類	Arum, 7: *XYDCYP*	1912, *ECSD*	11; Yuan 1
花點生鐵	Cast iron mottled, 8: *HXCL*. Mottled iron, 34: *QJ*	1904, *TT*	

37 *Yang danyang$_{wu}$* 養淡養$_{五}$ is annotated as part of translations in numerous lemmata within the *TVM*, invariably with meanings referring to nitrates.

TABLE 45 Neologisms in the *TVM*, MCST database, and the *Zhongguo jiben guji ku* database (*cont.*)

Chinese term	Attestation in the *TVM*	Year and attestation in the MCST database	*Zhongguo jiben guji ku*; number of attestations and the earliest one
阿魏	Assafœtida, 7: *XYDCYP*. Asafoetida, 4: *HXCL*	1912, *ECSD*	1740; Han (3)
小頭之白癬	Astragalus microcephalus 花成小頭阿司脫拉茄路司, 7: *XYDCYP*	1913, *A Dictionary of Philosophical Terms*	
齒形的	Barosma crenulata 細齒形邊巴路司瑪, Barosma serratifolia 齒形葉巴路司瑪, 8; Boswellia serrata 齒形蒲司韋里阿, 10; Emplernum serrulatum 微鋸齒形安蒲勒羅末, 22: *XYDCYP*	1910, *ECSD*	
鋇養	Baryta, 8: *XYDCYP*. Baryta, 5: *HXCL*. Baryta, 5: *JS*	1912, *ECSD*	87 (Qing)
漂白粉	Bleaching powder, 9; Lime chloride, Lime chlorinated, 35: *XYDCYP*. Chloride of lime, 9: *HXCL*	1886, *Nederlandsch-Chineesch Woordenboek*[38]	13 (Qing)
水銀丸	Blue pill, 10; Hydrargyri pilula, 29; Mercurial pill, 38; Pill blue, Pill mercury, 46; Pilula hydrargyri, 47: *XYDCYP*. Blue pill, 6: *HXCL*	1910, *ECSD*	50; Jin (2)
三葉	Bonplandia trifoliata 三葉蒲那普蘭地亞, Buckbean 三葉睡菜, 10; Marsh trefoil 三葉隰草, Menyanthes trifoliata 三葉米尼安替司, 38; Ptelea trifoliata 三葉替里亞, 49: *XYDCYP*	1910, *ECSD*	2617; Zhou (3); 0; 0; 0; 0; 0

38 Gustave Schlegel, *He-Hua wenyu leican—Nederlandsch-Chineesch Woordenboek met de Transcriptie der Chineesche Karakters in het Tsiang-Tsiu Dialekt* (Leiden: Brill, 1886).

TABLE 45 Neologisms in the *TVM*, MCST database, and the *Zhongguo jiben guji ku* database (*cont.*)

Chinese term	Attestation in the *TVM*	Year and attestation in the MCST database	*Zhongguo jiben guji ku*; number of attestations and the earliest one
白蘭地酒	Brandy, 10; Spirit French wine, Spiritus vini Gallici, 58: *XYDCYP*	1910, *ECSD*[39]	6 (Qing)
石灰水	Calcis liquor, 11; Lime water, 35; Liquor calcis, 36: *XYDCYP*	1911, *PBXDC*	150; Jin (1)
鎂養	Magnesia, Magnesic oxide, 37; Talc earth, 61: *XYDCYP*. Magnesia, Magnesia calcined, 21: *HXCL*. Magnesia, Magnesian limestone, 21: *JS*	1910, *ECSD*	123 (Qing)
穀類稅	Cereals 穀類, 14: *XYDCYP*	1927, *Vocabulaire Français-Chinois des sciences morales et politiques*[40]	0; 255; Han (3)
繖形花	Chimaphila umbellata 繖形奇瑪非拉, 14; Pyrola umbellata 繖形伯羅拉, 50; Umbelliferæ 繖形科, Umbellulæ 小繖形科, 65; Winter green 繖形奇瑪非拉, 67:[41] *XYDCYP*	1903, *Xin erya*	
長橢圓形葉	Cinchona oblongifolia 長橢圓葉金雞那, 15: *XYDCYP*	1903, *Xin erya*	
油膏	[42]	1935, *CEMMD*	117; Northern and Southern (4)

39 In Huang, *Jinxiandai ciyuan*, 14–15, an earlier attestation is only provided for the phonemic loan *Bailandi* 白蘭地, but not for the hybrid *Bailandi jiu* 白蘭地酒.

40 Jules Médard, *Fa-Han zhuanmen cidian—Vocabulaire Français-Chinois des sciences morales et politiques* (Tianjin: Société française de librairie et d'édition, 1927).

41 In this lemma, the Chinese rendering is a hybrid supplied as an alternative to *dong qing shu* 冬青樹; it refers to the Chimaphila umbellata, also known as Umbellate wintergreen.

42 *Yougao* 油膏 is also annotated as part of other translations in numerous other lemmata within the *TVM*, invariably with the meanings "ointment," "unguentum," "plaster," "emplastrum," and "amalgam."

TABLE 45 Neologisms in the *TVM*, MCST database, and the *Zhongguo jiben guji ku* database (*cont.*)

Chinese term	Attestation in the *TVM*	Year and attestation in the MCST database	*Zhongguo jiben guji ku*; number of attestations and the earliest one
魚鱗	Conyza squarrosa 鬆魚鱗形哥尼薩, 17: *XYDCYP*	1910, *ECSD*	5297; Zhou (1); 0
多瓣花冠	Apetalæ 無瓣花, 5; Corolliflor æ 瓣花屬, 18: *XYDCYP*	1903, *Xin erya*	
臍帶	Cotydon umbilicus 有臍帶苦替離屯, 18: *XYDCYP*	1910, *ECSD*	348; Tang (11); 0
冕號	Crown bark 冕號金雞那皮, 18: *XYDCYP*	1906, *Jinshi wulixue jiaokeshu* 近世物理學教科書 (A textbook of modern physics)[43]	21; Song (1); 0
暗生植物	Cryptogameæ, 18: *XYDCYP*	1912, *ECSD*	4 (Qing)
銀衰	Cyanide of silver, 19: *XYDCYP*	1901, "Xu xixue guanjian 續西學關鍵 (A sequel to Keys of Western learning)," *Huibao*	71 (Qing)[44]
鈍形	Dipterocarpus retusus 鈍形第不得露揩披, 21: *XYDCYP*	1903, *Xin erya*	23; Jin (1); 0
淡巴菰	Tabaci folia, Tabacum, 61; Tobacco, 64: *XYDCYP*	1890, *Riben guozhi* 日本國志 (A description of Japan)[45]	279; Ming (2)
審定	Erythrophlæum judicale 審定以里脫路甫里由末, 22: *XYDCYP*	1927, *Vocabulaire Français-Chinois des sciences morales et politiques*	3699; Spring and autumn—Warring States (7); 0
津液	Evacuants 令成津液藥, 23: *XYDCYP*. Excretions 棄津液, 14: *HXCL*	1910, *ECSD*	10481; Zhou (9); 0; 2 (Qing)

43 Nakamura Seiji, *Jinshi wulixue jiaokeshu*, trans. Xuebu bianyi tushuju (Beijing: Jingshi guanshuju, 1906).

44 Even though one attestation during the Ming dynasty can be found, the combination of the two characters has another unrelated meaning.

45 Huang Zunxian, *Riben guozhi* (Taipei: Wenhai chubanshe, 1968).

TABLE 45 Neologisms in the *TVM*, MCST database, and the *Zhongguo jiben guji ku* database (*cont.*)

Chinese term	Attestation in the *TVM*	Year and attestation in the MCST database	*Zhongguo jiben guji ku*; number of attestations and the earliest one
發芽	Extract of malt 發芽大麥膏, Extractum bynes 發芽大麥膏, 23; Secale cornutum 發芽麸麥, 55; Vinegar malt 發芽大麥醋, 66: *XYDCYP*. Wort 發芽大麥湳水, 37: *HXCL*	1912, *Tetsugaku jii* 哲學字彙 (Dictionary of philosophy)[46]	461; Song (7); 0; 0; 0; 0
鐵鏽	Ferri rubigo, Ferrugo, 24: *XYDCYP*	1902, "Xixue guanjian 西學關鍵 (Keys of Western learning)," *Huibao*, no. 211	156; Song (5)
普魯士藍	Ferri sesquiferrocyanidum, 24: *XYDCYP*. Blue Prussian, 6; Iron sesquiferrocyanide, 19; Prussian blue, 28: *HXCL*	1904, *TT*	10 (Qing)
乳香	Frankincense, 25; Olibanum, Olibanum emulsion, 43: *XYDCYP*. Frankincense, 15; Olibanum, 24: *HXCL*[47]	1910, *ECSD*	8247; Han (11)

46 Inoue Tetsujirō and Motora Yūjirō, *Tetsugaku jii*, 3rd ed. (Tokyo: Maruzen kabushiki kaisha, 1912).

47 The term *ruxiang* 乳香 in reality already appeared in bilingual glossaries at least since the *Dicionário Português-Chinês* by the two Italian Jesuit missionaries Michele Ruggieri (Luo Mingjian 羅明堅, 1543–1607) and Matteo Ricci (Li Madou 利瑪竇, 1552–1610), where it is listed under the lemma "Encenso," 89A. Refer to John W. Witek, ed., *Dicionário Português-Chinês* (Lisbon: Biblioteca Nacional Portugal; Macao: Instituto Português do Oriente; San Francisco: Ricci Institute for Chinese-Western Cultural History, University of San Francisco, 2001).

TABLE 45 Neologisms in the *TVM*, MCST database, and the *Zhongguo jiben guji ku* database (*cont.*)

Chinese term	Attestation in the *TVM*	Year and attestation in the MCST database	*Zhongguo jiben guji ku*; number of attestations and the earliest one
橢圓形	Garcinia elliptica 橢圓形加西尼亞, 26: *HXCL*. Elliptical boiler 橢圓形鍋爐, Elliptical tubes 橢圓形管, 18: *QJ*	1897, *Cehou congtan* 測候叢談 (Collected writings on astronomical and meteorological observations)[48]	73; Qing (67); 0; 0; 0
橢圓形葉	Xanthochymus ovalifolius 橢圓形葉散特扣密司, 67: *XYDCYP*	1903, *Xin erya*	
喜氣	Gas laughing, 26; Nitrogen oxyde, Nitrogen protoxyde, 41: *XYDCYP*	1904, *TT*	3066; Han (10)
金綠	Gold protochloride, 16: *HXCL*	1901, "Xu xixue guanjian," *Huibao*	213; Northern and Southern (1)[49]
金箔	Gold leaf, 27: *XYDCYP*. Gold leaf, 16: *HXCL*	1898, *Gewu zhixue*	1807; Han (1)
銳形	Gossypium acuminatum 銳形棉花, 27: *XYDCYP*	1903, *Xin erya*	68; Han (2); 0
筆鉛	Graphite, 27; Iron carburet, 32; Plumbago (graphite), 48: *XYDCYP*. Black lead, 6: *QJ*. Black lead, 6; Graphite, 16; Plumbago, 27: *HXCL*. Black lead, 6; Graphite, or Plumbago, 15; Plumbago, or Graphite, 26: *JS*	1887, *Dianxue xuzhi*	97; Tang (2)

48 In the MCST database, the text translated by Carl Traugott Kreyer and Hua Hengfang, *Cehou congtan*, is listed as included in Yuan Junde, ed., *Fuqiangzhai congshu* (Shanghai: Xiaocang shanfang, 1897); *Cehou congtan* was originally published in 1896 in Shanghai by Hongwen shuju 鴻文書局.

49 In the located text, the *Shenxian jiu dan jing* 神仙九丹經 [Scripture of the nine elixirs of the divine immortals], as well as in later attestations until the Qing dynasty, the meaning of *jinlü* 金綠 is basically not associated with chemistry, referring instead to the colours represented by the two characters.

TABLE 45 Neologisms in the *TVM*, MCST database, and the *Zhongguo jiben guji ku* database (*cont.*)

Chinese term	Attestation in the *TVM*	Year and attestation in the MCST database	*Zhongguo jiben guji ku*; number of attestations and the earliest one
棉花藥	Gun cotton, 27; Pyroxyline, 50: *XYDCYP*. Gun cotton, 17: *HXCL*	1904, *TT*	53; Ming (1)
裸子類	Gymnospermæ 裸子科, 27: *XYDCYP*	1916, *ECD*[50]	
汞養	Mercury oxide, 39: *XYDCYP*. Mercury oxide, 22; Red precipitate, 29: *HXCL*	1910, *ECSD*	87; Song (9)
紅汞養	Red oxide of mercury, 51: *XYDCYP*. Mercury red oxide, 22: *HXCL*	1910, *ECSD*	
膜翅類	Hymenoptera 膜翅族, 29: *XYDCYP*	1903, *Xin erya*	
月光石	Moonstone, (Orthoclase), 23: *JS*[51]	1912, *ECSD*	4 (Qing)
靛	Indigo, 30: *XYDCYP*. Indigo, 18: *HXCL*	1908, *Wulixue yuhui* 物理學語彙 (A glossary of physics)[52]	2840; Tang (1)
酸性	Infusion of roses, acid 玫瑰花酸性沖水, 30; Infusum rosæ acidum 玫瑰花酸性沖水, 31; Mistura chlorinii acida 綠氣酸性雜水, Mixture chlorine acid 綠氣酸性雜水, 39; Rosæ Gallicæ infusum acidum, Roses, acid infusion of 玫瑰花酸性沖水, 52: *XYDCYP*. Acidulous waters, 1; Salts acid 酸性鹽類, 30; Soda acid pyrophosphate 酸性二鈉養燐養$_{五}$, 32: *HXCL*	1897, *Cehou congtan*	207; Tang (2); 0; 0; 0; 0

50 In Huang, *Jinxiandai ciyuan*, 505, the first attestation for the term *luozi zhiwu* 裸子植物 is reported in 1906, in the *Bowuxue jiaokeshu*.

51 In this circumstance the term is listed within the second column of the *JS*, namely the "Terms used in Dr. Macgowan's translation," thus originally annotated within the *Jinshi shibie*, composed by Macgowan and Hua Hengfang.

52 Xuebu shendingke, *Wulixue yuhui* (Shanghai: Shangwu yinshuguan, 1908).

TABLE 45 Neologisms in the *TVM*, MCST database, and the *Zhongguo jiben guji ku* database (*cont.*)

Chinese term	Attestation in the *TVM*	Year and attestation in the MCST database	*Zhongguo jiben guji ku*; number of attestations and the earliest one
白芷	Iris florentina, 31: *XYDCYP*	1912, *ECSD*[53]	9785; Zhou (2)
長春藤	Ivy, 長春籐, 32: *XYDCYP*	1912, *ECSD*	26; Song (1)
形似鬚頭的[54]	Krameria triandra 三鬚頭架拉美刺阿, 33: *XYDCYP*	1912, *ECSD*	
落葉	Larix 落葉松族, 34: *XYDCYP*	1903, *Xin erya*	16520; Han (9); 0
鉛$_{\text{三}}$養$_{\text{四}}$	Lead red, Lead red oxyde, 34; Minium, 39; Plumbi oxydum rubrum, 48; Red lead, 51: *XYDCYP*. Minium, 23; Red lead, 29: *HXCL*. Lead phosphate, 19; Lead ore red, 20; Minium, or Red Lead, 23; Red lead, 28: *JS*	1910, *ECSD*	
鉛丹	Lead red, Lead red oxyde, 34; Minium, 39; Plumbi oxydum rubrum, 48: *HXCL*	1910, *ECSD*	776; Han (2)
水蛭	Leeches, 34: *XYDCYP*	1903, *Xin erya*	1146; Han (8)
石蕊類之植物	Lichens 石蕊類, 34: *XYDCYP*	1912, *ECSD*	
木精	Lignine, 34: *XYDCYP*	1910, *ECSD*	580; Zhou (1)

53 Within the *ECSD*, *bai zhi* 白芷 is listed under the lemma "Angelica," defined as "An aromatic plant."

54 *Xingsi xutou de* 形似鬚頭的 is the second translation of the lemma "Antheroid," with the meaning "similar to an anther"; "anther" is therefore rendered as *xutou* 鬚頭. From different Chinese renderings within the *TVM*, it is possible to infer that Fryer translates "stamen" with the term *xu* 鬚; the anther is actually the terminal part of the stamen, its *tou* 頭, head.

TABLE 45 Neologisms in the *TVM*, MCST database, and the *Zhongguo jiben guji ku* database (*cont.*)

Chinese term	Attestation in the TVM	Year and attestation in the MCST database	*Zhongguo jiben guji ku*; number of attestations and the earliest one
砒霜	Liquor arsenicalis 砒霜水, Liquor potassæ arsenitis 砒霜水, 36: *XYDCYP*. Arsenic white 白砒霜, 4; White arsenic 白砒霜, 36: *JS*[55]	1912, *ECSD*	878; Jin (2); 3 (Ming, 1); 4 (Yuan, 1)
胞衣	Mace 肉豆蔻子胞衣, 37: *XYDCYP*	1910, *ECSD*	808; Jin (1); 0
錳養	Manganesii binoxydum, 37: *XYDCYP*. Manganese oxide, Manganese protoxide 21: *HXCL*. Manganese oxide, 21: *JS*	1906, *Jiangsu shifan jiangyi—Wuli* 江蘇師範講義—物理 (Teaching notes for normal schools in Jiangsu—Physics)[56]	120 (Qing)
筆鉛形錳養二	Wad, 37: *HXCL*. Wad, 37: *JS*[57]	1910, *ECSD*	
楓樹	Maple 楓樹類, 38: *XYDCYP*. Sugar maple 楓樹糖, 33: *HXCL*	1912, *ECSD*	2430; Han (1); 0; 1 (Qing)
萬壽菊	Marigold 茼蒿萬壽菊等種, 38: *XYDCYP*	1912, *ECSD*	44; Ming (2); 0
金石藥	Materia mineral 金石藥品, 38: *XYDCYP*	1910, *ECSD*	240; Tang (9); 0
乳草	Milk worts 乳草科, 39: *XYDCYP*	1912, *ECSD*	33; Song (5); 0
含羞草類	Mimosads, 39: *XYDCYP*	1910, *ECSD*	
雌雄花	Monoecia 同樹有雌雄花, 39: *XYDCYP*	1912, *ECSD*	2; Ming (1); 0
鉤藤類	Nauclea Gambir 鉤籐, 41: *XYDCYP*	1910, *ECSD*	0; 1083; Tang (7)

55 In this circumstance the two terms are listed within the second column of the *JS*.
56 Nakamura Tamekuni, *Jiangsu shifan jiangyi—Wuli* (Jiangsu xuewuchu, 1906).
57 In this circumstance the term is listed within the third column of the *JS*; it indicates the principal element of the wad.

TABLE 45 Neologisms in the *TVM*, MCST database, and the *Zhongguo jiben guji ku* database (*cont.*)

Chinese term	Attestation in the *TVM*	Year and attestation in the MCST database	*Zhongguo jiben guji ku*; number of attestations and the earliest one
美麗[58]	Nelembium speciosum 美麗尼倫比由末, 41: *XYDCYP*	1912, *Tetsugaku jii*	1735; Spring and Autumn—Warring States (4); 0
臭養氣	Ozone, 44: *XYDCYP*. Ozone, 25: *HXCL*[59]	1901, "Xu xixue guanjian," *Huibao*	9 (Qing)
薑黄紙	Paper turmeric, 45: *XYDCYP*	1910, *ECSD*	
薑黄試驗紙	Turmeric paper 薑黄試紙, 64: *XYDCYP*	1910, *ECSD*	
蝴蝶舌門	Butterfly clack 蝶形舌門, Butterfly valve 蝴蝶形舌門, Butterfly valves of air pump 恒升車蝴蝶形舌門, 8; Valve butterfly 蝴蝶形門, 54: *QJ*	1904, *TT*	
多年植物	Perennial worm grass 多年殺蟲草, 45: *XYDCYP*	1908, *Bianxue*辨學 (Logic)[60]	

58 In the MCST database an earlier attestation is also provided for the term *meili zhi xueli* 美麗之學理, translated as "Aesthetic theory"; it is traced to *Xinlixue* 心理學 [Psychology] by the Chinese scholar Wang Guowei 王國維 (1877–1927), published in 1902 within the series *Zhexue congshu chuji* 哲學叢書初集 [Collectanea of philosophy: first collection] in Shanghai by Jiaoyu shijie chubanshe 教育世界出版社. The original text, *Shinrigaku* 心理学 [Psychology], was composed by the Japanese psychologist Motora Yūjirō 元良勇次郎 (1858–1912) and published in 1890 in Tokyo by Kinkōdō 金港堂.

59 In Huang, *Jinxiandai ciyuan*, 100, the first attestation of the contemporary term for ozone, *chou yang* 臭養, is also provided; it is listed in Zheng Zhenwen, *Ziran kexue cidian* (Shanghai: Huatong shuju, 1934).

60 Wang Guowei, trans., *Bianxue* (Beijing: Jingshi wudaomiao shoushuchu, 1908). The text is the translation of William Stanley Jevons, *Elementary Lessons in Logic: Deductive and Inductive; With Copious Questions and Examples, and a Vocabulary of Logical Terms* (London: Macmillan and Co., 1870).

TABLE 45 Neologisms in the *TVM*, MCST database, and the *Zhongguo jiben guji ku* database (*cont.*)

Chinese term	Attestation in the *TVM*	Year and attestation in the MCST database	*Zhongguo jiben guji ku*; number of attestations and the earliest one
變形	Phosphorus amorphous 變形燐, 46: *XYDCYP*. Phosphorus allotropic 變形燐, Phosphorus amorphous 變形燐, 27; Quinine amorphous 變形雞那以尼, Pyromorphite 火變形石, 29; Silica amorphous 變形矽養二, Silicon amorphous, 31: *HXCL*. Pyromorphite 火變形石, 27: *JS*	1888, *Dianxue yuanliu* 電學源流 (The origin of electricity)[61]	1151; Zhou (1); 2 (Qing); 0; 0; 0
紅色燐	Phosphorous red, 46: *XYDCYP*	1901, "Liuzhi hongbao 流質轟爆 (Detonation of liquid substances)," *Huibao*, no. 106	3 (Qing)
鉑絨	Platinum spongy, 47: *XYDCYP*. Platinum spongy, 27; Spongy platinum, 33: *HXCL*	1932, *Wulixue mingcihui*物理學名詞彙[62]	8 (Qing)
松香類	Resins, 51: *XYDCYP*. Resins, 29: *HXCL*	1912, *ECSD*	35 (Qing)
淡養	Protoxyde of nitrogen, 49: *XYDCYP*. Nitrogen oxide, Nitrogen protoxide, 24: *HXCL*	1908, *Bianxue*[63]	877; Northern and Southern[64]
無花莖的	Quercus sessiflora 無花莖苦爾苦司, 50: *XYDCYP*	1910, *ECSD*	
解熱藥	Refrigerants, 51: *XYDCYP*	1912, *ECSD*	5; Ming (3)

61 John Fryer and Xu Jianyin, trans., *Dianxue yuanliu* (Shanghai: Datong shuju, 1888).

62 Sa Bendong, ed., *Wulixue mingcihui—English-Chinese Physics Terminology* (Beiping: Zhonghua jiaoyu wenhua jijin dongshihui bianji weiyuanhui, 1932).

63 In the text, the term is listed as the translation of "Nitric acid."

64 In the text located, *Hongming ji* 弘明集 [Collection of propagation and clarification], composed by the monk Seng You僧祐 (455–518), as well as in the attestations until the Qing dynasty, the combination of the characters *dan* 淡 and *yang* 養 is not related to chemistry.

TABLE 45 Neologisms in the *TVM*, MCST database, and the *Zhongguo jiben guji ku* database (*cont.*)

Chinese term	Attestation in the *TVM*	Year and attestation in the MCST database	*Zhongguo jiben guji ku*; number of attestations and the earliest one
蔗糖	Saccharum officinarum, 53; Sugar cane, 59: *XYDCYP*. Cane sugar, 8: *HXCL*	1910, *ECSD*	195; Han (1)[65]
舉重螺絲器	Screw jack, 44: *QJ*	1904, *TT*	
起重汽機	Engine hoisting, 19: *QJ*	1904, *TT*	3 (Qing)
螺絲舉重器	Jack, 29: *QJ*	1904, *TT*	
浮表	Float gauge, 23; Gauge float, 24; Hydrometer, 28: *QJ*	1898, *Gewu zhixue*	81; Five Dynasties (1)[66]
火石玻璃	Glass flint, 16: *HXCL*	1906, *Jinshi wulixue jiaokeshu*	9 (Qing)
鈣鎂矽養二	Sahlite, (Pyroxene), 29; Spar tabular, (Wollastonite), 32; Tabular spar, 33; Wollastonite, or Tabular Spar, 36: *JS*[67]	1910, *ECSD*	
雖來脫	Sahlite, (Pyroxene), 29: *JS*[68]	1910, *ECSD*	

65 The first character of the attestation within the *Fangyan jianshu* 方言箋疏 [Commentary and annotation to dialects], composed by the Qing dynasty scholar Qian Yi 錢繹 (n.d.), is *tang* 餹, a graphical variant of *tang* 糖.

66 In the located text, *Taishang huanglu zhaiyi* 太上黃籙齋儀 [Protocols of the supreme yellow register retreat], compiled by the Taoist priest Du Guangting 杜光庭 (850–933), the characters *fu* 浮 and *biao* 表 should be read separately.

67 For the four lemmata, the term is listed within the third column of the *JS*; it indicates the principal element of the correspondent mineral.

68 The lemma, which corresponds to the first of the precedent line within table 45, is listed within the second column of the *JS*.

TABLE 45 Neologisms in the *TVM*, MCST database, and the *Zhongguo jiben guji ku* database (*cont.*)

Chinese term	Attestation in the *TVM*	Year and attestation in the MCST database	*Zhongguo jiben guji ku*; number of attestations and the earliest one
鋁鈣鈉矽養二	Brevicite, (Natrolite), 6; Chlorastrolite, 8; Dipyre, or Prehnitoid, 12; Faujasite, 13; Mesole, 22; Nuttallite, (Scapolite), 24; Orthoclase, or Felspar, Ozarkite, (Thomsonite), 25; Saccharite, (Andesine), 29; Sarcolite, 30; Stilbite, or Desmine, 32; Thomsonite, 33:*JS*[69]	1910, *ECSD*	
鋁鈣矽養二	Antrimolite, (Mesolite), 4; Canaanite, (Scapolite), 7; Carbuncle, or Garpet, Chabazite, 8; Comptonite, (Thompsonite), 10; Damourite, 11; Euphyllite, 13; Garnet, 14; Grossularite, (Garnet), Haydenite, (Chabasite), 15; Heulandite, 16; Idocrase, or Vesuvian, 17; Laumonite, 19; Leonhardite, (Laumontite), Lincolnite, (Heulandite), 20; Meionite, 22; Phacolite, (Chabasite), 25; Poonahlite, 26; Prehnite, or Chiltonite, 27; Scapolite, or Wernerite, Scolecite, 30; Skapolith, Skolecite, or Scolecite, 31; Thulite, (Epidote), 33:*JS*[70]	1910, *ECSD*	

69 For all the lemmata, the term is listed within the third column of the *JS*; it indicates the principal element of the correspondent mineral.

70 For all the lemmata, the term is listed within the third column of the *JS*; it indicates the principal element of the correspondent mineral.

TABLE 45 Neologisms in the *TVM*, MCST database, and the *Zhongguo jiben guji ku* database (*cont.*)

Chinese term	Attestation in the *TVM*	Year and attestation in the MCST database	*Zhongguo jiben guji ku*; number of attestations and the earliest one
膏藥	[71]	1912, *ECSD*	1457; Han (1)
松香膏藥	Emplastrum adhæsivum, Emplastrum resinæ, 22: *XYDCYP*	1912, *ECSD*	1 (Qing)
肥皂石	Soap stone, 56: *XYDCYP*. Soap stone, 32; Steatite, 33; Talc, 34: *HXCL*. Soapstone, or Steatite, 31; Steatite, or Soapstone, 32; Talc, 33: *JS*	1910, *ECSD*	3 (Qing)
鈉綠	Soda muriate, 56; Sodic chloride, Sodii chloridum, Sodium chloride, 57: *XYDCYP*. Chloride of sodium, 9; Soda chloride, Sodium chloride, 32: *HXCL*. Salt, common, 29; Sodium chloride, 31: *JS*[72]	1898, "Yan kao 鹽考 (An examination of salt)," *Gezhi yiwen huibao*, no. 1	62 (Qing)
木酒	Alcohol methylicum, 2; Spirit pyroligneous, Spirit wood, Spiritus pyroxilicus rectificatus, 58: *HXCL*	1904, *TT*	110; Northern and Southern (10)
成顆粒之水	Water of crystallization, 37: *HXCL*	1904, *TT*	2 (Qing)
木本	Tree ferns 木本背陰植物, 64: *XYDCYP*	1903, *Xin erya*	1818; Zhou (1); 0
蔓生	Triticum repens 蔓生麥, 64: *XYDCYP*	1912, *ECSD*	2618; Han (11); 1 (Qing)
蕁麻之類	Urticaceæ 蕁麻科, Urticeæ 蕁麻科, 65: *XYDCYP*. Nettles acid of 蕁麻酸, 24: *HXCL*	1910, *ECSD*	

71 *Gaoyao* 膏藥 is annotated as part of translations in numerous lemmata within the *XYDCYP*, invariably with the meanings "cerate," "ceratum," "emplastrum," "plaster," and "soap."

72 For these two lemmata, the term of common use and the principal chemical element within the third column of the *JS* both correspond to *na lü* 鈉綠.

TABLE 45 Neologisms in the *TVM*, MCST database, and the *Zhongguo jiben guji ku* database (*cont.*)

Chinese term	Attestation in the *TVM*	Year and attestation in the MCST database	*Zhongguo jiben guji ku*; number of attestations and the earliest one
鹼性的	Waters alkaline 鹼性水, Waters alkaline salt 鹼性鹽類水, 67: *XYDCYP*	1932, *Wulixue mingcihui*[73]	
鹼類	Waters alkaline saline 含鹼類與鹽類水, 67: *XYDCYP*	1888, *Dianxue yuanliu*	153 (Qing); 0
醋酸鹽類	Acetates, 1: *HXCL*	1912, *ECSD*	
土質	Waters earthy 含土質之水, 67: *XYDCYP*	1913, *Chinese New Terms and Expressions*[74]	378; Han (1); 0
全壓力	Absolute pressure, 1; Full pressure, 24; Gross pressure, 25; Pressure absolute, Pressure full, Pressure total, 39: *QJ*	1906, *Jinshi wulixue jiaokeshu*	
水壓力	Pressure hydraulic, 39: *QJ*	1899, *Tongwu dianguang* 通物電光 (The electric light piercing through all things)[75]	21 (Qing)
來往動	Motion reciprocating, 34: *QJ*	1904, *TT*	35; Tang (6)
往復動法	Reciprocating action, 42: *QJ*	1904, *TT*	
漸增速力	Accelerating effort, 1; Force accelerating, 23: *QJ*	1904, *TT*	
萍門	Valve safety, 55: *QJ*	1904, *TT*	67 (Qing)
空挺桿	Trunk, 53: *QJ*	1904, *TT*	3 (Qing)
升空挺	Trunk, 53: *QJ*	1904, *TT*	

73 The MCST database indicates that the term *fu jianxing* 附鹼性, with the meaning "alkalization," was already listed in Yan, *An English and Chinese Standard Dictionary* (1912).

74 The MCST database indicates the term *nian tuzhi zhi shi* 黏土質之石, with the meaning "argillaceous stone," was already listed in Yan, *An English and Chinese Standard Dictionary* (1912).

75 John Fryer and Wang Jilie, trans., *Tongwu dianguang* (Shanghai: Jiangnan zhizaoju, 1899).

TABLE 45 Neologisms in the *TVM*, MCST database, and the *Zhongguo jiben guji ku* database (*cont.*)

Chinese term	Attestation in the *TVM*	Year and attestation in the MCST database	*Zhongguo jiben guji ku*; number of attestations and the earliest one
增速率	Accelerate piston, pressure required 令韛韛增速所需壓力, Accelerating effort 漸增速力, Acceleration 漸增速, 1; Force accelerating 漸增速力, 23: *QJ*	1904, *TT*	7 (Qing); 0; 0; 2 (Qing)
螺輪翼	Propeller blade, 40: *QJ*	1904, *TT*	1 (Qing)
曲拐大軸	Crank shaft, 13: *QJ*	1904, *TT*	
相平曲拐	Balanced cranks, 4: *QJ*	1904, *TT*	
實馬力	Actual horse power, 1; Effective horse-power, 18; Horse power indicated, 27; Indicated horse power, 28: *QJ*	1904, *TT*	90 (Qing)
直立汽箭	Engine vertical, 19: *QJ*	1904, *TT*	
輪轂	Boss, Boss of wheel, 7: *QJ*	1904, *TT*	352; Han (3)
不傳力線	Adiabatic line 不傳力曲綫, 1: *QJ*	1916, *ECD*	
鍵輔	Guide, 25: *QJ*	1904, *TT*	7 (Qing)
襯管	Bilge pump neck blush 起船漏水箭頸襯管, 5; Brass bush 黃銅襯管, Brass bush of stuffing box 軟墊臼銅襯管, Brass bush of valve 門之襯管, Bush 銅襯管, Bush of piston rod stuffing box 挺桿軟墊臼底銅襯管, Bush of eccentric shaft 兩心輪軸銅襯管, 8; Circulating pump gland bush 冷水壓蓋襯管, Circulating pump neck bush 冷水箭頸襯管, 11; Gland tube 壓蓋襯管, 25; Gun metal bushes 礮銅襯管, 26: *QJ*	1904, *TT*	2 (Qing); 0; 0; 0; 0; 1 (Qing); 0; 0; 0; 0; 0; 0
襯管枕	Bush bearing, 8: *QJ*	1904, *TT*	
推路	Length of stroke, 30: *QJ*	1904, *TT*	29; Spring and Autumn—Warring States (1)

TABLE 45 Neologisms in the *TVM*, MCST database, and the *Zhongguo jiben guji ku* database (*cont.*)

Chinese term	Attestation in the *TVM*	Year and attestation in the MCST database	*Zhongguo jiben guji ku*; number of attestations and the earliest one
汽罨推路	Slide valve travel, 47: *QJ*	1904, *TT*	
出汽門	Eduction valve, 18; Exhaust valve, 20: *QJ*	1904, *TT*	2 (Qing)
進汽門	Induction valve, 28: *QJ*	1935, *CEMMD*[76]	2 (Qing)
進水門	Inlet valve, 28: *QJ*	1904, *TT*[77]	10; Ming (1)
船首	Stem of ship, 50: *QJ*	1904, *TT*	582; Han (4)
汽門桿	Bar slide, pressure on 汽門桿上壓力, 4; Guide valve rod 汽門桿鍵輔, 26; High pressure valve spindle 大抵力汽門桿, 27: *QJ*	1935, *CEMMD*[78]	
阿西他辣	Acetal, 1: *HXCL*	1912, *ECSD*	
阿西他阿美弟	Acetamide, 1: *HXCL*	1912, *ECSD*	
自漲圈	Expansion ring, 20: *QJ*	1904, *TT*	
活翼	Feathering paddle, 21: *QJ*	1904, *TT*	12 (Qing)
明輪活翼	Reefing paddle, 42: *QJ*	1904, *TT*	1 (Qing)
船底角	Bilge, 5: *QJ*	1904, *TT*	
停螺捎	Check belt, 10: *QJ*	1904, *TT*	
鉤螺捎	Hook bolt, 27: *QJ*	1904, *TT*	
相連之器具	Connecting gear, 12: *QJ*	1904, *TT*	
鍋爐座	Bed of boiler, 5, Boiler bedding, 6: *QJ*	1904, *TT*	
果全書鍋爐	Boiler Cornish 果臬書鍋爐, 6: *QJ*	1904, *TT*	0; 5 (Qing)
圓筒鍋爐	Cylindrical boiler 圓筩鍋爐, 14: *QJ*	1904, *TT*	2 (Qing)[79]
鍋爐聚汽膛	Dome of boiler, 16: *QJ*	1904, *TT*	

76 *Jinqi men* 進汽門 is listed in the text with the meaning "inlet valve."

77 *Jinshui men* 進水門 is listed in the text with the meaning "suction valve."

78 *Qimen gan* 汽門桿 is listed in the text with the meaning "valve stem."

79 The second character of the attestation within the *Shiwu tongkao* is *tong* 筩, the same variant commonly adopted in the *TVM*, and not *tong* 筒, adopted instead in Mateer, *Technical Terms*, 47.

TABLE 45 Neologisms in the *TVM*, MCST database, and the *Zhongguo jiben guji ku* database (*cont.*)

Chinese term	Attestation in the *TVM*	Year and attestation in the MCST database	*Zhongguo jiben guji ku*; number of attestations and the earliest one
附汽機鍋爐	Donkey boiler, 16: *QJ*	1904, *TT*	
鍋爐爆裂	Explosion of boiler, 20: *QJ*	1904, *TT*	
空筩鍋爐	Flue boiler, 23: *QJ*	1904, *TT*	1 (Qing)[80]
汽車鍋爐	Locomotive boiler, 31: *QJ*	1904, *TT*	2 (Qing)
回火路之鍋爐	Return flue boiler, 42: *QJ*	1904, *TT*	
分節鍋爐	Sectional boiler, 45: *QJ*	1904, *TT*	
汽鍋爐	Steam boiler, 49: *QJ*	1904, *TT*	1 (Qing)
鍋爐煙管	Boiler tube, 7: *QJ*[81]	1904, *TT*	5 (Qing)
鍋爐牽條	Boiler braces, 6; Boiler stays, 7: *QJ*	1904, *TT*	
阿古尼低尼	Aconitine, 1: *HXCL*	1912, *ECSD*	
阿各里哇里	Acroleine, 1: *XYDCYP*.	1912, *ECSD*	
以尼	Acroleine, 1: *HXCL*		
阿摩尼阿	Ammonia, 3: *HXCL*	1888, *Dianxue yuanliu*	25 (Qing)
阿大卡迷得	Atacamite, 5: *HXCL*. Atacamite, 5: *JS*[82]	1904, *TT*	
阿西低尼	Acetine, 1: *HXCL*	1912, *ECSD*	1 (Qing)
阿西台里	Acetyle, 1: *HXCL*	1912, *ECSD*	8 (Qing)[83]
阿西台里尼	Acetylene, 1; Klumene, 19: *HXCL*	1912, *ECSD*	1 (Qing)
那普塔	Naptha, 23: *JS*	1904, *TT*	44 (Qing)

80 In this case, the second character is *tong* 筩 within the *Shiwu tongkao*, the *TVM*, and Mateer, *Technical Terms*, 47.

81 The lemma "Tube boiler" on page 53 of the *QJ*, is in turn translated as *yanguan guolu* 烟管鍋爐; the latter will be listed in the *TT* as a rendering for "tubular boiler," which in the *TVM* is translated as *xiao yanguan guolu* 小烟管鍋爐. In any case, the character *yan* within the *TVM* is written in its simplified form.

82 In this circumstance, the term is listed within the second column of the *JS*.

83 Only seven should be considered, since one attestation is simply the incomplete form of the following item in the table.

TABLE 45 Neologisms in the TVM, MCST database, and the *Zhongguo jiben guji ku* database (*cont.*)

Chinese term	Attestation in the TVM	Year and attestation in the MCST database	*Zhongguo jiben guji ku*; number of attestations and the earliest one
乳皮	Cream, 12: *HXCL*	1911, *PBXDC*[84]	31; Northern and Southern (2)
迷以脫里醇	Alcohol methylated 迷以脫里醕, 2; Methylated spirit 迷以脫里醕, 22: *HXCL*	1911, *PBXDC*	
米以脫里醇	Spirit methylated 米以脫里醕, 58: *XYDCYP*	1911, *PBXDC*	
瓷器	Earthenware, 13; Porcelain, 27: *HXCL*	1911, *PBXDC*	1924; Jin (5)
黃紅石脂	Bole, 6: *JS*	1912, *ECSD*	
銀爆藥	Fulminate of silver, Fulminating silver, 15; Silver fulminate, 31: *HXCL*	1886, *Nederlandsch-Chineesch Woordenboek*	4 (Qing)
鍍鋅鐵	Galvanized iron 鍍鋅之鐵, 15; Iron galvanized 已鍍鋅之鐵, 19: *HXCL*	1932, *Wulixue mingcihui*	7 (Qing); 3 (Qing); 0
毛髮	Hair, 17: *HXCL*	1913, *A Dictionary of Philosophical Terms*	6450; Zhou (3)
鐵衣	Iron scales, 19: *HXCL*. Scale, 44: *QJ*	1904, *TT*	1735; Jin (1)
鏽節	Rust joint, 44: *QJ*	1904, *TT*	2 (Ming)
克虜伯	Krupp's steel 克虜伯氏鋼, 19; Steel Krupp's 克虜伯氏鋼, 33: *HXCL*	1916, *ECD*	292; Qing (290)

84 In the MCST database the first attestation of the term *niu rupi* 牛乳皮, with the same meaning "cream," is traced to the seventh edition of Kuang, *Hua-Ying zidian jicheng*, published in Hong Kong in 1923.

TABLE 45 Neologisms in the *TVM*, MCST database, and the *Zhongguo jiben guji ku* database (*cont.*)

Chinese term	Attestation in the *TVM*	Year and attestation in the MCST database	*Zhongguo jiben guji ku*; number of attestations and the earliest one
八面形	Magnetite 八面形鐵礦, 21: *HXCL*. Copper ore octahedral 八面形銅礦, 10; Iron ore octahedral 八面形鐵礦,[85] 18; Magnetite 八面形鐵礦, 21; Octahedrite, or Anatase 八面形粒礦, 24: *JS*	1911, *PBXDC*	26; Song (1); 1 (Qing); 0; 1 (Qing)
溺中之鹽	Microcosmic salt, 23: *HXCL*	1910, *ECSD*	
們子銅	Muntz metal, 34: *QJ*	1904, *TT*	1 (Qing)
非金類	Nonmetallic elements, 24: *HXCL*	1899, *Tongwu dianguang*	31 (Qing)
浸水	Photagraphic [sic] baths 照像浸水, 27: *HXCL*	1916, *ECD*	1169; Han (11); 0
泡面鋼	Blister steel, 6: *QJ*. Steel blistered, 33: *HXCL*	1904, *TT*	15 (Qing)
鉀綠	Chloride of potassium, 9; Potassium chloride, 28: *HXCL*. Potassium chloride, 27; Sylvine, 33: *JS*	1898, "Beizhi xinzhi 備製新質 (Preparing a new material)," *Gezhi yiwen huibao*, no. 1	38 (Qing)
掉成之鋼	Puddled steel, 28: *HXCL*. Puddled steel, 40: *QJ*	1904, *TT*	
負電性	Radicals negative 負電性本質, 29: *HXCL*	1932, *Wulixue mingcihui*	7 (Qing); 0
正電性	Radicals positive 正電性本質, 29: *HXCL*	1932, *Wulixue mingcihui*	4 (Qing); 0
中立性	Salts neutral 中立性鹽類, 30: *HXCL*	1887, *Dianxue xuzhi*	55; Song (10); 1 (Qing)
飽足之水	Saturated solutions, 30: *HXCL*	1904, *TT*	8 (Qing)
原動器	Prime mover, 39: *QJ*	1904, *TT*	
出泥孔	Mudholes, 34: *QJ*	1904, *TT*	

85 In this circumstance the Chinese term is also listed in exactly the same way within the second column of the *JS*.

TABLE 45 Neologisms in the TVM, MCST database, and the *Zhongguo jiben guji ku* database (*cont.*)

Chinese term	Attestation in the TVM	Year and attestation in the MCST database	*Zhongguo jiben guji ku*; number of attestations and the earliest one
剪鋼	Shear steel, 31; Steel shear, 33: *HXCL*. Shear steel, 45: *QJ*	1904, *TT*	14; Ming (2)
刀鋼	Shear steel, 45: *QJ*	1904, *TT*	19; Yuan (1)[86]
簧鋼	Shear steel, 45; Spring steel, 48: *QJ*	1904, *TT*	3 (Qing)
螺絲簧	Spring spiral, 48: *QJ*	1904, *TT*	8 (Qing)
無顆粒形	Antimony amorphous 無顆粒形銻 4; Boron amorphous 無顆粒形硼; 6, Sulphur amorphous 無顆粒形硫, 34: *HXCL*	1911, *DECF*[87]	
錫養二礦	Tin stone, 35: *HXCL*	1910, *ECSD*	
生硼砂	Tincal, 35: *HXCL*. Tincal, (Borax), 34: *JS*	1910, *ECSD*	
土耳末里尼	Tourmaline, 34: *JS*[88]	1910, *ECSD*	1 (Qing)
級形石	Trap rock, 35: *HXCL*. Trap, or Dolerite, 34: *JS*	1910, *ECSD*	
董斯敦	Tungsten, 36: *HXCL*	1910, *ECSD*	
由拉哇	Urao, (Trona), 35: *JS*	1910, *ECSD*	
尿酸	Uric acid, 36: *HXCL*	1910, *ECSD*	3 (Qing)
滑水	Water soft, 37: *HXCL*	1904, *TT*	275; Jin (4)
膠質	Wines, ropy 酒內已生膠質, 37: *HXCL*[89]	1908, *Wulixue yuhui*	25 (Qing); 0
光色鈷礦	Cobalt glance, 11: *HXCL*. Glance cobalt, 14: *JS*	1902, "Xixue guanjian," *Huibao*, no. 211	

86 The two characters of the attestation within the *Shui hu zhuan* 水滸傳 [Water margin], traditionally attributed to the novelist Shi Nai'an 施耐庵 (ca. 1296–1372), should indeed be read separately.

87 Wilhelm, *De-Ying-Han wen kexue zidian.*

88 In the HXCL the translation provided on page 35 is the hybrid *tuermolini shi* 土耳末里尼石.

89 In the MCST database, *jiao zhi* 膠質 is listed as the translation of "colloid."

TABLE 45 Neologisms in the *TVM*, MCST database, and the *Zhongguo jiben guji ku* database (*cont.*)

Chinese term	Attestation in the *TVM*	Year and attestation in the MCST database	*Zhongguo jiben guji ku*; number of attestations and the earliest one
外殼	Jacket, 29: *QJ*	1904, *TT*	134; Northern and Southern (1)
壓水舉重器	Jack hydraulic, 29: *QJ*	1904, *TT*	
同熱線	Isothermal lines, 29: *QJ*	1897, *Cehou congtan*	
鐵狗	Iron dog, 29: *QJ*	1904, *TT*	88; Northern and Southern (1)
叉頭鐵桿	Crow bar, 14: *QJ*	1904, *TT*	
兩心輪	Eccentric, 17: *QJ*	1889, *Zhongxue xuzhi*[90]	33 (Qing)
兩心輪動法	Eccentric metal of, 18: *QJ*	1889, *Zhongxue xuzhi*	
曲拐兩心距	Throw of crank, 52: *QJ*	1904, *TT*	
聯軸節	Coupling 聯軸節, 13: *QJ*	1904, *TT*	
聯軸	Crank shaft, 13; Intermediate shaft, 29: *QJ*	1904, *TT*	25; Song (2)
聯軸盤	Coupling, 13: *QJ*	1904, *TT*	
進水塞門	Inlet cock, 28: *QJ*	1904, *TT*	
尾舌門	Tail valve, 51: *QJ*	1904, *TT*	9 (Qing)
永靜性	Inertia, 28: *QJ*	1889, *Zhongxue xuzhi*	1 (Tang)
自記汽漲力表	Indicator, Indicator steam, 28: *QJ*	1904, *TT*	
磨阻力	Friction, 23: *QJ*	1904, *TT*	3 (Qing)
水重學	Hydraulics, 28: *QJ*	1911, *DECF*	
號馬力	Horse power nominal, 27; Nominal horse-power, 34; Power nominal, 39: *QJ*	1904, *TT*	90 (Qing)
推算之馬力	Horse power estimated, 27: *QJ*	1904, *TT*	
顯功用之馬力	Horse power effective, 27: *QJ*	1904, *TT*	
箍鐵	Hoop iron, 27: *QJ*	1904, *TT*	17; Ming (13)

90 John Fryer, ed., *Gezhi xuzhi er jiben* (Shanghai: Gezhi shushi, 1899).

TABLE 45 Neologisms in the *TVM*, MCST database, and the *Zhongguo jiben guji ku* database (*cont.*)

Chinese term	Attestation in the *TVM*	Year and attestation in the MCST database	*Zhongguo jiben guji ku*; number of attestations and the earliest one
大抵力機	High pressure engine, 27: *QJ*	1906, *Jiangsu shifan jiangyi—Wuli*	20 (Qing)
二心輪方捎膛[91]	Eccentric keyway, 18: *QJ*	1916, *ECD*	
二心輪推引杆	Eccentric rod 二心輪推引桿, 18: *QJ*	1904, *TT*	
螺蓋	Nut, 34: *QJ*	1889, *Zhongxue xuzhi*	81; Spring and Autumn—Warring States (1)[92]
上層螺蓋	Check nuts, 10; Lock nut, 31; Nuts lock, 34: *QJ*	1904, *TT*	
火切面	Fire surface, 22; Heating surface, 27; Surface heating, 51: *QJ*	1904, *TT*	140 (Qing)
熱度原數	Heat unit, 27: *QJ*	1904, *TT*	
容熱率	Heat specific, 27; Specific heat, 48: *QJ*	1904, *TT*	5 (Qing)
外散之熱	Radiating heat, 41 *QJ*	1904, *TT*	2 (Qing)
虎頭鉗	Vice, 56: *QJ*	1904, *TT*	
手桿	Hand spike, 26: *QJ*	1904, *TT*	5; Ming (1)
扶手欄杆	Hand rail, 26: *QJ*	1904, *TT*	4; Ming (1)
附汽機起水筩	Donkey pump, 16; Pump donkey, 41: *QJ*	1904, *TT*	
風車起水筩	Pump windmill, 41; Windmill pump, 57: *QJ*	1904, *TT*	

91 Within the MCST database, the term *er xinlun* 二心輪 is listed as an independent lemma attested for the first time in Parker and Xie, *Gewu zhixue*, with the meaning "eccentric engine." In addition to this lemma and the following one in table 45, *er xinlun* is also annotated as part of other translations in numerous other lemmata within the *QJ*, in all the cases except one with the meaning "eccentric."

92 The two characters of the attestation within *Xunzi* 荀子, the very well-known philosophical book of the Warring States period (475–221 BCE), should indeed be read separately.

TABLE 45 Neologisms in the *TVM*, MCST database, and the *Zhongguo jiben guji ku* database (*cont.*)

Chinese term	Attestation in the *TVM*	Year and attestation in the MCST database	*Zhongguo jiben guji ku*; number of attestations and the earliest one
手動	Hand gear 手動器具, 或手動齒輪, 26: *QJ*	1935, *CEMMD*	428; Jin (2);[93] 0; 0; 0
橫割紋銼	Cross cut file, 13: *QJ*	1904, *TT*	
細紋銼	Smooth file, 47: *QJ*	1904, *TT*	
旋轉法器	Gyroscope, 26: *QJ*	1904, *TT*	
牽條	Stays, 48: *QJ*	1904, *TT*	57; Northern and Southern (5)
鍋爐牽條	Boiler braces, 6; Boiler stays, 7: *QJ*	1904, *TT*	
摺邊節	Flange joint, 22: *QJ*	1904, *TT*	
護條	Guard rail, 25: *QJ*	1904, *TT*	10; Song (1)[94]
爐柵	Bars, grate, 4; Fire grate, 22; Furnace grate, 24; Grate, 25: *QJ*	1904, *TT*	169 (Qing)
弧口鑿子	Gouge, 25: *QJ*	1904, *TT*	
貨車	Freight car, 23; Goods truck, Goods wagon, 25: *QJ*	1904, *TT*[95]	290; Tang (1)
退行杆	Lever revesing, 30; Revesing lever, 42: *QJ*	1904, *TT*	
油膛	Oil chamber, 34: *QJ*	1904, *TT*	

93 The two characters of the attestation within the *Mo he seng qi lü* 摩訶僧祇律 [Mahāsāṃghikavinaya], translated by the Indian Buddhist monk Buddhabhadra (359–429) and the Chinese monk and translator Faxian 法顯 (340–418), should indeed be read separately.

94 The two characters of the attestation within the *Huangshi richao* 黄氏日鈔 [Daily notes by Mister Huang], by the Song dynasty scholar Huang Zhen 黄震 (1213–1281), should indeed be read separately.

95 In the MCST database the first attestation of the term *yunhuo che* 運貨車 is also listed, specifically within Schlegel, *Nederlandsch-Chineesch Woordenboek*.

TABLE 45 Neologisms in the *TVM*, MCST database, and the *Zhongguo jiben guji ku* database (*cont.*)

Chinese term	Attestation in the *TVM*	Year and attestation in the MCST database	*Zhongguo jiben guji ku*; number of attestations and the earliest one
空樞	Bearings of trunnion 空樞頸, 5; Brasses for trunnions 空樞頸銅襯, 8; Cast iron trunnion caps 生鐵空樞枕蓋, 9; Flange of trunnion 空樞摺邊, 22; Gland of trunnions 空樞壓蓋, 25; Trunnion, bearing of 空樞頸, Trunnion brasses 空樞頸銅襯, Trunnion brasses supports 空樞銅襯枕座, Trunnion cap bolts 空樞相連枕蓋螺釘, Trunnion cap, cast iron 空樞生鐵枕蓋, Trunnion pipe 空樞管, Trunnion plummer blocks 空樞之枕, Trunnions 搖汽筩空樞, Trunnions steam opening 空樞內汽管, Trunnions steam passages 空樞汽路, Trunnions stuffing box 空樞軟墊臼, 53: *QJ*	1910, *ECSD*	35; Song;[96] 0; 0; 0; 0; 0; 0; 0; 0; 0; 0; 0; 0; 3 (Qing); 0; 0; 0
坐榫	Gib, 25: *QJ*	1904, *TT*	
扁栓	Gib, 25; Key, 30; Wedge, 56: *QJ*	1904, *TT*	1 (Qing)
方梢	Key, 30: *QJ*	1904, *TT*	3; Ming (1)
長髈	Gib, 25: *QJ*	1904, *TT*	16; Song (1)
小抵力汽機	Engine low pressure, 19; Low pressure engine, 32: *QJ*	1898, *Gewu zhixue*	
起行機	Gear starting 令起行機, 25: *QJ*	1904, *TT*	
退行機	Gear reversing, 24: *QJ*	1904, *TT*	

96 The two characters of the attestation within *Xu Zizhi tongjian changbian* 續資治通鑑長編 [Complete version of the continuation of the Comprehensive mirror for aid in governance], completed in 1182 by the Song dynasty scholar Li Tao 李燾 (1115–1184), *Taiping zhiji tonglei* 太平治蹟統類 [A record of various topics pertaining to politics in the Taiping period], by Peng Baichuan 彭百川 (late 12th century), and *Huang Song tongjian jishi benmo* 皇宋通鑑長編紀事本末 [The complete version of the comprehensive mirror of the Song dynasty: a full account of events], by the scholar Yang Zhongliang 楊仲良 (n.d.) of the Southern Song dynasty (1127–1279), should indeed be read separately.

TABLE 45 Neologisms in the *TVM*, MCST database, and the *Zhongguo jiben guji ku* database (*cont.*)

Chinese term	Attestation in the *TVM*	Year and attestation in the MCST database	*Zhongguo jiben guji ku*; number of attestations and the earliest one
看水表	Gauge water, 24: *QJ*	1904, *TT*	
縮力表	Gauge vacuum, 24: *QJ*	1904, *TT*	
漲力表	Gauge pressure, 24: *QJ*	1904, *TT*	
煤氣機器	Engine gas, 19; Gas engine, 24: *QJ*	1904, *TT*	1 (Qing)
凹節	Gab, 24; Swivel or socket, 51: *QJ*	1904, *TT*	1 (Ming)
能鎔化之性	Fusibility, 24: *QJ*	1904, *TT*	
套節煙通	Funnel telescopic 套節烟通, 24: *QJ*	1904, *TT*	
附汽機	Donkey engine, 16: *QJ*	1916, *ECD*	7 (Qing)
根基	Foundation, 23: *QJ*	1916, *ECD*	1434; Zhou (1)
壓水筩	Forcing pump, 23: *QJ*	1904, *TT*	7 (Qing)
常力	Constant force, 12; Force constant, 23: *QJ*	1904, *TT*	351; Han (3)
彎力	Bending force, 5; Force bending, 23: *QJ*	1898, *Gewu zhixue*	2; Tang (1)
底門	Foot valve, 23; Valve foot, 55: *QJ*	1904, *TT*	84; Song (12)
飛輪汽制	Fly governor, 23: *QJ*	1904, *TT*	
刮器	Tool scraping, 52: *QJ*	1904, *TT*	2 (Qing)
刷子	Boiler tube brush 烟管刷子, 7; Flue brush 火路刷子, 23: *QJ*	1904, *TT*	229; Tang (2); 0; 0
明輪翼	Floats of paddle wheel, 23; Paddle floats, Paddle wheel float, 36: *QJ*	1904, *TT*	1 (Qing)
輪翼	Floats, 23: *QJ*	1904, *TT*	25; Song (1)
滅火水龍	Fire engine, 22: *QJ*	1904, *TT*	1 (Qing)
亂紋銼	Flat bastard file 平面亂紋銼, 23: *QJ*	1904, *TT*	
扇門	Flap, 22; Flap valves, 23: *QJ*	1898, *Gewu zhixue*	409; Qin (1)[97]

97 The two characters of the attestation within the *Lüshi chunqiu* 呂氏春秋 [Mister Lü's spring and autumn annals], by the Qin dynasty prime minister Lü Buwei 呂不韋 (292–235 BCE), should indeed be read separately.

TABLE 45 Neologisms in the *TVM*, MCST database, and the *Zhongguo jiben guji ku* database (*cont.*)

Chinese term	Attestation in the *TVM*	Year and attestation in the MCST database	*Zhongguo jiben guji ku*; number of attestations and the earliest one
軸枕	Axle boxes, 3; Bearing, Bearing of shaft, 4: *QJ*	1895, *Zhongxue tushuo* 重學圖說 (An illustrated exposition of mechanics)[98]	20; Song (2)
軸頸	Bearing, 4; Journal, 29: *QJ*	1904, *TT*	13; Ming (1)
凸圈	Fillet, 22: *QJ*	1904, *TT*	3 (Qing)
陰螺絲	Female screw, 21: *QJ*	1889, *Zhongxue xuzhi*	2 (Qing)
添水筩	Boiler feed pump 鍋爐添水筩, 6; Cast iron feed pump 生鐵添水筩, 10; Donkey engine feed pump 附汽機添水筩, Donkey engine pump 添水筩附汽機, 16; Feed pump 鍋爐添水筩, 21; Flange of feed pump 添水筩口摺邊, 22: *QJ*	1904, *TT*	16 (Qing); 0; 0; 0; 0; 0; 0
添水塞門	Feed cock, 21: *QJ*	1916, *ECD*	4 (Qing)
鐵柱	Column, facing of 鐵柱刨平面, Column feet of 鐵柱座, 11; Facings of columns 鐵柱刨平之面, 21: *QJ*	1904, *TT*	2691; Han (5); 0; 0; 0
壓環	Face ring, 21; Packing ring, 35: *QJ*	1904, *TT*	3 (Qing)
劈圈	Face ring, 21: *QJ*	1904, *TT*	
吹汽舌門	Blow valve, 6: *QJ*	1904, *TT*	
漲縮節	Expansion joint, 20; Joint expansion, 29: *QJ*	1904, *TT*	
餘汽門	Escape valve, 20; Relief valve, 42: *QJ*	1916, *ECD*	1 (Qing)
餘水	Waste water, 56: *QJ*	1904, *TT*[99]	2414; Han (8)
放出之汽	Exhaust steam, 20: *QJ*	1911, *DECF*	1 (Qing)

98 John Fryer, *Zhongxue tushuo*, collected in Wang Xiqing and Lu Tiqing, ed., *Xixue dacheng* (Shanghai: Zui liutang shufang, 1885).

99 In the MCST database the first attestation of the term *yushui zha* 餘水閘, with the meaning "soupape de trop-plein," is also listed, specifically within *Dictionnaire Technique Français-Chinois*, in Lemaire and Giquel, *Dictionnaire de poche français-chinois*, 408.

TABLE 45 Neologisms in the *TVM*, MCST database, and the *Zhongguo jiben guji ku* database (*cont.*)

Chinese term	Attestation in the *TVM*	Year and attestation in the MCST database	*Zhongguo jiben guji ku*; number of attestations and the earliest one
汽罨出汽孔	Slide valve exhaust port of, 46: *QJ*	1904, *TT*	
核算	Estimated horse power 核算馬力, 20: *QJ*	1904, *TT*	1875; Ming (18); 0
門簧	Escape valve spring 放汽門簧, 20: *QJ*	1904, *TT*	1 (Ming);[100] 0
漏洩	Escape of steam 漏洩汽, 20: *QJ*	1913, *Han yi Riben falü jingji cidian* 漢譯日本法律經濟辭典 (A Japanese dictionary of law and economics translated into Chinese)[101]	4491; Zhou (1); 0
熱之值數	Equivalent of heat, 20: *QJ*	1904, *TT*	
物料	Engineer's stores 司汽機日用物料, 20; Fittings 配合物料, 22; Store tanks 存物料鐵箱, 50: *QJ*	1904, *TT*	6968; Jin 1;[102] 0; 0; 0
司汽機之人	Engineer, 20: *QJ*	1904, *TT*	
司汽機事	Engineering, 20: *QJ*	1904, *TT*	
工程事	Engineering, 20: *QJ*	1916, *ECD*	104; Ming (15)
轉行汽機	Engine rotary, 19; Rotatory engines, 43: *QJ*	1904, *TT*	1 (Qing)

100 The two characters of the attestation within the *Wanli xinhai jingcha jishi shimo* 萬曆辛亥京察記事始末 [Evaluation of the capital during the year xinhai—1611—of Wanli reign: a full account of events], composed by Zhou Nianzu 周念祖 (n.d.) during the Ming dynasty, should indeed be read separately.

101 Tanabe Keiya and Wang Wozang, *Han yi Riben falü jingji cidian* (Shanghai: Shangwu yinshuguan, 1913).

102 The two characters of the attestation within the *Zhou hou beiji fang* 肘後備急方 [A handbook of prescriptions for emergencies], by the Eastern Jin dynasty scholar Ge Hong 葛洪 (283–343?), should indeed be read separately.

TABLE 45 Neologisms in the *TVM*, MCST database, and the *Zhongguo jiben guji ku* database (*cont.*)

Chinese term	Attestation in the *TVM*	Year and attestation in the MCST database	*Zhongguo jiben guji ku*; number of attestations and the earliest one
天窗	Engine room skylight 汽機房天牕, 19; Stokehole skylight 火艙天牕, 50: *QJ*	1911, *PBXDC*	1004; Han (1); 0; 0
平臥汽機	Engine horizontal, 19: *QJ*	1904, *TT*	
漲力汽機	Engine expansion, 19: *QJ*	1904, *TT*	
電氣運動機器	Electro motor, 18: *QJ*	1904, *TT*	
電氣運動力	Electro motive force, 18: *QJ*	1904, *TT*	
電力機器	Electro dynamic engine, 18: *QJ*	1904, *TT*	
號令報	Telegraph, 51: *QJ*	1904, *TT*	1 (Minguo)
印字電報器	Electric telegraph 電報器, 18: *QJ*	1910, *Deutsch-Chinesisches Verzeichnis von Fachausdrücken*[103]	1 (Qing); 7 (Qing)[104]
顯電表	Electric indicator, 18: *QJ*[105]	1904, *TT*	
電氣時辰鐘	Electric clock, 18: *QJ*[106]	1914, *Vocabulaire français-chinois*[107]	3 (Qing)
出凝水器	Ejector, 18: *QJ*	1904, *TT*	

103 Hans Wirtz, *Deutsch-Chinesisches Verzeichnis von Fachausdrücken aus dem Gebiete der Physik und Verwandten Gebieten* (Qingdao: Deutsch-Chinesische Hochschule, Übersetzungsanstalt, 1910); within the text the term is annotated as a translation of "Drucktelegraph." The first occurrence of the term *dianbaoqi* 電報器 is in turn listed in Mateer, *Technical Terms*, 366, in the entry *dianhua dianbaoqi* 點畫電報器, with the meaning "recorder."

104 Only six should be considered, since one is included within the previous item on the same line of the table.

105 In Mateer, *Technical Terms*, 147, the term is listed as a translation of "electroscopes."

106 The first attestation of *shichenzhong* 時辰鐘 as an independent term dates back to Wilhelm Lobscheid, *Ying-Hua zidian—English and Chinese Dictionary, with the Punti and Mandarin Pronunciation* (Hong Kong: Daily Press Office, 1866–1869), I, 406.

107 Charles Taranzano, *Vocabulaire français-chinois des sciences mathématiques, physiques et naturelles suivi d'un index anglais-français* (Sien-hsien: Imprimerie de la Mission Catholique, 1914). See also Idem, *Supplément au vocabulaire français-chinois des sciences: classifications—Tableaux synoptiques* (Sien-hsien: Imprimerie de la Mission Catholique, 1923).

TABLE 45 Neologisms in the *TVM*, MCST database, and the *Zhongguo jiben guji ku* database (*cont.*)

Chinese term	Attestation in the *TVM*	Year and attestation in the MCST database	*Zhongguo jiben guji ku*; number of attestations and the earliest one
間隙	Clearance, 11: *QJ*	1916, *ECD*	1851; Qin (1)
汽隙	Clearance, 11: *QJ*	1916, *ECD*	
稱力器	Dynameter, 17: *QJ*	1904, *TT*	5 (Qing)
火膛	Combustion chamber, 12; Fire chamber, 22: *QJ*	1904, *TT*	20 (Qing)
鼓形滑車	Drum, 17: *QJ*	1904, *TT*	
鬆連盤	Loose coupling, 32: *QJ*	1904, *TT*	
發動輪	Driver, 17: *QJ*	1904, *TT*	2 (Qing)
洩水管	Drain pipe, 17: *QJ*	1904, *TT*	
連汽車之鈎	Draw bolt, 17: *QJ*	1904, *TT*	
炭養$_{二}$	Acidum carbonicum, 1; Carbonic acid, 12: *XYDCYP*. Carbonic acid, 8: *HXCL*. Acid carbonic, 2: *JS*	1913, *Chinese New Terms and Expressions*[108]	13 (Qing)

TABLE 46 *JS*, page 1
Vocabulary of Mineralogical Terms Occurring in the Manual by J. D. Dana A. M., page 1

Terms used in J. D. Dana's manual	Terms used in Dr. Macgowan's translation	Terms in general use and chief elements
Acadiolite (Chabasite)	哀開台育來脫	阿卡弟哇來得 鋁鈣鈉鉀矽
Achmite	鴨克每脫	阿格迷得 鐵鈉矽養$_{二}$
Acid, arsenious	少砒酸	鉀養$_{三}$
Acid boracic	布而倫酸	硼養$_{三}$

108 The first attestation within the MCST database provides only the date, 1898, without specifying the text in which the term was listed. The first attestation of *tanyang*$_{er}$ *qi* 炭養$_{二}$氣 is in turn annotated within John Fryer and Xu Jianyin, trans., *Shengxue* (Shanghai: Jiangnan zhizaoju 1874). The same term is listed on page 26 of the *XYDCYP* as a rendering of "Gas, carbonic acid," within the *HXCL* as part of the translations of two different lemmata, and on page 8 of the *JS* as the translation "Carbonic acid."

TABLE 46 *JS*, page 1 (*cont.*)

Terms used in J. D. Dana's manual	Terms used in Dr. Macgowan's translation	Terms in general use and chief elements
Acid carbonic	炭酸	炭養$_{二}$
Acid hydrochloric	鹽酸氣	輕綠氣
Acid muriatic	鹽酸氣	輕綠氣
Acid sulphuric	硫磺酸	硫養$_{三}$
Acid sulphurous	硫酸氣	硫養$_{二}$
Acid tungstic	東斯天酸	鎢養$_{三}$
Acmite	鴨克每脫	阿格迷得 鐵鈉矽養$_{二}$
Actinolite (Hornblende)	阿克低摩兒愛脫	光線石 鎂鈣矽養$_{二}$
Actinolite asbestiform	哀斯倍斯得斯阿克低摩兒愛脫	不灰木形光線石
Actinolite glassy	玻璃阿克低摩兒愛脫	玻璃形光線石
Actinolite massive	塊阿克低摩兒愛脫	成塊光線石
Actinolite radiated	星阿克低摩兒愛脫	星形光線石
Adamant, or Diamond	金剛石	金剛石 明炭
Adamantine spar	阿德蠻淡斯罷	金剛性光石 鋁$_{二}$養$_{三}$
Adularia, (Orthoclase)	愛度琉璃耶	阿度拉里阿石 鋁矽養$_{二}$鉀
Aeschynite	曷斯間奈脫	依喜奈得 鍇鐵鐠鈮
Agalmatolite, or Pagodite	像石	像石 鋁鉀矽養$_{二}$
Agaric mineral, (Calcite)	灰拓發	蕈形石 鈣養炭養$_{二}$
Agate, (Quartz)	鴨呆脫	黑白瑪瑙 矽養$_{二}$
Agate moss	莫斯鴨呆脫	苔紋瑪瑙 矽養$_{二}$
Alabandine		阿拉班弟尼礦 錳硫
Alabaster, (Gypsum)	阿拉罷斯登	阿拉巴司得 鈣養硫養$_{三}$
Alalite, (Pyroxene)	哀來來脫	阿拉來得 鎂矽養$_{二}$
Albite, (Cleavelandite)	阿兒倍脫	阿勒倍得 鋁鉀鈉矽養$_{二}$
Albite granite	阿兒倍脫合拉尼脫	含阿勒倍得之花剛石
Alexandrite, (Chrysoberyl)	哀來刻殘奪來脫	亞勒散[109] 得來得 鋊鋁鐵
Allagite, (Rhodonite)	鴨拉呆脫	阿拉蓋得 錳矽養$_{二}$
Allanite, or Cerine	俺蘭奈脫	阿蘭奈得 鍇鋁鐵鈣
Allophane, or Reimannite	哀盧非能	阿陸法尼石 鋁鐵銅矽養$_{二}$
Allophane opal	阿背爾哀盧非能	阿陸法尼哇巴勒石
Alluaudite	哀盧哀得愛脫	阿魯哇台得 鐵錳燐養$_{五}$
Almandine, (Garnet)	鴨兒蠻定	阿利滿的尼石 鋁鐵矽養$_{二}$

109 The character *san* 散 presents *ri* 日 in place of *yue* 月, as almost always occurs in both the manuscripts and the *TVM*.

TABLE 46 *JS*, page 1 (*cont.*)

Terms used in J. D. Dana's manual	Terms used in Dr. Macgowan's translation	Terms in general use and chief elements
Alum	明礬	白礬
Alum ammonia	阿摩尼阿礬	淡輕$_{四}$礬
Alum iron	鐵礬	鐵礬
Alum magnesia	美合尼西養礬	鎂礬
Alum manganese	孟葛尼斯礬	錳礬
Alum potash	卜對斯礬	鉀礬
Alum soda	素特礬	鈉礬
Alumina	哀盧彌那	鋁$_{二}$養$_{三}$
Alumina dihydrate	二水哀盧彌那	鋁$_{二}$養$_{三}$二輕養
Alumina fluate	夫羅而林酸哀盧彌那	鋁$_{二}$弗$_{三}$

TABLE 47 *HXCL*, page 1
Vocabulary of Names of Chemical Substances, page 1

Acetal 阿西他辣
Acetamide 阿西他阿美弟
Acetates 醋酸鹽類
Acetic acid 醋酸
Acetic acid, glacial 冰形醋酸
Acetic anhydride 無水醋酸
Acetic ether 醋酸以脫
Acetic oxychloride 醋酸綠養$_{二}$
Acetic peroxide 醋酸養$_{二}$
Acetine 阿西低尼
Acetone 阿西多尼
Acetone diethylated 二以脫里阿西多尼
Acetone dimethylated 二迷以脫里阿西多尼
Acetone ethylated 以脫里阿西多尼
Acetone methylated 迷以脫里阿西多尼
Acetones 阿西多尼類
Acetonic acid 阿西多尼克酸
Acetonitrile 阿西多內脫來里
Acetyle 阿西台里
Acetyle binoxide 阿西台里養$_{二}$
Acetyle chloride 阿西台里綠
Acids, polybasic 多本之酸質
Acids tribasic 三本之酸質
Acids vegetable 植物酸質
Acids volatile 能自散酸質
Acidulous waters 酸性水
Aconitic acid 阿古尼低克酸
Aconitine 阿古尼低尼
Acrinyl 阿格里乃里
Acroleine 阿各里哇里以尼
Acrylic acid 阿各里里克酸
Actinic rays 光內化物線
Actinolite 阿格替奴來得 即光線石
Adipic acid 阿弟比克酸
Adipocere 阿弟布昔里
Adularia 阿度拉里阿石
Aeschylic alcohol 依喜里克酯
Aerated bread 壓空氣發鬆之饅頭
After dump 煤硐燒後變成炭養$_{二}$氣
Agalmatolite 阿軋辣馬土來得 即像石
Agate 黑白瑪瑙
Aich-metal 愛止銅 又名苟之銅

TABLE 47 *HXCL*, page 1 (*cont.*)

Acetyle urea 阿西台里由里阿
Acetylene 阿西台里尼
Acetylene silver precipitate 銀阿西台里尼
Acetylide of copper 銅阿西台里弟
Acetylide of potassium 鉀阿西台里弟
Acetylide of sodium 鈉阿西台里弟
Acid 酸質 又配[110] 質
Acid of sugar 糖酸類
Acid radicals 酸底子 或配底子
Acids, acetic series 醋酸級
Acids acrylic series 阿各里里克酸級
Acids anhydrous 無水配質 或酸質
Acids aromatic 香酸類
Acids bibasic 二本之酸類
Acids dibasic 二本之酸類
Acids hydrated 含輕養之酸質
Acids lactic series 拉格的克酸級
Acids non-volatile 不自散酸質
Acids organic 生物酸類
Air 空氣
Air benzolized 含偏蘇里之空氣
Alabaster 阿拉巴司得
Alabaster, oriental 東方所產阿拉巴司得
Albanum 阿立白奴末
Albite 阿勒倍得
Albumen 阿立白門 即蛋白質
Alcarsin 阿里卡耳新
Alcohol 酯
Alcohol absolute 無水酯
Alcohol allylic 阿來里克酯
Alcohol amylic 阿美里克酯
Alcohol anisic 阿尼西克酯
Alcohol benzoic 偏蘇以克酯
Alcohol caprylic 加布里里克酯
Alcohol cerylic 西立里克酯

TABLE 48 *HXCL* and VCT, pages 1, 2, and 3
Synoptic table including page 1 of the *Vocabulary of Names of Chemical Substances*, corresponding to pages 1, 2, and 3 of the "Vocabulary of chemical terms"[111]

Original term	*HXCL*	VCT, first version	VCT, second version
[Absinthin]			
Abels' fuze composition		吸銕[112] 亜比勒電引史料	電引史料
" gum cotton pulp		棉花藥漿	亜比勒棉花藥漿

110 All occurrences of the character *pei* 配 listed in this work present the graphical variant, with *yi* 已 in place of *ji* 己 in the *TVM*, but not in the manuscripts.

111 The entries Fryer annotated in the manuscript in brackets are also listed in brackets within table 48; these entries have been almost always excluded from the *HXCL*. Entries that have already been cancelled in the manuscript, though still readable, are signalled within table 48 in square brackets.

112 In the manuscript, this version of the character *tie* 鐵 is frequently used; *tie* 鐵 is in turn employed in the *HXCL*.

TABLE 48 *HXCL* and VCT, pages 1, 2, and 3 (*cont.*)

Original term	*HXCL*	VCT, first version	VCT, second version
Acetal	阿西他辣	阿西台辣	阿西他辣
Acetamide	阿西他阿美弟	阿西的美弟	阿西他阿美弟
Acetates[113]	醋酸鹽類	醋酸鹽類	
1[114] Acetic acid	醋酸	醋酸	
1 ” ”, glacial	冰形醋酸	冰形醋酸	
1 ” anhydride	無水醋酸	無水醋酸	
1 ” ether	醋酸以脫	醋酸以脫	
” oxychloride	醋酸綠養二	阿西他(多) 尼綠[115]	醋酸綠養二
” peroxyde	醋酸養二	醋酸養二	
(Acetification)		成醋酸	
1 Acetine	阿西低尼	阿西低尼	
1 Acetone	阿西多尼	阿西多尼	
” diethylated	二以脫里阿西多尼	二以脫里阿西多尼	
” dimethylated	二迷以脫里阿西多尼	二米以脫里阿西多米[116]	二迷以脫里阿西多尼
” ethylated	以脫里阿西多尼	以脫里阿西多尼	
” methylated	迷以脫里阿西多尼	迷以脫里阿西多尼	
Acetones	阿西多尼類	阿西多尼類	
Acetonic acid	阿西多尼克酸	阿西多尼克	阿西多尼克酸
Page 2			
Acetonitrile	阿西多內脫來里	醋酸內脫來里	阿西多內脫來里

113 The entry has presumably been added in a second phase, between the precedent, “Acetamide,” and the following, “Acetic acid.” The Chinese characters of its translation are to be read from left to right, not from top to bottom.

114 Some entries are distinguished in the manuscript by a graphical sign similar to a checkmark; it has been impossible to clarify the meaning of the sign. In order to avoid adding a footnote for every single entry, terms characterised by this sign are preceded in table 48 by “1.” In the table, only this graphical peculiarity is annotated, together with the sign “(b),” whose meaning could not be ascertained.

115 Another phase of correction can be noted: the first version corresponds to *axiduoni lü* 阿西多尼綠, later modified in *axitani lü* 阿西他尼綠, finally changed to the version annotated in table 48.

116 This was presumably a mere oversight by Fryer, since the character *ni* 尼 has apparently been added with the same ink immediately after the draft. In this and the following tables, in addition to the evident cases of erasure, all entries in which a character, a part of a character or other changes have been carried out with ink patently different from that of the first draft, have been considered second versions.

TABLE 48 *HXCL* and VCT, pages 1, 2, and 3 (*cont.*)

Original term	*HXCL*	VCT, first version	VCT, second version
1 Acetyle	阿西台里	阿西台里	
" binoxide	阿西台里養二	阿西台里綠[117]	阿西台里養二
" chloride	阿西台里綠	阿西台里綠	
" urea	阿西台里由里阿	阿西台里由里阿	
Acetylene	阿西台里尼	阿西里尼	阿西台里尼
" silver precipitate	銀阿西台里尼	銀阿西台里弟[118]	銀阿西台里尼
Acetylide of copper	銅阿西台里弟	銅阿西台里弟	
" potassium	鉀阿西台里弟	鉀阿西台里弟	
" sodium	鈉阿西台里弟	鈉阿西台里弟	
Acid	酸質 又配質	配質即酸質	酸質又配質
" of sugar	糖酸類	糖酸類	
" radicals	酸底子 或配底子	酸底子	酸底子或配底子
Acid anhydrous	無水配質 或酸質	無水配質	無水配質或酸質
" acrylic series	阿各里里克酸級	阿克羅以里酸類	阿各里里克酸級
" aromatic	香酸類	香酸類	
" bibasic	二本之酸類	二本之酸類	
" dibasic	二本之酸類	二本之酸類	
" hydrated	含輕養之酸質	含輕養之配質	含輕養之酸質
" monobasic		一本之配質	一本之酸質
" acetic series[119]	醋酸級	醋酸類	醋酸級
" lactic series	拉格的克酸級	拉克的克	拉各的克酸級

117 This is possibly a mere writing mistake, rather than an actual reconsideration on the translation itself. Lemmata from "Acetonic acid" to "Acetyle urea" are slightly misaligned as compared with the lines pre-printed in the volume. This might have misled Fryer, leading him to originally translate the term "Acetyle chloride" and not "Acetyle binoxide," also considering that the modification from *lü* 綠 to $yang_{er}$ 養二 has been realised with the same ink, therefore implying an immediate change.

118 This as well is presumably an oversight by Fryer. Since all the subsequent lemmata show *di* 弟 as the last character in the phonetic representation of the term "Acetylide," Fryer might simply have been mistaken, originally translating "Acetylene" with the phonemic loan for "Acetylide," to wit *axitailidi* 阿西台里弟, all the more so considering that the modification has been realised with the same ink, therefore implying an immediate change.

119 In table 48, the original sequence of lemmata within the manuscript has been maintained. Whenever an entry with a different order compared with the printed version is annotated in the volume, one or more arrows indicating the correct alphabetical order are annotated in the manuscript. Moreover, numbers and other graphical signs are also annotated in pencil in the volume, beside and in between these lemmata, at times also

TABLE 48 *HXCL* and VCT, pages 1, 2, and 3 (*cont.*)

Original term	*HXCL*	VCT, first version	VCT, second version
" organic	生物酸類	生物酸類	
" polybasic	多本之酸質	多本之配質	多本之酸質
" tribasic	三本之酸質	三本之配質	三本之酸質
" vegetable	植物酸質	植物酸質	
" volatile	能自散酸質	能散配質	能自散酸質
" non-volatile	不自散酸質	不化散配質	自不化散酸質
Acidulous waters	酸性水	酸性水	
Aconitic acid	阿古尼低克酸	阿古尼低克	阿古尼低克酸
Aconitine	阿古尼低尼	阿古尼低尼	
1 Acroleine	阿各里哇里以尼	阿克羅阿里以尼	阿各里哇里以尼
1 Acrinyl	阿格里乃里	阿格里乃里	
[Adapters]		接管	
[Agalmatolite]		阿軋辣馬土來得	
Acrylic acid	阿各里里克酸	阿克羅以里酸	阿各里里克酸
Page 3			
Actinic rays[120]	光内化物線	光内化物氣	
Actinolite	阿格替奴來得 即光線石	阿克替奴來得	阿各替奴來得
(Adapters)[121]		接管	
Adipic acid	阿弟比克酸	阿弟比克	阿弟比克酸
Adipocere	阿弟布昔里	阿弟布西里	阿弟布昔里
Adularia	阿度拉里阿石	阿度拉里阿	
Aeschylic alcohol	依喜里克醕	依喜里克醕	
Aerated bread	壓空氣發鬆之饅頭	壓空氣發鬆之饅頭	
After dump	煤硐燒後變成炭養二氣	煤硐燒後變成炭養二氣	

beside the relevant Chinese translations. Nevertheless, as many of these signs are cancelled or unclear, they will not as a rule be listed in this work.

120 After "Actinic rays," a graphical sign corresponding to "(b)," annotated in pencil, can be read, even if cancelled.

121 The Chinese translation of the lemma has to be read from left to right, not from top to bottom. It is plausible that the namesake preceding entry, listed a few entries above in table 48, had been cancelled due to a mistake in the alphabetical order; in any case, "Adapters" will not be listed in the *HXCL*.

TABLE 48 *HXCL* and VCT, pages 1, 2, and 3 (*cont.*)

Original term	*HXCL*	VCT, first version	VCT, second version
Agalmatolite[122]	阿軋辣馬土來得 即像石	阿軋辣馬土來得	
Agate	黑白瑪瑙	黑瑪瑙	
Aich-metal	愛止銅 又名苟之銅	愛止銅又名苟之銅	
Air	空氣	空氣	
" benzolized	含偏蘇里之空氣	含偏蘇里之空氣	
Alabaster[123]	阿拉巴司得	阿拉巴司打	
" oriental	東方所產阿拉巴司得	東方所產阿拉巴司打	
Albanum	阿立白奴末	阿勒白奴末	
Albite	阿勒倍得	阿勒倍得	阿勒比得[124]
Albumen	阿立白門 即蛋白質	阿勒布門	
Alcarsin	阿里卡耳新	阿勒卡耳新	阿里卡耳新[125]
Alcohol	醕	醕	
" absolute	無水醕	無水醕	
" allylic	阿來里克醕	阿來里克醕	
" amylic	阿美里克醕	阿美里克醕	
" anisic	阿尼西克醕	阿呢西克醕	
" benzoic	偏蘇以克醕	偏蘇以克醕	
" caprylic	加布里里克醕	加布里里克醕	
" cerylic	西立里克醕	昔立里克醕	西立里克醕

122 The Chinese rendering has to be read from left to right, not from top to bottom. As in the case of "Adapters," it is plausible that the namesake preceding entry, listed a few entries above in table 48, had been cancelled due to a mistake in the alphabetical order. Nevertheless, in contrast with "Adapters," "Agalmatolite" will be listed on page 1 of the *HXCL*.

123 Lemmata from "Alabaster" to "Alcarsin" are connected by a graphical sign, above which the characters *a yin* 阿音 are annotated. It is presumably a note that Fryer himself considered necessary in relation to the theories endorsed about phonemic loans.

124 The character *bi* 比 has been added close to *bei* 倍, which in turn has been cancelled; the modification indeed has not been accepted in the *HXCL*, in which "Albite" is translated on page 1 as *alebeide* 阿勒倍得.

125 Also in this case the character *li* 里 has been simply added beside *le* 勒. Nevertheless, contrary to the translation for "Albite," the modification has been finally accepted in the *HXCL*, in which "Alcarsin" is translated on page 1 as *alikaerxin* 阿里卡耳新.

TABLE 49 *HXCL*, page 8

Carbonates 炭養$_{二}$鹽類
Carbonates alkaline 炭養$_{二}$鹻性鹽類
Carbonates normal 炭養$_{二}$合法鹽類
Carbonic acid 炭養$_{二}$
Carbonic anhydride 無水炭養$_{二}$
Carbonic ether 炭養$_{二}$以脫
Carbonic oxide 炭養
Carbotriamine 加耳波三阿美尼
Carbovinate of potash 鉀養加耳波費尼克
Carburetted hydrogen 炭$_{二}$輕$_{四}$
Carbuncle 卡朋可勒石
Carmine 卡耳迷尼
Carmine red, or lake 卡耳迷尼紅料
Carminic acid 卡耳迷尼克酸
Carnallite 卡耳那來得
Carnelian 卡耳尼里恩 即紅瑪瑙
Carraway oil 茴香油 即揩羅衣油
Carron oil 卡倫油
Carthamine 卡而大迷尼
Cartilage 脆骨
Caseine 加西以尼
Caseine vegetable 植物加西以尼
Cassiterite 加西特來得
Cast iron 生鐵
Cast iron grey 灰色生鐵
Cast iron malleable 韌性生鐵
Cast iron mottled 花點生鐵
Cast iron containing phosphorus 含燐生鐵
Cast iron containing silicon 含矽生鐵
Cast iron containing sulphur 含硫生鐵
Castor 楷司土爾 即海騾
Castor oil 蓖麻油
Castor oil, cold drawn 冷壓蓖麻油
Castorine 楷司土爾以尼
Cast steel 生鋼 即鑄鋼
Catechu 加的主 即兒茶
Catechuic acid 加的主以克酸
Cat's eye 貓兒眼石
Caustic alkali 烙炙鹻類 即無炭養$_{二}$者

Calc spar 丐克司巴耳
Calomel 汞$_{二}$綠
Cameos 刻陽紋花寶石
Camomile 揩暮迷拉 即野菊花
Camphiline 加暮非里尼
Camphic acid 加暮非以克酸
Camphine 加暮非尼
Camphol 加暮非哇里
Campholic acid 加暮非哇里克酸
Camphor 樟腦
Camphoric acid 樟腦酸 即加暮夫里克酸
Camphorimide 加暮夫衣美弟
Candles, composite 洋燭
Cane sugar 蔗糖
Cannel coal 干尼里煤 即燭煤
Caoutchine 古得止以尼
Caoutchouc 古得止格
Caoutchouc solvents 化古得止格之料
Cap composition 銅帽藥料
Capillary pyrites 毛形鎳硫$_{二}$礦
Capric acid 加布里克酸
Caprine 加布里尼
Caproic acid 加布路以克酸
Caproic alcohol 加布路以克醕
Caproine 加布路以尼
Caproyle 加布路以里
Caproylene 加布路以里尼
Caprylene 加布里里尼
Caprylic acid 加布里里克酸
Caprylic alcohol 加布里里克醕
Capsicine 揩普西幾尼
Caramel 卡拉末勒
Carbazotic acid 加耳波阿蘇的克酸
Carbolic acid 加耳波力克酸
Carbon 炭 西名加耳波
Carbon bichloride 炭綠$_{二}$
Carbon bisulphide 炭硫$_{二}$
Carbon chloride 炭綠
Carbon iodide 炭碘

TABLE 49 *HXCL*, page 8 (*cont.*)

Caustic lunar 銀養淡養$_{五}$	Carbon oxide 炭養
Caustic potash 鉀養	Carbon oxychloride 炭$_{二}$養$_{二}$綠$_{二}$
Caustic soda 鈉養	Carbon protochloride 炭$_{二}$綠
Cawk 夸克 鋇養礦之類	Carbon sesquichloride 炭$_{二}$綠$_{三}$
Cedar wood 片柏木	Carbon sesquichloride liquid 炭$_{二}$綠$_{三}$流質
Cedrene 西特里尼	Carbon subchloride 炭$_{二}$綠
	Carbonate of baryta and lime 鋇養炭養$_{二}$鈣養炭養$_{二}$
	Carbonate of lime and soda 鈣養炭養$_{二}$鈉養炭養$_{二}$

TABLE 50 *HXCL* and VCT, pages 20, 21, and 22
Synoptic table including page 8 of the *Vocabulary of Names of Chemical Substances* corresponding to pages 20, 21, and 22 of the "Vocabulary of chemical terms"

Original term	***HXCL***	**VCT, first version**	**VCT, second version**
Calc spar x	丐克司巴耳	鈣克司怕而	丐克[126] 司巴耳
Calomel	汞$_{二}$綠	汞$_{二}$綠[127]	
Cameos	刻陽紋花寶石	刻陽文花寶石	
Camomile	揩暮迷拉 即野菊花	野菊花	
Camphilene[128]	加暮非里尼	加末非里尼	加暮非里尼
Camphic acid	加暮非以克酸	嵌末夫以克酸	

126 An intermediate phase of correction can be noted in the second version of the translation: the character *ke* 克 had been replaced by *ge* 各.

127 Beside this and the following Chinese translation, *ke yang wenhua baoshi* 刻陽文花寶石, the terms "Calumbine" and "Calumbic" are respectively annotated. The two terms have not been included in the *HXCL*, but are instead listed on page 11 of the *XYDCYP*, translated respectively as *gaolinbuyike suan* 高林布以克酸 and *gaolinbuyini* 高林布以尼. In the manuscript, other terms can be read beside the Chinese translations; the great majority of these English lemmata, added in pencil beside Chinese renderings annotated in pen, has been later inserted in the *XYDCYP*. In turn, terms added in pencil in the space between two lines, or preceded by numbers put close to the Chinese or English lemma—for example, those reported in the first page of the manuscript—are seldom if ever found in the *TVM*.

128 Lemmata from "Camphilene" to "Camphorimide" are connected by a graphical sign, above which the characters *mu jia* 暮加 are annotated; Fryer was referring to a correction in the phonetic rendition necessary in all the terms.

TABLE 50 *HXCL* and VCT, pages 20, 21, and 22 (*cont.*)

Original term	*HXCL*	VCT, first version	VCT, second version
Camphine	加暮非尼	加暮非尼	
Camphol	加暮非哇里	嵌末夫勒	加暮[129] 夫哇里
Campholic acid	加暮非哇里克酸	嵌末夫里克	加暮夫里克
Camphor	樟腦	樟腦	
Camphoric acid	樟腦酸 即加暮夫里克酸	樟腦酸即嵌末夫拉以克酸	樟腦酸即加暮夫拉以克酸
Camphorimide	加暮夫衣美弟	嵌末夫以米弟	加暮夫衣美弟
Candles, composite[130]	洋燭	洋燭	
Cane sugar	蔗糖	蔗[131] 糖	
Cannel coal x	干尼里煤 即燭煤	干尼里煤 即燭煤	
Caoutchine	古得止以尼	古得止以尼	
Caoutchouc	古得止格	古得止格	
" solvents	化古得止格之料	化古得止格之料	
Cap composition	銅帽藥料	銅冒藥之料	銅冒藥料
Capillary pyrites x	毛形鎳硫二礦	微管形鎳礦	微管形鎳礦毛形鎳硫二礦
Capric acid	加布里克酸	加布里克	
Caprine	加布里尼	加布里尼	
Caproic acid	加布路以克酸	加布路以克	
Caproic alcohol	加布路以克醕	加布路以克醕	
Caproine	加布路以尼	加布路以尼	
Caproyle	加布路以里	加布路愛里	加布路以里
Caproylene	加布路以里尼	加布路以里尼	
Caprylene	加布里里尼	加布里里尼	
Caprylic acid	加布里里克酸	加布里里克	
Capsicine[132]	揩普西幾尼	加布西格尼	揩普西幾尼
Caramel	卡拉末勒	卡拉末辣	

129 The two characters *qian* 嵌 and *mo* 末 have been cancelled in the manuscript, without being replaced. Nevertheless, the need to replace them with *mu* 暮 and *jia* 加 is implied. The same applies, later on, to "Campholic acid," "Camphoric acid," and "Camphorimide."

130 In this and the following lemma, beside the Chinese translations, the entry "Cannabene" and "Cannellin" can be read. The first will later merge in the *XYDCYP*, translated on page 12 as *kannabini* 看拿比尼, while the second has not been included in the *TVM*.

131 In place of the radical *huo* 灬, the character *zhe* 蔗 presents the graphical variant composed of two *ren* 人.

132 "Capsicina" can be read between "Capsicine" and the subsequent English lemma, "Caramel"; "Capsicina" will be listed on page 12 of the *XYDCYP*, translated as *kaipuxiji'na* 楷普西幾那.

TABLE 50 *HXCL* and VCT, pages 20, 21, and 22 (*cont.*)

Original term	*HXCL*	VCT, first version	VCT, second version
Carbazotic acid	加耳波阿蘇的克酸	加貝所的酸	加耳波阿蘇的克酸
Carbolic acid	加耳波力克酸	加波力酸	加波力克酸
Page 20			
Carbon	炭 西名加耳波	炭	
" bichloride	炭綠二	炭綠二	
" chloride	炭綠	炭綠	
Carbon oxide	炭養	炭養	
" sesquichloride	炭二綠三	炭二綠三	
" subchloride	炭二綠	炭二綠	
Carbonate of baryta & lime	鋇養炭養二鈣養炭養二	鋇養炭養二鈣養炭養二	
" lime and soda	鈣養炭養二鈉養炭養二	鈣養炭養二鈉養炭養二	
Carbonates	炭養二鹽類	含炭養二之質	
" alkaline	炭養二鹻性鹽類	鹻類[133]含炭養二之質	鹻類含炭養二之質
" normal[134]	炭養二合法鹽類	合法含炭養二之質	
Carbon bisulphide	炭硫二	炭硫二	
" chloride	炭綠	炭綠	
" iodide	炭碘	炭碘	
" liquid sesquichloride	炭二綠三流質	炭二綠三流質	
" oxychloride	炭二養二綠二	炭二養二綠二	
" protochloride	炭二綠	炭綠	
Caprylic alcohol[135]	加布里里克醕	加布里里克醕	
Carbonic acid	炭養二	炭養二	
" anhydride	無水炭養二	無水炭養二	
" ether	炭養二以脫	炭養二以脫	
" oxide	炭養	炭養	
(Carbonization) (b)		變成炭之事	

133 The right part of the character *lei* 類 is simplified.

134 After "Carbonates normal," seven entries follow: "Caprylic alcohol," "Carbon bisulphide," "Carbon chloride," "Carbon iodide," "Carbon liquid sesquichloride," "Carbon oxychloride," and "Carbon protochloride." Nevertheless, these entries have been later moved to another position within the *HXCL* for reasons related to the alphabetical order.

135 In the manuscript, "Caprylic alcohol" was originally annotated on the facing page; the term is inserted in the correct alphabetical order through an arrow.

TABLE 50 *HXCL* and VCT, pages 20, 21, and 22 (*cont.*)

Original term	*HXCL*	VCT, first version	VCT, second version
(Carboniying fermentation)		發炭酵	
Carbotriamine	加耳波三阿美尼	三阿米尼炭	三阿美尼炭
Carbovinate of potash	鉀養加耳波費尼克	鋰[136]養以養二炭養$_{二}$	鉀養輕養二炭養$_{二}$
Carburetted hydrogen	炭$_{二}$輕$_{四}$	炭$_{二}$輕$_{四}$	
Carbuncle x[137]	卡朋可勒石	卡朋可辣	卡朋可勒
Carmine	卡耳迷尼	卡耳米尼	卡耳迷尼
" red or lake	卡耳迷尼紅料	卡耳米尼紅料	卡耳迷尼紅料
Carminic acid	卡耳迷尼克酸	卡耳米尼克酸	卡耳迷尼克[138]
Carnallite x	卡耳那來得	卡那來得	卡耳那來得[139]
Carnelian x	卡耳尼里恩 即紅瑪瑙	卡尼里恩 即紅寶石	卡耳尼里恩 即紅寶石
(Carre's freezing apparatus) (b)?)		卡來成冰器	卡來造冰器
Page 21			
Carraway oil	茴香油 即揩羅衣油	茴香油 即楷羅衣油	
Carron oil[140]	卡倫油	卡倫油	

136 Since the two characters *li* 鋰 and *jia* 鉀, meaning respectively lithium and potash, are graphically similar, it can be assumed that the mistake in the first version for the Chinese translation of "Carbovinate of potash" is simply an oversight by Fryer; the correction of the second version has been realised with the same ink, plausibly at the time of the first draft.

137 The sign "x" is added in pencil beside some lemmata; it was impossible to ascertain its purpose.

138 In the Chinese translation, the character *suan* 酸 is simply signalled with a circle written in pencil. It is not clear whether, as will be the case for other entries, Fryer initially wanted to cancel it. The alternation of translations with and without the character *suan* 酸 as the right-hand component of the word to indicate acids is a constant of the manuscript and, partially, also of the *HXCL*.

139 An intermediate version between the first and the second can also be inferred as *ke'ernalaide* 客而那來得.

140 Beside this and the two following English terms, "Carthamine" and "Cartilage," the entries "Carthartine," "Carvene," and "Carvol" are respectively annotated. The three lemmata can be read on page 12 of the *XYDCYP*, rendered respectively as *kaiertatini* 楷爾他替尼, *kaierfuyini* 揩爾夫以尼, and *kaierfuwali* 揩爾夫哇里. Moreover, beside the Chinese translations for "Carthamine" and "Cartilage," the entries "Caryopthyllin" and "Caryopthyllum" can be found; they can be identified with the lemmata "Caryophyllin"

TABLE 50 *HXCL* and VCT, pages 20, 21, and 22 (*cont.*)

Original term	*HXCL*	VCT, first version	VCT, second version
Carthamine	卡而大迷尼	卡耳太迷尼	卡而太迷尼
Cartilage	脆骨	脆骨	
(Case hardening) (b)[141]		銕面變鋼	
Caseine	加西以尼	加西衣尼	加西以尼
" vegetable	植物加西以尼	植物加西衣尼	植物加西以尼
Cassia[142]		粗桂皮	
Cassiterite x	加西特來得	卡西脫來得	卡西特來得
Cast iron	生鐵	生銕	
" " grey	灰色生鐵	灰色生銕	
" " malleable	韌性生鐵	韌性生銕	
" " mottled	花點生鐵	花点生銕	
" " containing phosphorus[143]	含燐生鐵	含燐生銕	
" " " silicon	含矽生鐵	含矽生銕	
" " " sulphur	含硫生鐵	含硫生銕	
Castorine	楷司土爾以尼	楷司土里尼	楷司土尔以尼
Castor oil	蓖麻油	蓖麻油	
" " cold drawn	冷壓蓖麻油	冷壓蓖麻油	
Castor	楷司土爾 即海騾	楷司土尔	
Cast steel[144]	生鋼 即鑄鋼	生鋼即鑄鋼	
(Catalan process) (b)		卡太閑取銕法	
(Catalysis) (b)		在別質之旁變化	
Catechu[145]	加的主 即兒茶	加的主 即兒茶	
Catechuic acid	加的主以克酸	加的主以克酸	

and "Caryophyllum," rendered with *jialiyoufeilini* 加里由非里尼 and *jialiyoufeilumo* 加里由非路末 also on page 12 of the *XYDCYP*.

141 Below "Case hardening" the entry "Cascarilline" is annotated; "Cascarilline" will be translated on page 13 of the *XYDCYP* as *kaisikaililini* 揩思揩里里尼.

142 "Cassia" is not signalled in any way compared with other lemmata; the term, though, will not be listed in the *HXCL*, while in the *XYDCYP* it can be found on page 13, translated as *jiaxiye* 加西耶.

143 An arrow indicates the need to insert "Cast iron containing phosphorus" after "Cast iron," but the repositioning has not been realised in the *HXCL*.

144 Beside "Cast steel," "Cathartine" can be read; the latter is not found in the *TVM*.

145 Beside "Catechu," "Catechine" can be read; moreover, next to the Chinese translation, "Catechuic" is annotated. Both "Catechine" and "Catechuic," though, are not found in the *TVM*.

TABLE 50 *HXCL* and VCT, pages 20, 21, and 22 (*cont.*)

Original term	*HXCL*	VCT, first version	VCT, second version
Cats' eye x[146]	貓兒眼石	貓鬼[147] 眼寶石	
Caustic alkali	烙炙鹻類 即無炭養二者	烙性鹻類[148] 即無炭養二者	烙炙鹻類 即無炭養二者
" lunar	銀養淡養五	銀養淡養五	
" potash	鉀養	鉀養	
" soda	鈉養	鈉養	
Cawk x	夸克 鋇養礦之類	夸克 鋇養礦之類	
Cedar wood	片柏木	片柏木	
Cedrene	西特里尼	西特里尼	

TABLE 51 *HXCL*, page 14

Ethylene binoxide 以脫里尼養二	Enysite 衣尼歲得
Ethylene diamine 以脫里尼二阿美尼	Epidote 以比佗得
Ethylene hexethyl-diphosphonium-hydrate 以脫里尼六以脫里二燐輕養	Epsom salts 鎂養硫養三
Ethylformate of soda 鈉養以脫里福耳密克	Erbium 鉺
Ethylglucose 以脫里哥路哥司	Erucic acid 衣魯西克酸
Ethylic alcohol 以脫里克醕	Erucine 衣魯西尼
Ethylic bromide 以脫里克溴	Erytherite 以里特來得
Ethylic chloride 以脫里克綠	Erythric acid 以里特里克酸
Ethylic ether 以脫里克以脫	Erythrine 以里特里尼礦 即紅鈷礦
Ethylic iodide 以脫里克碘	Erythrosic acid 以里特魯西克酸
Ethyl-methyl-phenylamine 以脫里迷以脫里非尼里阿美尼	Esculetine 愛思古里低尼
Ethyl-methyl-urea 以脫里迷以脫里由里阿	Esculine 愛思古里尼
Ethylo-platammonium hydrate 以脫里鉑淡輕四養輕養	Eserina 以西里那

146 "Catechurentine" can be read beside the Chinese translation of "Cat's eye," annotated as "Cats' eye," as per usual for Fryer. "Catechurentine" can be plausibly identified with "Catechuretine," translated as *jiadezhulideni* 加的主里的尼 on page 13 of the *XYDCYP*.

147 Apparently, the character is *gui* 鬼, with a stroke *dian* 丶 in place of *si* 厶.

148 The right part of both characters *lei* 類 is simplified.

TABLE 51 *HXCL*, page 14 (*cont.*)

Ethylo-toluidine 以脫里朵路以弟尼
Ethyloxamide 以脫里哇格司阿美弟
Ethylurea 以脫里由里阿
Euchlorine 由綠氣
Euclase 由可賴斯石 即易斷石
Eudoic acid 由度以克酸
Eugenic acid 由幾尼克酸
Eugenine 由幾尼尼
Euperion matches 由比里恩自來火
Euphorbium 由福比由末 即大戟香
Eupione 由比阿尼
Evernic acid 以分尼克酸
Excretions 棄津液
Exitile 愛格西台里
Fahlerz 發勒士 即灰色銅礦
Faroelite 發里來得
Fast colours 定質顏料
Fats 定質油
Fatty acid series 油酸類
Feldspar 非勒司巴耳

Feldspar potash 鉀養鋁$_{二}$養$_{三}$六矽養$_{二}$
Feldspar soda 鈉養鋁$_{二}$養$_{三}$六矽養$_{二}$
Fennel 茴香
Ferric acid 鐵養$_{三}$
Ferric chloride 鐵綠
Ferric oxide 鐵$_{二}$養$_{三}$
Ferric sulphate 鐵養硫養$_{三}$

Essence of almonds 杏仁露
Essence of turpentine 松香露
Essential oils 能蒸油類
Essonite 愛蘇奈得 即桂皮色寶石
Ethal 以脫勒
Ethalic acid 以脫勒克酸
Ether 以脫
Ethereal oil 以脫油
Etherine 以脫以尼
Ethers 以脫類
Ethers double 雙以脫類
Ethers perfuming 香以脫類
Ethers flavouring 加味以脫類
Ethylamine 以脫里阿美尼
Ethylammonia 以脫里淡輕$_{三}$
Ethylaniline 以脫里阿尼里尼
Ethylate of potash 鉀養以脫里
Ethylate of soda 鈉養以脫里
Ethylate of zinc 以脫里鋅
Ethyl-codeyl-ammonium hydrate 以脫里苛弟里淡輕$_{四}$養輕養
Ethyle 以脫里
Ethyle amyle 以脫里阿埋里
Ethyle butyle 以脫里布低里
Ethyle cyanide 以脫里衰
Ethyle hydride 以脫里輕
Ethyle iodide 以脫里碘
Ethyle kakodyle 以脫里卡可待里
Ethyle nicotyl-ammonium-hydrate 以脫里尼古弟里淡輕$_{四}$養輕養
Ethyle orthcarbonate 以脫里炭養$_{二}$
Ethyle subcarbonate 二以脫里炭養$_{二}$
Ethyle sulphide 以脫里硫
Ethylene 以脫里尼
Ethylene bibromide 以脫里尼溴$_{二}$

TABLE 52 *HXCL* and VCT, pages 36, 37, 38, and 39
Synoptic table including page 14 of the *Vocabulary of Names of Chemical Substances*, corresponding to pages 36, 37, 38, and 39 of the "Vocabulary of chemical terms"

Original term	*HXCL*	VCT, first version	VCT, second version
Enysite x	衣尼歲得	衣尼歲得	
Enamel glass	發郎玻璃	發郎玻璃	
Endosmose (b)	通膜和匀	通膜和匀	
Epidote x	以比佗得	以比佗得	
Erythrosic acid	以里特魯西克酸	以里脫魯西克酸	以里特魯西克酸[150]
Page 36[149]			
Epsom salts	鎂養硫養$_{三}$	鎂養硫養$_{三}$	
(Equivalent) (b)		分劑數	
(” formula) (b)		分劑式	
Erbium	鉺	鉺	
Erucic acid[151]	衣魯西克酸	衣魯西克酸[152]	
Erucine[153]	衣魯西尼	衣魯西尼	
Erythrine[154]	以里特里尼礦 即紅鈷礦	以里脫里尼	以里特里尼
Erythric acid[155]	以里特里克酸	以里脫里酸	以里特里克酸

149 The numbering annotated in the glossary ends on page 29; it has been added in table 52 for the reader's convenience.

150 On page 14 of the *HXCL* the translation of "Erythrosic acid" is *yiliteluxike suan* 以里特魯西克酸, while on page 22 of the *XYDCYP* it corresponds to *yilituoluxike suan* 以里脫魯西克酸, reflecting the same alternation of the manuscript.

151 "Erucic acid" is placed side by side with the term "Ergotina"; "Ergotina" will not be listed in the *HXCL*, but can be located instead on page 22 of the *XYDCYP*, translated as *erwodayi'na* 耳臥達以那.

152 The character *suan* 酸 is apparently separated from the remaining part of the Chinese term; it is difficult to affirm with certainty whether it was added at a later phase, also considering the strong resemblance of the two varieties of ink used.

153 "Erucine" is placed side by side with the term "Ergotic acid"; "Ergotic acid" will not be listed in the *HXCL*, but can be located instead on page 22 of the *XYDCYP*, translated as *erwodake suan* 耳臥達克酸.

154 "Erythrine" is placed side by side with the term "Ergotine"; "Ergotine" will not be listed in the *HXCL*, but can be located instead on page 22 of the *XYDCYP*, translated as *erwodayini* 耳臥達以尼.

155 "Erythric acid" is placed side by side with the term "Ericinol"; "Ericinol" will not be listed in the *HXCL*, but can be located instead on page 22 of the *XYDCYP*, translated as *yilixinuli* 以里西奴里. Moreover, its Chinese translation is juxtaposed with the term

TABLE 52 *HXCL* and VCT, pages 36, 37, 38, and 39 (*cont.*)

Original term	*HXCL*	VCT, first version	VCT, second version
Erytherite[156]	以里特來得	以里脫來脫	以里特來脫
Esculetine	愛思古里低尼	愛思古里低尼	
Esculine	愛思古里尼	愛思古里尼	
Eserina	以西里那	以西里那	
Essence of almonds	杏仁露	杏仁滷	杏仁露
" turpentine	松香露	松仁滷	松香露
Essential oils	能蒸油類	能蒸油類	
Essonite x	愛蘇奈得 即桂皮色寶石	愛蘇內得	愛蘇奈得
Ethal	以脫勒	以脫辣	
Ethalic acid	以脫勒克酸	以脫辣克	
Ether	以脫	以脫	
Etherine	以脫以尼	以脫以尼	
(Etherification continuous) (b)		連成以脫之工	
Ethereal oil	以脫油	以脫油	
Ethers	以脫類	以脫類	
" double	雙以脫類	雙以脫	
" perfuming[157]	香以脫類	香以脫	
" flavouring	加味以脫類	加味以脫	
Ethylamine	以脫里阿美尼	以脫里阿米尼	以脫里阿美尼
Ethylammonia	以脫里淡輕三	以脫里淡輕三	
Ethylaniline	以脫里阿尼里尼	以脫阿尼里尼	以脫里阿尼里尼
Ethylate of potash	鉀養以脫里	鉀養以脫里	
" soda	鈉養以脫里	鈉養以脫里	
" zinc	以脫里鋅	以脫里鋅	
Ethyl-codeyl-ammonium hydrate	以脫里苟弟里淡輕四養輕養	以脫里苟弟里淡輕四養輕養	

"Erythroretine"; "Erythroretine" is also attested on page 22 of the *XYDCYP*, rendered as *yilituoliwalitini* 以里脫里哇里替尼.

156 "Erytherite" is placed side by side with the term "Ericolin," preceded by an encircled "x"; "Ericolin" will not be listed in the *HXCL*, but can be located instead on page 22 of the *XYDCYP*, translated as *yiligelini* 以里哥里尼.

157 Three terms are annotated on the left of this and the following two lemmata, not on the right, as in the case of terms analysed in the previous footnotes, which later merged in the *XYDCYP*: they are "Ethionic," "Ethine," and "Ethide." The first two are not found in the *XYDCYP*, and the third is only annotated in the glossary as part of other lemmata.

TABLE 52 *HXCL* and VCT, pages 36, 37, 38, and 39 (*cont.*)

Original term	*HXCL*	VCT, first version	VCT, second version
Ethyle	以脱里	以脱里	
Ethyle amyle	以脱里阿埋里	以脱里阿美里	以脱里阿埋里
" butyle	以脱里布低里	以脱里布低里	
" cyanide	以脱里衰	以脱里衰	
Euclase x	由可賴斯石 即易斷石	由可賴斯礦	
Page 37			
Ethyle hydride	以脱里輕	以脱里輕	
Ethyle-nicotyl-ammonium-hydrate	以脱里尼古弟里淡輕$_{四}$養輕養	以脱里尼古弟里淡輕$_{四}$養輕養	
Ethyle iodide	以脱里碘	以脱里碘	
" kakodyle	以脱里卡可待里	以脱里卡苦待里	以脱里卡可待里
" orthcarbonate	以脱里炭養$_{二}$	以脱里炭養$_{二}$	
" subcarbonate	二以脱里炭養$_{二}$	二以脱里炭養$_{二}$	
" sulphide	以脱里硫	以脱里硫	
Ethylene	以脱里尼	以脱里尼	
" bibromide	以脱里尼溴$_{二}$	以脱里尼溴$_{二}$	
" binoxide	以脱里尼養$_{二}$	以脱里尼養$_{二}$	
" diamine	以脱里尼二阿美尼	以脱里尼二阿米尼	以脱里尼二阿美尼
" hexethyl-diphosphonium-hydrate	以脱里尼六以脱里二燐輕養	以脱里尼六以脱里二燐輕養	
Ethylformate of soda	鈉養以脱里福耳密克	鈉養以脱里福密克	
Ethylglucose	以脱里哥路哥司	以脱里哥路哥司	
Ethylic alcohol	以脱里克醕	以脱里克醕	
" bromide	以脱里克溴	以脱里克溴	
" chloride	以脱里克綠	以脱里克綠	
" ether	以脱里克以脱	以脱里克以脱	
" iodide	以脱里克碘	以脱里克碘	
Ethyl-methyl-phenylamine	以脱里迷以脱里非尼里阿美尼	以脱里迷以脱里非內里阿米尼	以脱里迷以脱里非尼里阿美尼
" " urea	以脱里迷以脱里由里阿	以脱里迷以脱里由里阿	
Ethylo-platammonium hydrate[158]	以脱里鉑淡輕$_{四}$養輕養	以脱里鉑淡輕$_{四}$輕養	

158 Beside "Ethylo-platammonium hydrate," in the manuscript the characters *yituolini gudini danqing$_{si}$ qingyang* 以脱里尼古低尼淡輕$_{四}$輕養, can be read.

TABLE 52 *HXCL* and VCT, pages 36, 37, 38, and 39 (*cont.*)

Original term	*HXCL*	VCT, first version	VCT, second version
" toluidine	以脱里朶路以弟尼	以脱里多路以弟尼	以脱里朶路以弟尼
Ethyloxamide	以脱里哇格司阿美弟	以脱里哇克司美弟	以脱里哇格司阿美弟
Ethyl-urea[159]	以脱里由里阿	以脱里由里阿	
Fahlerz[160]	發勒士 即灰色銅礦	發勒士 即灰色銅礦	
Euchlorine	由綠氣	由綠氣	
Eudoic acid[161]	由度以克酸	由阿弟酸	由度以克酸
(Eudiometer) (b)		通電化合氣質管	
(" Cavendishes'?) (b)		卡分弟詩通電管	
(" siphon) (b)		虹吸形通電管	
(" Ure's[162]) (b)		幼而通電管	
Euphorbium[163]	由福比由末 即大戟香	大戟香	由福比由末 即大戟香
Eugenic acid	由幾尼克酸	由幾尼克	
Eugenine[164]	由幾尼尼	由幾尼尼	
Eupione	由比阿尼	由比阿尼	
Page 38			
Euperion matches (b)[165]	由比里恩自來火	由比里恩自來火	
Evernic acid	以分尼克酸	以分尼酸	以分尼克酸
Excretions	棄津液	身内放出之質	棄津液
Exitile x	愛格西台里	愛克西台里	
(Fagothing) (b)		發里來得	銕條成束軋連

159 "Ethyl-urea" will be annotated as "Ethylurea" on page 14 of the *HXCL*.

160 In the manuscript, "Fahlerz" is listed before "Euchlorine"; "Fahlerz" has been later inserted through an arrow in the suitable alphabetical order.

161 In the manuscript, an arrow indicates the need to move "Eudoic acid" after the preceding four entries, annotated in brackets. This indicates that, at least initially, the four entries, later not included in the *TVM*, were destined to be inserted into the printed version.

162 Annotated by Fryer without leaving a space.

163 "Euphorbium" is placed side by side with the term "Euphorbon"; "Euphorbon" will not be listed in the *HXCL*, but can be located instead on page 23 of the *XYDCYP*, translated as *youfubi'en* 由福比恩.

164 "Eugenine" is placed side by side with the term "Eugenin"; "Eugenin" will not be listed in the *HXCL*, but can be located instead on page 22 of the *XYDCYP*, translated as *youjinini* 由幾尼尼.

165 "Euperion matches" appears after "Eupione," but contrary to similar cases, no arrow to reposition it in the correct alphabetical order is present.

TABLE 52 *HXCL* and VCT, pages 36, 37, 38, and 39 (*cont.*)

Original term	*HXCL*	VCT, first version	VCT, second version
Faroelite x	發里來得	發里來得	
Fast colours (b)	定質顏料	定質顏料	
Fats	定質油	定質油	
Fatty acid series	油酸類	油酸類	
Feldspar x	非勒司巴耳	非勒特司怕耳	非勒司巴耳
" potash	鉀養鋁$_{二}$養$_{三}$六矽養$_{二}$	鉀養鋁$_{二}$養$_{三}$六矽養$_{二}$	
" soda	鈉養鋁$_{二}$養$_{三}$六矽養$_{二}$	鈉養鋁$_{二}$養$_{三}$六矽養$_{二}$	
Fennel	茴香	茴[166] 香	
(Fermentation) (b)		發酵	
(" acetous) (b)		發醋酸酵	
(" alcoholic) (b)		發醕酵	
(" viscous) (b)		發膠形之酵	
Ferric acid[167]	鐵養$_{三}$	銕養$_{三}$	
" chloride	鐵綠	銕綠	
" oxide	鐵$_{二}$養$_{三}$	銕養	
" sulphate	鐵養硫養$_{三}$	銕養硫養$_{三}$	

TABLE 53 *HXCL*, page 17

Heavy oil of wine 重酒油 即以脫油	Guaicol 古以阿哥里
Hepatic waters 益肝之水	Guanidine 古阿尼弟尼
Hexamincobaltic chloride 六阿美尼鈷綠	Guanite 古阿奴愛得
Hippuric acid 希布由里克酸	Guano 古阿奴
Homogeneous metal 勻質之銅	Guelder rose 繡毬花
Heliotrope 希里倭特路巴石 即血點石	Gum Arabic 阿喇伯樹膠
Heulandite 許蘭台得	Gum anime 阿你迷膠
Hircine 希爾西尼	Gum British 英國膠

166 In *hui* 茴, below the radical *cao* 艹, the graphical variant *hui* 回 is to be found.

167 In the manuscript, close to this and the following three lemmata, as well as others not included in table 52, a graphical sign similar to an "a" is annotated. The most plausible hypothesis is that it represents a first version of the work, as opposed to the sign "(b)" appearing in other lemmata.

TABLE 53 *HXCL*, page 17 (*cont.*)

Honey 蜜
Hoofs, material 獸蹄料
Hops 霍布花
Hornblende 河拿布侖得
Horn lead 鉛綠 即明角形鉛礦
Horn material 獸角料
Horn silver 銀綠 即明角形銀礦
Horse chesnut bark 馬栗樹皮
Horseradish 辣根
Hot blast iron 熱風鐵
Humic acid 呼迷克酸
Humine 呼迷尼
Humus 呼莫司
Humus coal of 呼莫司煤
Hyacinth 海耶辛得
Hyalite 海耶來得 即玻璃色石
Hydrargyrum cum creta 汞含白石粉
Hydrated bases 含水之本質
Hydrated oxides 含水之配質
Hydrate of lime 鈣養輕養
Hydrate of potash 鉀養輕養
Hydrates 含輕養之質
Hydraulic cements 水內結成之灰
Hydrides of alcohol radicals 含水醋本質
Hydriodate of potash 鉀養輕碘
Hydriodic acid 輕碘
Hydriodic ether 輕碘以脫
Hydroboracite 海得路硼拉歲得 即水硼養$_{三}$礦
Hydrobenzamide 海得路偏蘇阿美弟
Hydrobromic acid 輕溴
Hydrobromic ether 輕溴以脫
Hydrocarbons 輕炭類質
Hydrochloric acid 輕綠
Hydrochloric ether 輕綠以脫
Gum Senegal 歲尼加勒膠
Gum tragacanth 特拉茄看得膠
Gum resins 松香膠類
Gums 樹膠類
Gun cotton 棉花藥
Gun cotton pulp, Abel's 亞比勒氏棉花藥漿
Gun metal 砲銅
Gun paper 紙藥
Gun powder 火藥
Gutta percha 格搭伯查
Gutta, pure 淨格搭
Gypsum 紀[168] 布速末 即石膏
Haemateine 喜瑪替以尼
Haematine 喜瑪替尼
Haematite brown 椶色喜瑪台得 即椶色鐵礦
Haematite red 紅色喜瑪台得 即紅色鐵礦
Haematosine 喜瑪托西尼
Haematoxyline 喜瑪托客西里尼
Haemoglobin 喜瑪格路比尼
Hair 毛髮
Hair dye 染髮料
Halite 哈來得
Halogens 成鹽類之底質
Haloid salts 似鹽之鹽類
Hammer slag 打鐵所出之滓
Hard metal 硬銅
Hard water 濇水
Harmatome 哈馬土密 即劈節石 又名十字石
Harrogate water 哈路該得水
Hartshorn, spirit 淡輕$_{三}$水
Hausmannite 何司曼內得 即錳黑礦
Haüyne 何以尼石
Hayesine 亥以西尼石 即鈣養硼養$_{三}$
Heath's steel 希德氏鋼 即加錳之鋼

168 The character presents the graphical variant with *yi* 已 in place of *ji* 己.

TABLE 53 *HXCL*, page 17 (*cont.*)

Hydrocotarnia 海得路哥他而尼阿
Hydrocyanic acid 輕衰
Hydrocyanic ether 輕衰以脫
Heavy lead ore 鉛養二礦
Heavy spar 鋇養硫養三礦 又名重光石

TABLE 54 *HXCL* and VCT, pages 45, 46, 47, and 48
Synoptic table including page 17 of the *Vocabulary of Names of Chemical Substances*, corresponding to pages 45, 46, 47, and 48 of the "Vocabulary of chemical terms"

Original term	*TVM*	VCT, first version	VCT, second version
Guaicol[169]	古以阿哥里	古以阿哥里	
Guaicum resin	古阿以古末	古阿以苦末	
Guanidine	古阿尼弟尼	古阿尼弟尼	
Guanite x	古阿奴愛得	古阿內得	古阿奴愛得
Guano	古阿奴	古阿奴	
Guelder rose	繡毬花	繡毬花	
Gum Arabic	阿喇伯樹膠	阿拉伯樹膠	阿喇伯樹膠
Gum anime	阿你迷膠	安尼米膠	阿你迷膠
Page 45			
" British	英國膠	英國膠	
" Senegal	歲尼加勒膠	歲尼加勒膠	
" tragacanth	特拉茄看得膠	脫辣茄嵌得膠	脫辣茄看得膠
Gum resins	松香膠類	松香膠類[170]	
Gums	樹膠類	樹膠類	
Gun cotton	棉花藥	棉花藥	
Gun cotton pulp, Abel's	亞比勒氏棉花藥漿		
Gun metal	砲銅	砲銅	
Gun paper	紙藥	紙藥	
Gun powder	火藥	火藥	
Gutta percha	格搭伯查	格搭伯查	
Gutta, pure	淨格搭	净格搭	
Gypsum x	紀布速末 即石膏	紀布速末	紀布速末 即石膏

169 In the manuscript, "Guaicol" appears before "Guaicum resin," with an arrow indicating the need to insert the entry in the correct alphabetical order.

170 The right part of the character *lei* 類 is simplified.

TABLE 54 *HXCL* and VCT, pages 45, 46, 47, and 48 (*cont.*)

Original term	*TVM*	VCT, first version	VCT, second version
Hayesine x	亥以西尼石 即鈣養硴養$_{三}$	海以西尼石	亥以西尼 即鈣養硴養$_{三}$礦
Hæmateine	喜瑪替以尼	喜瑪替以尼	
Hæmatine	喜瑪替尼	喜瑪替尼	
Hæmatite brown x	椶色喜瑪台得 即椶色鐵礦	喜瑪台得棕色者	棕色喜瑪台得
" red x	紅色喜瑪台得 即紅色鐵礦	喜瑪台得紅色者	紅色喜瑪台得
Hæmatosine	喜瑪托西尼	喜瑪托西尼	
Hæmatoxyline	喜瑪托客西里尼	喜瑪托客西里尼	
Haemoglobin	喜瑪格路比尼	喜瑪格路比尼	
Hair dye	染髮料	染髮料	
Hair	毛髮	毛髮	
Halite x	哈來得	哈來得	
Halogens	成鹽類之底質	成塩類之本質	成塩類之底質
Haloid salts	似鹽之鹽類	似塩之塩類	
Hammer slag	打鐵所出之滓	打銕所出之滓	
(Hardness, degrees of) (b)		石質軟硬率	
Hard metal	硬銅	硬銅	
Hard water x	濇水	濇水	
Harmatome	哈馬土密 即劈節石又名十字石	哈馬土密	
Harrogate water x	哈路該得水	哈路該得水	
Hartshorn, spirit	淡輕$_{三}$水	淡輕$_{三}$水	
Hausmannite x	何司曼內得 即錳黑礦	何司曼內得	
Haüyne x	何以尼石	好哇尼石	何以尼石
(Heat atomic) (b)		質点熱水[171]	質点熱率
Hexamincobaltic chloride	六阿美尼鈷綠	六阿米尼鈷綠	
Page 46			
Heath's steel	希德氏鋼 即加錳之鋼	希德鋼 即加錳之鋼	

171 Due to confusing handwriting, it is not possible to state with absolute certainty whether the cancelled character is *shui* 水.

TABLE 54 *HXCL* and VCT, pages 45, 46, 47, and 48 (*cont.*)

Original term	*TVM*	VCT, first version	VCT, second version
Heat units) (b)		熱率	
Heavy lead ore x	鉛養二礦	鉛養二礦	
" spar x	鋇養硫養三礦 又名重光石	鋇養硫養三礦	
Heavy oil of wine	重酒油 即以脫油	重酒油 即以脫油	
Hemihedral crystals) (b)		六面形顆粒	
(Hemmings' safety jet) (b)		海明斯輕養燈嘴	
Hepatic waters x	益肝之水	益肝之水	
Hippuric acid	希布由里克酸	希布由里酸	希布由里克酸
Homogeneous metal	勻質之銅	勻質之銅	
Heliotrope x	希里倭特路巴石 即血點石	希里倭脫路巴	希里倭脫路巴石 血點石[172]
Heulandite x	許蘭台得	許蘭[173] 台得	
Hircine	希爾西尼	希爾西尼	
(Homologous series) (b)		級數類	
(Homologue) (b)		同級數[?] 之质[?] 灰[?]	
Honey	蜜	蜜	
Hoofs, material	獸蹄料	獸蹄料	
Hops	霍布花	霍布花	
Hornblende x	河拿布侖得	河拿布侖特	河拿布侖得
Horn lead x	鉛綠 即明角形鉛礦	鉛綠 即牛角質	鉛綠 即明角形之鉛礦
Horn material	獸角料	獸角料	
" silver x	銀綠 即明角形銀礦	銀綠 即牛角質	銀綠 即明角形之銀礦
Horse chesnut bark	馬栗樹皮	馬栗樹皮	
Horseradish	辣根	辣根	
Hot blast iron	熱風鐵	熱風鐵	

172 The characters of *xuedian shi* 血點石 should be read from left to right, not from top to bottom, as is usual in the manuscript.

173 The radical *cao* 艹 within *lan* 蘭 has apparently been added at a later phase.

TABLE 54 *HXCL* and VCT, pages 45, 46, 47, and 48 (*cont.*)

Original term	*TVM*	VCT, first version	VCT, second version
Humic acid	呼迷克酸	呼米克	呼迷克
Humine	呼迷尼	呼米尼	
Humus x	呼莫司	呼莫司	
" coal of x	呼莫司煤	呼莫斯煤	
Hyacinth x	海耶辛得	海耶辛得	
Hyalite x	海耶來得 即玻璃色石	海耶來得	
Hydrargyrum cum creta	汞含白石粉	汞含白石粉	
Hydrated oxides	含水之配質	含水之配質	
" bases	含水之本質	含水之本質	
Hydrate of lime	鈣養輕養	鈣養軽養	
" " potash	鉀養輕養	鉀養軽養	
Hydrates	含輕養之質	含軽養之質	
Hydraulic cements	水內結成之灰	水內結成之灰	
Hydrides of alcohol radicals	含水醕本質	含水醕本質	
Hydriodate of potash	鉀養輕碘	鉀養軽碘	
Page 47			
Hydriodic acid	輕碘	軽碘	
" ether	輕碘以脫	軽碘以脫	
Hydroboracite x	海得路硑拉歲得 即水硑養$_{三}$礦	海脫路硑拉歲得	海得路硑拉歲得
Hydrobenzamide	海得路偏蘇阿美弟	海得路偏蘇阿美弟	
Hydrobromic acid	輕溴	軽溴	
" ether	輕溴以脫	軽溴以脫	
Hydrocarbons	輕炭類質	軽炭類質	
Hydrochloric acid	輕綠	軽綠	
" ether	輕綠以脫	軽綠以脫	
Hydrocotarnia	海得路哥他而尼阿	海得路哥他而尼阿[174]	
Hydrocyanic acid	輕衰	軽衰	
" ether	輕衰以脫	軽衰以脫	

174 The characters are to be read from left to right.

TABLE 55 *HXCL*, page 33

Strontia carbonate 鍶養炭養$_{二}$
Strontia nitrate 鍶養淡養$_{五}$
Strontia sulphate 鍶養硫養$_{三}$
Strontianite 鍶礦 卽鍶養炭養$_{二}$礦
Strontium 鍶
Strontium sulphide 鍶硫
Struvite 司特路非得
Strychnine 司脫立格尼尼
Stucco 司得苟
Styracine 司土辣西尼
Styrole 司土辣哇里
Styrone 司土辣哇尼
Suberic acid 蘇毗里克酸
Sublimate, corrosive 汞硫$_{二}$
Substitution products 替代法變成之質
Succinic acid 瑟格西尼克酸
Sucrose 蘇格羅司
Suet 生牛羊等定質油
Sugar, beet root 胡蘿蔔糖
Sugar candy 冰糖
Sugar cane 甘蔗
Sugar flesh 肉糖
Sugar fruit 果糖
Sugar grape 葡萄糖
Sugar lime 含鈣養糖
Sugar loaf 糖塔
Sugar maple 楓樹糖
Sugar manna 瑪內糖
Sugar milk 乳糖
Sulphamylic acid 硫阿埋里克酸
Sulphantimoniate 含銻硫$_{五}$合鹻類與硫之質
Sulphantimonite 含銻硫$_{三}$合鹻類與硫之質
Sulpharsenic acid 鉮硫$_{五}$

Spongio-piline 象皮面絨毡
Spongy platinum 鉑絨
Stalactites 上成石鍾乳
Stalagmites 下成石鍾乳
Stannates 含錫養$_{二}$質
Stannic acid 錫養$_{二}$
Stannic chloride 錫綠$_{二}$
Stannic oxide 錫養$_{二}$
Stannic sulphide 錫硫$_{二}$
Stannine 司塔尼尼
Star antimony 星形面銻
Stannous chloride 錫綠
Stannous oxide 錫養
Stannous sulphide 錫硫
Starch 小粉
Staurolite 司托路來得
Staurodite 司托路台得 即十字形石
Steam 汽[175]
Stearic acid 司替阿里克酸
Stearic glucose 司替阿里克哥路哥司
Stearine 司替阿里尼
Steatite 司替阿台得 即肥皂石
Steel 鋼
Steel Bessemer's 別色麻氏鋼
Steel cast 鑄鋼
Steel blistered 泡面鋼
Steel Krupp's 克虜伯氏鋼
Steel mild 輭性鋼
Steel shear 剪鋼
Steel puddled 掉鋼
Steel tempered 退火之鋼
Steel titanic 含鍇之鋼
Stereochromy 含玻璃釉色

175 *Qi* 汽 presents a graphical variant similar to *qi* 汔, with another horizontal stroke above *yi* 乙.

TABLE 55 *HXCL*, page 33 (*cont.*)

Sulpharsenious acid 鉮硫三
Sulphate of soda and lime 鈉養鈣養硫養三
Sulphates 含硫養三質
Sulphethylic acid 含硫之質
Sulphides 含硫質
Sulphindigotic acid 硫養三靛
Sulphindylic acid 硫養三奄的里克酸
Sulphites 硫養二鹽類
Sulphobenzolic acid 硫養三偏蘇里克酸
Sulphocarbolates 硫養三加波力克鹽類
Sulphocarbonates 硫合炭養二之鹽類
Sulphocarbonic acid 硫養炭養二
Sulphocyanide of ammonium 淡輕三輕衰硫二

Sterro-metal 固鋼
Stibethyle 銻以脫里
Stibio triethyle 銻三以脫里
Stibio trimethyle 銻三迷以脫里
Stibnite 司弟伯奈得 即灰色銻礦
Stilbite 司替勒倍得 即光亮石
Stockholm tar 木黑油
Stone ware 上釉之瓦器
Storax 司土辣克司
Stout 黑苦酒
Straits tin 彭加錫
Strassfurthite 斯太司富太得
Stream tin ore 河底聚之錫礦

TABLE 56 *HXCL* and VCT, pages 89, 90, and 91
Synoptic table including page 33 of the *Vocabulary of Names of Chemical Substances*, corresponding to pages 89, 90, and 91 of the "Vocabulary of chemical terms"

Original term	*TVM*	VCT, first version	VCT, second version
Spongio-piline	象皮面絨毡[176]	像皮面絨毡	
Spongy platinum	鉑絨	鉑絨	
(Spontaneous combustion) (b)		自生火	
Stalactites x	上成石鐘乳	上成石鍾乳	
Stalagmites x	下成石鐘乳	下成石鍾乳	
Stannates	含錫養二質	含錫養二質	
Stannic acid	錫養二	錫養二	
" chloride	錫綠二	錫綠二	
" oxide	錫養二	錫養二	
" sulphide	錫硫二	錫硫二	
Stannine	司塔尼尼	司塔尼尼	

176 The characters are to be read from left to right.

TABLE 56 *HXCL* and VCT, pages 89, 90, and 91 (*cont.*)

Original term	*TVM*	VCT, first version	VCT, second version
Star antimony[177]	星形面銻	星形面銻	
Stannous chloride	錫綠	錫綠	
" oxide	錫養	錫養	
" sulphide	錫硫	錫硫	
Starch	小粉	小粉	
Strassfurthite x	斯太司富太得	斯太司富太得	
Staurolite x	司托路來得	司托路來得	
Staurodite x	司托路台得 即十字形石	司托路台得	
Steam	汽	汽	
Stearic acid	司替阿里克酸	司替阿里酸	司替阿里克酸
Stearic glucose	司替阿里克哥路哥司	司替阿里酸古路哥司	司替阿里克哥路哥司
Stearine	司替阿里尼	司替阿里酸	司替阿里尼
Steatite x	司替阿台得 即肥皂石	司替阿台得	
Steel	鋼	鋼	
(" annealing) (b)		鋼退火之工	
" Bessemer's	別色麻氏鋼	別色麻鋼	
" cast	鑄鋼	鑄鋼	
Stibnite x	司弟伯奈得 即灰色	司弟伯内得即灰色	
Page 89	銻礦	銻礦	
Steel blistered	泡面鋼	泡面鋼	
" Krupp's	克虜伯氏鋼	克虜伯鋼	
" mild	輭性鋼	軟性鋼	
" shear[178]	剪鋼	剪鋼	
" puddled	掉鋼	掉鋼	
" tempered	退火之鋼	退火之鋼	
" titanic	含鐟之鋼	含鐟之鋼	
Stereochromy	含玻璃釉色	含玻璃釉色	

177 In the manuscript, the Chinese translations of "Star antimony" and of the following three entries cross one another; the renderings have been connected to their relevant English terms through dashed arrows. Therefore, "Star antimony" was originally annotated on the same line as *xi lü* 錫綠, "Stannous chloride" on that of *xi yang* 錫養, "Stannous oxide" on that of *xi liu* 錫硫, and "Stannous sulphide" on that of *xing xing mian ti* 星形面銻.

178 The alphabetical order of this entry is erroneous in both the manuscript and the *HXCL*.

TABLE 56 *HXCL* and VCT, pages 89, 90, and 91 (*cont.*)

Original term	*TVM*	VCT, first version	VCT, second version
Sterro metal[179]	固鋼	固鋼	
Stilbite x[180]	司替勒倍得 即光亮石	司替勒倍得	
Stibethyle	銻以脫里	銻以脫里	
Stibio triethyle	銻三以脫里	銻三以脫里	
Stibio trimethyle	銻三迷以脫里	銻三米以脫里	銻三迷[181] 以脫里
(Still) (b)		蒸物之器	
Stockholm tar	木黑油	木黑油	
Stoneware (b)	上釉之瓦器	上釉之瓦器	
Storax	司土辣克司	司土辣克司	
Stout	黑苦酒	黑苦酒	
Straits tin	彭加錫	彭加錫	
Stream tin ore x	河底聚之錫礦	河底聚之錫礦	
Strontia carbonate	鍶養炭養$_{二}$	鍶養炭養$_{二}$	
" nitrate	鍶養淡養$_{五}$	鍶養淡養$_{五}$	
" sulphate	鍶養硫養$_{三}$	鍶養硫養$_{三}$	
Strontianite x	鍶礦 即鍶養炭養$_{二}$礦	鍶礦	
Strontium	鍶	鍶	
" sulphide	鍶硫	鍶硫	
Struvite x	司特路非得	司脫路韋得	司特路非得
Strychnine	司脫立格尼尼	司脫立格尼尼	
Stucco	司得苟	司得苟	
Styracine	司土辣西尼	司台辣西尼	司土辣西尼
Styrole	司土辣哇里	司台路里	司土辣哇里
Suberic acid	蘇毗里克酸	蘇昆里克酸	
Sublimate corrosive[182]	汞硫$_{二}$	汞硫$_{二}$	

179 In the *HXCL*, "Sterro metal" presents a dash between the two words.

180 "Stilbite" had been originally inserted before "Stibethyle"; an arrow repositions it in the correct alphabetical order.

181 It has been difficult at times to establish whether the radical *chuo* 辶 has been added at a later phase of the manuscript's draft. In this work, its addition has been considered certain only in cases where it appeared to be patently realised in a subsequent stage, for example when annotated with different ink.

182 In the *HXCL*, "Sublimate corrosive" presents a comma between the two words.

TABLE 56 *HXCL* and VCT, pages 89, 90, and 91 (*cont.*)

Original term	*TVM*	VCT, first version	VCT, second version
Styrone	司土辣哇尼	司土辣哇尼	
Sucrose	蘇格羅司	蘇格羅司	
Page 90			
Substitution products	替代法變成之質	司類[183]	替代法變成之質
(Sublimation) (b)		乾蒸	
Succinic acid	瑟格西尼克酸	色克西尼酸	瑟格西尼克酸
(Succussion) (b)		猛滚	
Suet	生牛羊等定質油	生牛羊等定質油	
Sugar, beet root	胡蘿葡糖	胡蘿葡糖	
Sugar candy	冰糖	冰糖	
Sugar cane	甘蔗	甘蔗[184]	
Sugar lime	含鈣養糖	含鈣養糖	
Sugar loaf	糖塔	糖塔	
Sugar maple	楓樹糖	楓樹糖	
Sugar flesh	肉糖	肉糖	
Sugar fruit	果糖	果糖	
Sugar grape	葡萄糖	葡萄糖	
Sugar milk	乳糖	乳糖	
Sugar manna	瑪內糖	瑪那糖	瑪內糖
Sulphamylic acid	硫阿埋里克酸	硫阿美里克酸	硫阿埋里克酸[185]
Sulphantimoniate	含銻硫$_{\text{五}}$合鹻類與硫之質	含銻硫$_{\text{五}}$合鹻類與硫之質	
Sulphantimonite	含銻硫$_{\text{三}}$合鹻類與硫之質	含銻硫$_{\text{三}}$合鹻類與硫之質	
Sulpharsenic acid	鉮硫$_{\text{五}}$	鉮硫$_{\text{五}}$	
Sulpharsenious acid	鉮硫$_{\text{三}}$	鉮硫$_{\text{三}}$	

183 The two characters are difficult to interpret, particularly the second one. The hypothesis is that it is an oversight by Fryer or that, at least initially, he also wanted to provide a phonemic loan for "Substitution products."

184 In place of the radical *huo* 灬 the character *zhe* 蔗 presents the graphical variant composed of two *ren* 人.

185 In the manuscript, the character *suan* 酸, used as the right-hand component of the word to indicate acids, is frequently separated from the remaining part of the Chinese translation, apparently denoting a later addition. As it is impossible to define it with certainty, it is difficult to affirm whether the renderings provided are two different drafts or the same.

TABLE 56 *HXCL* and VCT, pages 89, 90, and 91 (*cont.*)

Original term	*TVM*	VCT, first version	VCT, second version
Sulphate of soda and lime	鈉養鈣養硫養三	鈉養鈣養硫養三	
Sulphates	含硫養三質	含硫養三質	
Sulphethylic acid	含硫之質	含硫之質	
Sulphides	含硫質	含硫質	
Sulphindigotic acid	硫養三靛	硫養三靛	
Sulphindylic acid	硫養三奄的里克酸	硫養三奄的里克酸	
Sulphites	硫養二鹽類	含硫養二之質	硫養二之塩類
Sulphobenzolic acid	硫養三徧蘇里克酸	硫養三徧蘇里克酸	
Sulphocarbolates[186]	硫養三加波力克鹽類	硫養三加波力克塩類	
Sulphocarbonates	硫合炭養二之鹽類	含硫合炭養二之質	硫合炭養二之塩類
Sulphocarbonic acid	硫養炭養二	硫養炭養二之質	硫養炭養二
Sulphocyanide of ammonium	淡輕三輕衰硫二	淡軽三軽衰硫二	

TABLE 57 *XYDCYP*, page 1
Vocabulary of Names of Materia Medica, page 1

Acetate lead 鉛養醋酸
Acetate morphia 嗼啡啞醋酸
Acetate potash 鉀養醋酸
Acetate soda 鈉養醋酸
Acetate zinc 鋅養醋酸
Acetic acid 醋酸
Acetic acid glacial 冰形醋酸
Acetic aldehyde 醋酸阿勒弟海特
Acetic ether 醋酸以脫
Acetone 阿西多尼
Acetosella 阿西多薩拉
Abies 阿比司族 杉族
Abies balsamea 波勒殺末阿比司
Abies Canadensis 加拿大阿比司
Abies excelsa 高品阿比司 高品杉
Abies nigra 黑色阿比司 黑杉
Abies picea 產柏油阿比司 白松
Abietinic 阿比以弟尼克
Abietis resina 杉香
Abrus precatorius 伏地阿勃路司
Absinthin 阿蒲星弟尼
Absinthium 阿蒲星弟由末

186 In the manuscript, "Sulphocarbolates" and its Chinese translation were annotated on the upper part of the sheet of paper facing the one in which "Sulphocarbonates" is written; "Sulphocarbolates" has been later inserted in the correct alphabetical order through an arrow.

TABLE 57 *XYDCYP*, page 1 (*cont.*)

Acetum 醋
Acetum Britannicum 英國醋
Acetum cantharidis 斑蝥醋
Acetum colchici 哥勒枝噤醋
Acetum destillatum 蒸醋
Acetum ipecacuanhae 叱嗶格醋
Acetum Gallicum 法國醋
Acetum opii 鴉片醋
Acetum scillae 士哇盧醋
Achras mammosa 乳頭形阿格拉司
Achras sapota 薩布他阿格拉司
Acidum aceticum 醋酸
Acidum aceticum dilutum 淡醋酸
Acidum aceticum glaciale 冰形醋酸
Acidum arsenicum 鉮養五
Acidum arseniosum 鉮養三 信石
Acidum benzoicum 徧蘇以克酸 安息酸
Acidum boracicum 硼養三
Acidum carbolicum 加波力克酸
Acidum carbonicum 炭養二
Acidum citricum 檸檬酸[187]
Acidum gallicum 沒石子酸

Acidum hydrochloricum 輕綠水 鹽强水
Acidum hydrochor: dil: 淡鹽强水
Acidum hydrocyanicum 輕衰水
Acidum hydrocyan: dil: 淡性輕衰水
Acidum hydrosulphuric 輕硫水
Acidum hyposulphurous 硫二養二水
Acidum nitricum 硝[188] 强水

Abuta rufescens 略紅阿蒲他
Acacia 阿揩西耶
Acacia Adamsonii 阿但孫阿揩西耶
Acacia albida 略白色阿揩西耶
Acacia Arabica 阿喇伯阿揩西耶
Acacia catechu 兒茶阿揩西耶
Acacia decurrens 蔓形阿揩西耶
Acacia Ehrenbergii 依倫白辯阿揩西耶
Acacia fistula 管形阿揩西耶
Acacia giraffae 其拉夫阿揩西耶
Acacia gum 阿揩西耶樹膠
Acacia gummi 阿揩西耶樹膠
Acacia gummifera 成膠阿揩西耶
Acacia horrida 粗惡阿揩西耶
Acacia Karroo 楷羅阿揩西耶
Acacia mucilage 阿揩西耶樹膠水
Acacia pycnantha厚鬚頭阿揩西耶
Acacia Senegal 塞內加阿揩西耶
Acacia serissa 色里薩阿揩西耶
Acacia Seyal 西亞勒阿揩西耶
Acacia tortilis 絞形阿揩西耶
Acacia vachelia furnesiana 伐扣里亞伐尼西亞捺阿揩西耶
Acacia vera 眞阿揩西耶
Acacia verec 羞性阿揩西耶
Aceraceae 阿西拉西依
Acetate, ammonia 淡輕四養醋酸
Acetate copper 銅養醋酸
Acetate iron 鐵養醋酸

187 All occurrences of the character *ning* 檸 in the printed version of the TVM listed in this work present the graphical variant with *han* 丆 in place of *xin* 心.

188 All occurrences of the character *xiao* 硝 in the printed version of the TVM listed in this work present the graphical variant with *guang* ⺌ overturned.

TABLE 58 *XYDCYP* manuscript and its printed version
Synoptic table including page 1 of the *Vocabulary of Names of Materia Medica* and the corresponding entries of the Vocabulary of names of Materia Medica

Original term	*TVM*	*XYDCYP* manuscript, first version
Abies	阿比司族 杉族	阿比司族 杉族
Abies balsamea	波勒殺末阿比司	波勒殺末阿比司
Abies Canadensis	加拿大阿比司	加拿大阿比司
Abies excelsa	高品阿比司 高品杉	高品阿比司 高品杉
Abies nigra	黑色阿比司 黑杉	黑色阿比司 黑杉
Abies picea	產柏油阿比司 白松	產柏油阿比司 白松
Abietinic	阿比以弟尼克	阿比以弟尼克
Abietis resina	杉香	杉香
Abrus precatorius	伏地阿勃路司	伏地阿勃路司
Absinthin	阿蒲星弟尼	阿蒲星弟尼
Absinthium	阿蒲星弟由末	阿蒲星弟由末
Abuta rufescens	略紅阿蒲他	略紅阿蒲他
Acacia	阿揩西耶	阿揩西耶
Acacia Adamsonii	阿但孫阿揩西耶	阿但孫阿揩西耶
Acacia albida	略白色阿揩西耶	略白色阿揩西耶
Acacia Arabica	阿喇伯阿揩西耶	阿喇伯阿揩西耶
Acacia catechu	兒茶阿揩西耶	兒茶阿揩西耶
Acacia decurrens	蔓形阿揩西耶	蔓形阿揩西耶
Acacia Ehrenbergii	依倫白豣阿揩西耶	依倫白豣阿揩西耶
Acacia fistula	管形阿揩西耶	管形阿揩西耶
Acacia giraffae	其拉夫阿揩西耶	其拉夫阿揩西耶
Acacia gum	阿揩西耶樹膠	阿揩西耶樹膠
Acacia gummi	阿揩西耶樹膠	阿揩西耶樹膠
Acacia gummifera	成膠阿揩西耶	成膠阿揩西耶
Acacia horrida	粗惡阿揩西耶	粗惡阿揩西耶
Acacia Karroo	楷羅阿揩西耶	楷羅阿揩西耶
Acacia mucilage	阿揩西耶樹膠水	阿揩西耶樹膠水
Acacia pycnantha	厚鬚頭阿揩西耶	厚鬚頭阿揩西耶
Acacia Senegal	塞內加阿揩西耶	塞內加阿揩西耶
Acacia serissa	色里薩阿揩西耶	色里薩阿揩西耶
Acacia Seyal	西亞勒阿揩西耶	西亞勒阿揩西耶
Acacia tortilis	絞形阿揩西耶	絞形阿揩西耶

TABLE 58 *XYDCYP* manuscript and its printed version (*cont.*)

Original term	*TVM*	*XYDCYP* manuscript, first version
Acacia vachelia furnesiana	伐扣里亞伐尼西亞捺阿揩西耶	伐扣里亞伐尼西亞捺阿揩西耶
Acacia vera	眞阿揩西耶	真阿揩西耶
Acacia verec	羞性阿揩西耶	羞性阿揩西耶
Aceraceae	阿西拉西依	阿西拉西依
Acetate, ammonia[189]	淡輕四養醋酸	淡輕四養醋酸
Acetate copper	銅養醋酸	銅養醋酸
Acetate iron	鐵養醋酸	銕養醋酸
Acetate lead	鉛養醋酸	鉛養醋酸
Acetate morphia	嗼啡啞醋酸	嗼啡啞醋酸
Acetate potash	鉀養醋酸	鉀養醋酸
Acetate soda	鈉養醋酸	鈉養醋酸
Acetate zinc	鋅養醋酸	鋅養醋酸
Acetic acid[190]	醋酸	醋酸
Acetic acid glacial	冰形醋酸	冰形醋酸
Acetic aldehyde	醋酸阿勒弟海特	醋酸阿勒弟海特
Acetic ether	醋酸以脫	醋酸以脫
Acetone	阿西多尼	阿西多尼
Acetosella	阿西多薩拉	阿西多薩拉
Acetum	醋	醋
Acetum Britannicum	英國醋	英國醋
Acetum cantharidis	斑蝥醋	斑蝥醋
Acetum colchici	哥勒枝噤醋	哥勒枝噤醋
Acetum destillatum	蒸醋	蒸醋
Acetum ipecacuanhae	叱嗶格醋	叱嗶喀醋
Acetum Gallicum	法國醋	法國醋
Acetum opii	鴉片醋	鴉片醋
Acetum scillae	士哇盧醋	士哇盧醋
Achras mammosa	乳頭形阿格拉司	乳頭形阿格拉司
Achras sapota	薩布他阿格拉司	薩布他阿格拉司

189 In the manuscript, "Acetate, ammonia" does not present the comma between the two words.

190 In the manuscript, "Acetic acid" is preceded by "10" written in pencil.

TABLE 58 *XYDCYP* manuscript and its printed version (*cont.*)

Original term	*TVM*	*XYDCYP* manuscript, first version
Acidum aceticum	醋酸	醋酸
Acidum aceticum dilutum	淡醋酸	淡醋酸
Acidum aceticum glaciale	冰形醋酸	冰形醋酸
Acidum arsenicum	鉮養$_{五}$	鉮養$_{五}$
Acidum arseniosum	鉮養$_{三}$ 信石	鉮養$_{三}$ 信石
Acidum benzoicum	偏蘇以克酸 安息酸	偏蘇以克酸 安息酸
Acidum boracicum	硑養$_{三}$	硑養$_{三}$
Acidum carbolicum	加波力克酸	加波力克酸
Acidum carbonicum	炭養$_{二}$	炭養$_{二}$
Acidum citricum	檸檬酸	檽檬酸
Acidum gallicum	沒石子酸	沒石子酸
Acidum hydrochloricum	輕綠水 鹽强水	輕綠水 塩強水
Acidum hydrochor: dil:	淡鹽强水	淡塩強水
Acidum hydrocyanicum	輕衰水	輕衰水
Acidum hydrocyan: dil:	淡性輕衰水	淡性輕衰水
Acidum hydrosulphuric[191]	輕硫水	輕硫水
Acidum hyposulphurous	硫$_{二}$養$_{二}$水	硫$_{二}$養$_{二}$水
Acidum nitricum	硝强水	硝強水

TABLE 59 *XYDCYP* manuscript and its printed version: discrepancies
Discrepancies between the manuscript and the printed version of the *Vocabulary of Names of Materia Medica*[192]

Page	Original term	Chinese translation
A 12	Anemone coronaria	冕形阿尼磨尼
B 8	Bryonia dioica	弟啞以楷布來啞尼亞

191 In the manuscript, this and the two following entries, "Acidum hyposulphurous" and "Acidum nitricum," are closed by a colon, even though they do not include any abbreviation.

192 The manuscript count does not abide by a regular numerical succession—it restarts with every change of the starting letter, with some exceptions. Therefore, the letter indicates the first of the terms listed, while the digit corresponds to the progressive number of the pages including words starting with that letter; exceptions are reported in footnotes.

TABLE 59 *XYDCYP* manuscript and its printed version: discrepancies (*cont.*)

Page	Original term	Chinese translation
C 2	Caltha palustris	卑溼地卡勒他
C 5	Caper spurge	開潑司普尔知
C 10	Celandine	西蘭弟尼
D 1	Daffodil	大福弟拉
D 4	Diascoriaceae	代阿斯苦里亜西依
E 5	Euphorbia lathyris	拉替里司由福比阿
	Euphorbia officinarum	藥品由福比阿
	Euphorbia peplus	帆形由福比阿
F 1	Felon-grass	疔瘡草
G 6	Gratiola officinalis	藥品格拉弟由勒
H 1	Helleborus orientalis	東方赫勒蒲爾
H 3	Hiera picra	海以辣配克拉, 即聖苦藥[193]
H 4	Hyacinthus nonscriptus	未寫海耶新拖司
J 1	Jatropha curcas	苦尔楷司壓脫羅法[194]
M 2[195]	Manchineel	曼西尼勒
M 5	Mercurialis perennis	多年墨古里阿里司, 水銀草
N 1	Narcisus poeticus	詩家那而西蘇司
	Narcisus pseudo	假那而西蘇司
	Nectandria	奈克坦特里亞[196]

193 Below this entry, another one can be read, but it has been cut and only the second part can be inferred. It conceivably is "Hippomane Mancinella," translated as *Manxinala xibumani* 曼西納拉希布瑪尼. Furthermore, transversally to the list of voices, the characters *feigelantuoluosi xibumani* 非格蘭佗羅司希布瑪尼 can be read, without the original term; it may be a reference to the "Hippomane biglandulosa." Other characters can be inferred as *youci xibumani* 有刺希布瑪尼, probably a rendering of "Hippomane spinosa" but the characters are cut and the original term is not indicated.

194 After this entry, another can be read; the original English, "Juice hamlock," and the Chinese rendering, *heimulake zhi* 黑暮拉克汁 have not been included in the printed version.

195 As for the entries starting with the letter "M," it should be noted that the count starts twice; on page 6 numeration starts again from 1 to arrive once more at 6. This and the following entry are included in the first count.

196 The entry has not been added, as was the case with others, directly onto the strips of paper, but has been inserted among the strips of paper glued in the volume. Nevertheless, it has been cancelled from the manuscript. In the *XYDCYP* its correct version, "Nectandra," can be read, translated as *naiketantela* 奈克坦特拉, as confirmed by the correction in the manuscript of the subsequent entry, from "Nectandria Rodiei" to "Nectandra Rodiei."

TABLE 59 *XYDCYP* manuscript and its printed version: discrepancies (*cont.*)

Page	Original term	Chinese translation
O 3	Oil phosphorated	含燐之油[197]
P 2	Parsnip	怕司尼鉢
	Paris quadrifolia	四葉巴黎斯
P 4	Phellandrium aquaticum	水產非蘭[198] 特里

TABLE 60 *XYDCYP* and VCT
Synoptic table including page 1 and 2 of the "Vocabulary of chemical terms," botanical section, and the corresponding entries of the *Vocabulary of Names of Materia Medica*

Original term	*XYDCYP*	VCT, first version	VCT, second version
Abies[199]	阿比司族 杉族	阿比司族 即杉族	

197 The entry has been added transversally to the reading order of the strips of papers glued onto the manuscript; nevertheless, on the manuscript the term is cut and it is not easily understandable. In the *XYDCYP* only the English entry will be listed, without translation; this is a one-off in the whole *TVM*, together with the one indicated in section 4.6, footnote no. 93.

198 The actual character in the manuscript does not exist in standard Chinese dictionaries; it is composed of the radical *cao* 艹, with *men* 门 below and *nü* 女 inside the latter. It can be supposed as being an unconventional simplified form of *lan* 蘭, by analogy with other phonemic loans.

199 The actual *XYDCYP* manuscript is in reality included in another volume. The one found in the final section of the *HXCL* manuscript is a partial version, with fewer entries as the volume approaches the end. From the very first page, it appears much less organised from an alphabetical point of view compared with the *HXCL* manuscript, especially in the first two pages. Some points should be taken into account while analysing table 60. First, the different signs Fryer used have been included; most are difficult to interpret, but they can provide an idea of the complexity of interpreting the manuscript. Second, some entries reported here are not listed in the *TVM*, for example "Acumen" and "Aculeate," rendered respectively as *jian* 尖 and *pi sheng xiao ci* 皮生小刺, "Acrospice," *zhongzi fa xiao gen* 種子發小根, "Abortive," *bu chengshu* 不成熟, "Acrogens" *shang chang* 上長, "Acotyledones" *wu ziban bu* 無子瓣部, "Acrogenæ" *shang ding chang shu* 上頂長屬, "Achenia" *duzi zai qiao nei bu xianglian* 獨子在殼內不相連, "Acuminate" *jian xing zhe* 尖形者, and "Acuminated," *rui xing* 銳形. Third, other entries, in turn annotated in the *TVM*, cannot be located in the manuscript, for instance some in the chemical field. Finally, in table 60, lemmata are listed according to the order of the printed version. Hereafter the actual order in the manuscript of the original terms included in the alphabetical range of the page is reported. Some lemmata are annotated in different subsequent pages,

TABLE 60 *XYDCYP* and *VCT* (*cont.*)

Original term	*XYDCYP*	VCT, first version	VCT, second version
Abies balsamea	波勒殺末阿比司	阿比司[200] 波勒殺米阿	波勒殺末阿比司
Abies Canadensis	加拿大阿比司	阿比司加拿大	加拿大阿比司
Abies excelsa	高品阿比司 高品杉	阿比司愛克色撒 即高品杉	高品阿比司 即高品杉
Abies nigra	黑色阿比司 黑杉	阿比司尼哥賴 即黑杉	黑色阿比司 即黑杉
Abies picea	虀柏油阿比司	阿比司比西阿	虀柏油阿比司 即白松
Abietinic	阿比以弟尼克	阿比以弟尼克	
Abietis resina	杉香	杉香	

but they are not listed here: "Abietis resina," "Abietinic," "Absinthum," "Acorus calamus," "Agaricus campestris," "Agaric," "Ammanita Muscaria," "Achœnia," "Abies," "Abies eccelsa," "Abies picea," "Abies balsamea," "Abies canadendis," "Abies nigra," "1 Allium," "2 Allium olativum," "3 Aloine (a)," "4 Aloineae," "5 Allium Porrum," "6 Alllium cepa," "7 Aloesic acid (a)," "- + Acumen," "+ Aculeate," "8 Aloes," "9 Aloe Barbadensis," "10 Aloe Socotrina," "11 Aloe hepatica," "12 Aloe vulgaris," "13 Aloe Abyssinea," "14 Aloes rubescens," "15 Aloe Arabica," "16 Aloe Africana," "17 Aloe phicatilis," "18 Aloe spicata," "19 Aloe variegata," "20 Aloe linguaformis," "21 Aloe commelina," "21 Aloe ferox," "x Acrospia," "x Acrospice," "1 Abuta rufescens," "2 x Abortive," "3 Acacia," "4 x Acrogens," "5 Aconitia (a)," "Aconella," "6 Acetic acid (a)," "Acraconitine," "7 Aconite," "x Acotyledones," "7 x—Acotyledons," "8 x Acrogenæ," "9 x Achenia," "10 Acoraceae," "Aconitum heterophyllum," "Aconitum intermedium," "12 Actæa spicata," "12 Actæa racemosa," "Aconitum ferox," "Aconitum Stœrkianum," "15 Aconitic acid (a)," "Aconitum decorum," "16 Aconitina (a)," "Aconitum paniculatum," "Aconitum neomontanum," "17 Aconita (a)," "Aconitum napellus," "18 x Acuminate," "19 x Acuminated," "20 Acacia Seyal," "21? Acacia Ehrenbergii," "22 Acacia tortilis," "23 Acacia vera," "24 Acacia arabica," "25 Acacia albida," "26 Acacia adamsonii," "27 Acacia serissa," "28 Acacia vachelia furnesiana," "29 Acacia Karroo," "30 Acacia horrida," "31 Acacia giraffæ," "32 Acacia decurrens," "33 Acacia catechu," "34 Acacia Senegal," "Acacia fistula." Other lemmata, as "Aceraceae," "Acacia pycnantha," "35 Acacia verec," "36 Acacia gummifera," are annotated after numerous pages, therefore indicating the numbering continues in the manuscript, but with terms relevant to botany instead, for instance "37 x Adelphous" *xu hecheng ruogan cong* 鬚合成若干叢 and "38 Adenanthera pavonia," not with terms pertaining to chemistry, which are not found.

200 Fryer frequently left a considerable space between two different Chinese translations for the same original term; he left a smaller space among the different words of a single Chinese rendering, in the event that such words are a phonemic loan for an original term composed of more than one word. In other circumstances, for instance with "Abrus precatorius," Fryer added a small dot between the two words. In table 60, a space is left only in case two different translations for the same lemma were provided.

TABLE 60 *XYDCYP* and VCT (*cont.*)

Original term	*XYDCYP*	VCT, first version	VCT, second version
Abrus precatorius[201]	伏地阿勃路司	伏地阿勃路司	
Absinthin	阿蒲星弟尼	阿蒲星弟尼	
Absinthium[202]	阿蒲星弟由末		
Abuta rufescens	略紅阿蒲他	阿蒲他羅非孫司	略紅阿蒲他
Acacia	阿揩西耶	阿揩西耶	
Acacia Adamsonii	阿但孫阿揩西耶	阿楷西耶阿但孫	阿但孫阿楷西耶
Acacia albida	略白色阿揩西耶	阿楷西耶阿勒比達	略白色阿楷西耶
Acacia Arabica	阿喇伯阿揩西耶	阿楷西耶阿喇伯	阿喇伯阿楷西耶
Acacia catechu	兒茶阿揩西耶	阿楷西耶加的主 即兒茶	兒茶阿楷西耶
Acacia decurrens	蔓形阿揩西耶	阿楷西耶弟克倫司	蔓形阿楷西耶
Acacia Ehrenbergii	依倫白辯阿揩西耶	阿楷西耶依倫白辯	依倫白辯阿楷西耶
Acacia fistula	管形阿揩西耶	管形阿楷西耶	
Acacia giraffae	其拉夫阿揩西耶	阿楷西耶其拉夫	其拉夫阿楷西耶
Acacia gum	阿揩西耶樹膠		
Acacia gummi	阿揩西耶樹膠		
Acacia gummifera	成膠阿揩西耶	阿楷西耶格米菲勒	成膠阿楷西耶
Acacia horrida	粗惡阿揩西耶	阿楷西耶喝里達	粗惡阿楷西耶
Acacia Karroo	楷羅阿揩西耶	阿楷西耶楷羅	楷羅阿楷西耶
Acacia mucilage	阿揩西耶樹膠水		
Acacia pycnantha	厚鬚頭阿揩西耶	厚鬚頭阿揩西耶	
Acacia Senegal	塞內加阿揩西耶	阿楷西耶生[203] 尼加勒	塞內加阿楷西耶
Acacia serissa	色里薩阿揩西耶		
Acacia Seyal	西亞勒阿揩西耶	阿楷西耶西亞勒	西亞勒阿楷西耶
Acacia tortilis	絞形阿揩西耶	阿楷西耶託替里斯	絞形阿楷西耶
Acacia vachelia furnesiana	伐扣里亞伐尼西亞捺阿揩西耶	阿楷西耶伐扣里亞又名伐尼西亞捺	阿楷西耶伐扣里亞伐尼西亞捺
Acacia vera	眞阿揩西耶	阿楷西耶菲拉	真阿楷西耶

201 "Abrus precatorius" is listed on page 7 of the manuscript, in a different alphabetical order. Some entries may have escaped the author's attention, given their extremely complicated arrangement in the manuscript, and also because at least two pages are glued together within the volume.

202 The entry has not been located in the *XYDCYP* manuscript; in every other analogous case, the corresponding cell in table 60 has been left blank.

203 *Saineijia* 塞內加 has been added in place of *shengnijiale* 生尼加勒; in the latter, in turn, the replacement of the character *sheng* 生 with *sun* 孫 could already be noted.

TABLE 60 *XYDCYP* and VCT (*cont.*)

Original term	*XYDCYP*	VCT, first version	VCT, second version
Acacia verec	羞性阿揩西耶	阿楷西耶菲里克	羞性阿楷西耶
Aceraceae	阿西拉西依	阿西拉西依	
Acetate, ammonia[204]	淡輕四養醋酸		
Acetate copper	銅養醋酸		
Acetate iron	鐵養醋酸		
Acetate lead	鉛養醋酸		
Acetate morphia	嗼啡啞醋酸		
Acetate potash	鉀養醋酸		
Acetate soda	鈉養醋酸		
Acetate zinc	鋅養醋酸		
Acetic acid	醋酸		
Acetic acid glacial	冰形醋酸		
Acetic aldehyde	醋酸阿勒弟海特		
Acetic ether	醋酸以脫		
Acetone	阿西多尼		
Acetosella	阿西多薩拉		
Acetum	醋		
Acetum Britannicum	英國醋		
Acetum cantharidis	斑蝥醋		
Acetum colchici	哥勒枝嗦醋		
Acetum destillatum	蒸醋		
Acetum ipecacuanhae	吐嗶格醋		
Acetum Gallicum	法國醋		
Acetum opii	鴉片醋		
Acetum scillae	士哇盧醋		
Achras mammosa[205]	乳頭形阿格拉司	乳頭形阿格拉司	
Achras sapota	薩布他阿格拉司	薩布他阿格拉司	
Acidum aceticum	醋酸		

204 As the *XYDCYP* is a glossary of "materia medica," terms listed refer to different branches of knowledge, such as botany and chemistry. Nevertheless, many of the terms pertaining to chemistry and, partially, to medicine, are not found within the *XYDCYP* manuscript which, as already stated, is rather confusing and incomplete, as is the case for botanical nomenclature. In any case, some of the terms pertaining to chemistry listed in the manuscript are not found in the *HXCL*, some only partially, while others are faithfully reported.

205 "Achras mammosa" and the subsequent entry, "Achras sapota," are listed in an alphabetical order extremely different from the relevant one.

TABLE 60 *XYDCYP* and VCT (*cont.*)

Original term	*XYDCYP*	VCT, first version	VCT, second version
Acidum aceticum dilutum	淡醋酸		
Acidum aceticum glaciale	冰形醋酸		
Acidum arsenicum	鉀養$_{五}$		
Acidum arseniosum	鉀養$_{三}$ 信石		
Acidum benzoicum	徧蘇以克酸 安息酸		
Acidum boracicum	硼養$_{三}$		
Acidum carbolicum	加波力克酸		
Acidum carbonicum	炭養$_{二}$		
Acidum citricum	檸檬酸		
Acidum gallicum	沒石子酸		
Acidum hydrochloricum	輕綠水 鹽强水		
Acidum hydrochor: dil:	淡鹽强水		
Acidum hydrocyanicum	輕衰水		
Acidum hydrocyan: dil:	淡性輕衰水		
Acidum hydrosulphuric	輕硫水		
Acidum hyposulphurous	硫$_{二}$養$_{二}$水		
Acidum nitricum	硝强水		

TABLE 61 *QJ*, page 1
Vocabulary of Terms Relating to the Steam Engine, page 1

Absolute pressure 全壓力
Absolute zero 准零熱度
Absorption of heat 収熱
Accelerate piston, pressure required 令鞴鞴增速所需壓力
Accelerated motion 漸加動速
Advance angle 前進角
Aether engines 以脫機器
After crank shaft 後曲拐
After water ballast tank 後水壓儎箱
Aggregate combinations 積合

TABLE 61 *QJ*, page 1 (*cont.*)

Accelerating effort 漸增速力	Aggregate paths 積路
Accelerating force 漸加速力	Agricultural engine 農汽機
Acceleration 漸增速	Ahead motion 前行動
Accumulation of pressure 聚壓力	Air 風， 或氣
Acting, perpendicularly 直加力	Air casing of funnel 路通空氣壳
Action and reaction 施力復力	Air cock 空氣塞門
Action of propeller 螺輪施力	Air engine 空氣機
Actual energy 實能力	Air expansion of 氣之漲
Actual horizontal pressure on crank 曲拐平受實力	Air extractor 抽空氣機
Actual horse power 實馬力	Air friction of 氣之磨阻力
Actual mean pressure 實中壓力	Air gun 風鎗
Actual power 實能力	Air passages 通氣咽[206]
Actual pressure 實壓力	Air pipe 空氣管
Adamson's vertical blowing engine 阿担生直立進風器	Air pressure 空氣壓力
Addendum circle 齒輪增長之界圈	Air pressure engine 氣抵力機器
Addendum of a tooth 輪齒增配之長	Air pump 恒升車， 或抽氣筩
Addendum line 齒輪增長之架綫	Air pump arm 動恒升車之臂
Adequate strength 足用之力	Air pump back link 恒升車後連桿
Adhesion of locomotive 車輪與鐵路之滯力	Air pump barrel 恒升車筩
Adiabatic line 不傳力曲綫	Air pump bucket 恒升車鞲鞴
Adiabatic line for air 氣之不傳力曲綫	Air pump bucket ring 恒升車鞲鞴圈
Adiabatic line for steam 汽之不傳力曲綫	Air pump bucket valve 恒升車鞲鞴舌門
Adjusting 配准	Air pump chamber 恒升車空膛
Adjusting bolt of brasses 配准枕蓋螺釘	Air pump connecting rod 恒升車搖桿
Adjusting part of shoe 配准鍵輔活襯	Air pump connecting rod socket 升搖桿樞凹
Adjustment for motion blocks 配准行動塊器	Air pump cover 恒升車蓋
Adjustment of mechanism 配准機器内各件法	Air pump cross head 恒升車横擔
Adjustment of speed 配准速法	Air pump cross head journal 恒升車横擔樞
Adjustment of stroke 配准推路法	Air pump cylinder 恒升車筩

206 The last character of the Chinese translation is illegible in all the TVM editions consulted; *yan* 咽 is the author's deduction.

TABLE 61 *QJ*, page 1 (*cont.*)

Admiralty bronze 英國戰船部礮銅	Air pump cylinder, single 單行恒升車箭
Admiralty knot 英國戰船部所定之海里	Air pump delivery valve 恒升車進水門
Admiralty rules 英國戰船部章程	Air pump delivery valve guard 恒升車進水門擋
Admiralty rules for horse power 英國戰船部馬力法	Air pump delivery valve seating 恒升車進水門座
Admission port of slide valve 汽門進汽孔	Air pump delivery valve spindle 恒升車進水門桿
	Air pump discharge pipe 恒升車出水管
	Air pump discharge valve 恒升車放水門
	Air pump double acting 雙行恒升車
	Air pump efficiency of 恒升車功力
	Air pump foot valve 恒升車底門

TABLE 62 *QJ* manuscript and its printed version
Synoptic table including page 28 and 29 of the *Vocabulary of Terms Relating to the Steam Engine*, corresponding to page 38 of the Vocabulary of terms relating to the steam engine[207]

Original term	*QJ*	*QJ* manuscript, first version	*QJ* manuscript, second version
一 Ice cock (added)	冰塞门	冰塞门	
Ice locomotive (red)	行冰汽車	行冰汽車	
Immersed section (red)	船入水體横剖面	船入水中横剖面	船入水體横剖面
Impact (sky blue)	擊力	擊力	

207 The choice of the pages presented stems in this circumstance from the material conditions of the volume; the manuscript is extremely fragile and numerous strips of paper are at risk of coming away from the pages. The page shown here is the one preserved in the best condition. In table 62, after the original term, the colour of the legend is annotated in brackets, while "added" indicates that the entry was annotated directly on the corresponding facing left-hand page. Numbers written in characters, close to the respective terms, and in Arabic numerals, written in pencil, annotated beside the single entries, are also reported in table 62, if present. Finally, whenever there are discrepancies in the annotation of singulars or plurals of the terms listed, or merely mistakes, the manuscript version has been maintained.

TABLE 62 *QJ* manuscript and its printed version (*cont.*)

Original term	***QJ***	***QJ* manuscript, first version**	***QJ* manuscript, second version**
Imperfect elasticity (sky blue)	不全凹凸力	不全凸力	不全凹凸力
Imperial gallon (blue)	准加倫	准加倫	
Impermeater (blue)	汽筩內進油器	汽筩內進油器[208]	
Impinge (sky blue)	聚擊	擊聚	聚擊
Impressed force (sky blue)	收壓力	收壓力	
Impulse of fluids (sky blue)	流質動力	流質流力	流質動力
Inclined engine (red)	斜置汽筩汽機	斜置汽筩汽机	
Inclined plane (sky blue)	斜面	斜面	
二 Increase of pressure (added)	增壓力	增壓力	
二 Increased temperature (added)	增熱度	增熱度	
二 Increasing motion (added)	增動	增動	
Increasing pitch (red)	漸增螺距	漸增螺距	
8 Incrustation (blue)	鍋爐內面生衣	鍋[209] 炉內面生衣	
India rubber (sky blue)	象皮	象皮	
India rubber disc valve (green)	象皮圓板門	象皮圓板門	象皮圓門板[210]
India rubber disk (green)	象皮圓板	象皮圓板	

208 It is also possible to read an intermediate version, *ling qi you zhi* 令汽油質, with an additional character inserted between the other four, but is has been completely cancelled and is therefore illegible.

209 The right part of the character corresponds to that of the simplified version *guo* 锅.

210 This is one of the rare cases in which the Chinese translation listed in the *QJ* is not the one corrected in Fryer's manuscript, but is instead the version before the correction; the same applies to "India rubber square valve."

TABLE 62 *QJ* manuscript and its printed version (*cont.*)

Original term	*QJ*	*QJ* manuscript, first version	*QJ* manuscript, second version
三 India rubber join (added)	象皮節	象皮成節	象皮所做節
三 India rubber packing (added)	象皮軟墊	象皮軟墊	
India rubber ring (green)	象皮圈	象皮圈	
India rubber springs (red)	象皮簧	象皮簧	
India rubber squares valves (green)	象皮方板門	象皮方板	象皮方門板
9 India rubber valve (red)	象皮門	象皮門	
四 Indicated horse power (added)	實馬力	實馬力	
四 Indicated power (added)	所顯之力	所顯之力	
四 Indicated thrust of propeller (added)	螺輪所顯之推力	螺輪所顯之推力	
o Indicator (blue)	自記汽漲力表	汽試漲力表	自記汽漲力表
1 Indicator barrel (blue)	自記汽漲力表筩	自記汽漲力表筩	
2 Indicator card (blue)	自記汽漲力表紙面	自記汽漲力表紙面	
3 Indicator cock (blue)	自記汽漲力表塞門	自記汽漲力表塞門	
Indicator, continuous (blue)	連自記汽漲力表器	連自記漲[211] 力表器	
4 Indicator cylinder (blue)	自記汽漲力表筩	自記漲力表筩	
5 Indicator diagrams (red)	自記汽漲力表圖	均力圖	自記漲力表圖

211 In the manuscript, in numerous subsequent entries the character *qi* 汽 is not annotated.

TABLE 62 *QJ* manuscript and its printed version (*cont.*)

Original term	*QJ*	*QJ* manuscript, first version	*QJ* manuscript, second version
Indicator diagrams of air pump	恒升車自記縮力表圖	恒升車自記縮力表	
Indicator driving (blue)	動自記汽漲力表繩	動自記漲力表繩	
Indicator friction	自記汽漲力表磨阻力	自記漲力表磨阻力	
6 Indicator pencil (blue)	自記汽漲力表筆	自記漲力表筆	
7 Indicator pipe (blue)	自記汽漲力表管	自記漲力表管	
8 Indicator piston (blue)	自記汽漲力表韝韛	自記漲力表韝韛	
9 Indicator piston rod (blue)	自記汽漲力表挺桿	自記漲力表挺桿	
o Indicator spring (blue)	自記汽漲力表管	自記漲力表簧[212]	
Indicator, steam (red)	指汽力器, 或自記汽漲力表	指力器	指汽力器
Induction valves (red)	進汽門	進汽門	
Inelastic (sky blue)	無凹凸力	無凸力	無凹凸力
Inertia (red)	永靜性, 或不肯動之性	永靜性	永靜性又名不肯動之性
Ingot iron[213]	鐵錠鍊成之熟鐵	錠成之熟銕	

212 The character *huang* 簧 has been added beside another, which is in turn cancelled and impossible to establish with certainty; it may possibly be *gang* 鋼. In any case, in this entry as well, the translation provided in the manuscript is different from the one listed in the *QJ*.

213 The entry has been inserted between the previous and the subsequent strip of paper, to wit "Inertia," red, and "Inertia," sky blue, is cancelled. In reality, the second rendering of "Inertia," red, namely *bu ken dong zhi xing* 不肯動之性, was included in the namesake entry of the subjacent strip, "Inertia," sky blue; this included precisely the translation *bu ken dong zhi xing* 不肯動之性. Further down, another namesake entry is listed, marked

TABLE 62 *QJ* manuscript and its printed version (*cont.*)

Original term	*QJ*	*QJ* manuscript, first version	*QJ* manuscript, second version
五 Initial absolute pressure (added)	原全壓力	原全壓力	
五 Initial pressure (added)	原壓力	遠壓力	原壓力
2 Injection (blue)	噴冷水凝水法	噴凝水法	噴冷水凝水法
Injection cock (red)	噴冷水門, 或噴冷水塞門	噴冷門	噴冷水門, 又名噴冷水塞門[214]
Injection condensers (red)	用噴水法之凝水櫃	噴水凝水法	用噴水法之凝水櫃
4 Injection handle (blue)	噴凝水門柄	噴凝水柄	噴凝水門柄
5 Injection lever (blue)	噴凝水門桿	噴凝水桿	噴凝水門桿
Injection orifice (red)	噴冷水孔	噴水孔	噴冷水孔
6 Injection pipe (red)	噴冷水管	噴水管	噴冷水管
Injection pipe (blue)[215]		噴進管	噴冷水管
Injection snifting valve (red)	噴冷水尾舌門	尾舌門	噴冷水尾舌門
Injection spray pipes (red)	噴冷水多小孔管	噴水多小孔管	噴冷水多小孔管
7 Injection valve (green)	噴冷水門	噴水門	噴冷水門
Injection valve hole (green)	噴冷水門孔	噴水門孔	噴冷水門孔

with blue, for which the Chinese translation is *yong jing xing* 永靜性; this has also been cancelled and covered by the translation of "Ingot iron."

214 In this case as well, the second translation, *you ming pen lengshui saimen* 又名噴冷水塞門, was included in the namesake entry "(Injection cock)" of the subjacent strip, preceded by "3" and marked with sky blue; the original translation on the latter strip was *pen jin saimen* 噴進塞門.

215 The entry has been cancelled with a line, written in pencil, already in the manuscript; it is therefore not listed in the *QJ*.

TABLE 62 *QJ* manuscript and its printed version (*cont.*)

Original term	*QJ*	*QJ* manuscript, first version	*QJ* manuscript, second version
Injection valve rod (green)	噴冷水門桿	噴水門桿	噴冷水門桿
六 Injection valve spindle (added)	噴冷水門桿	噴冷水門之桿	
Injector, Giffard's (red)	格法得鍋爐進水器	格法得噴水器	格法得鍋爐進水器
8 Injectors (red)	噴水器	噴水器	
七 Inlet cock (added)	進水塞門	進水塞門	
七 Inlet pipe (added)	進水管	進水管	
七 Inlet valve (added)	進水門	進水門	
Inner end of piston rod (green)	挺桿內端	挺桿內端	
Inner switch (red)	鐵路內移條	內移條[216]	銕路內移條
八 Insertion cloth joint (added)	布嵌成之節	塞布成節	布嵌成之節
八 Inside bearing of paddle shaft (added)	明輪軸內襯	明輪軸內襯	
八 Inside callipers (added)	比內徑規	比內径規	
Inside crank pin (red)	曲拐內軸徑	內曲拐軸頸	曲拐內軸頸
Inside cut off slide valve	內斷汽之汽門	內斷汽之汽罨	
Inside cylinder locomotives (red)	內汽箭汽車	內汽箭汽車	
Inside lap of valve (green)	汽門內餘面	汽罨內餘面	

216 Within this entry another translation appears, on the top right, in which four characters can be distinguished: the first and third are *tie* 鐵 and *ju* 距, while the remaining two are impossible to interpret.

TABLE 62 *QJ* manuscript and its printed version (*cont.*)

Original term	*QJ*	*QJ* manuscript, first version	*QJ* manuscript, second version
Inside lap to slide valves (blue)	汽門內餘面	汽鼂內餘面	
Instant (sky blue)	霎時, 或微分時	霎時	霎時 又名微分時
九 Insufficient pressure (added)	壓力不足	壓力不足	
九 Intense heat (added)	極大之熱	極大之熱	
九 Intensity of draught (added)	爐內風力率	炉風力率	
Intercepter	丁字形受水管	丁字形受水管	
Intermediate posts (red)	木柵間桂	木柵間桂	
十 Intermediate pressure engine (added)	中壓力汽機	中壓力汽机	

TABLE 63 CT, page 1
"Vocabulary of Terms in Naval Architecture," page 1

Absolute-pressure
Accelerating-force 漸加力
Accelerated motion
- Accomodation ladder 船邊便梯
Accumulation of pressure
- Action of propeller 螺輪顯力
Actual-pressure
- Adequate-strength 合宜之堅固
- Adjoining-strake 船皮相切板層
- Adjusting 配准
Admission-port (of slide valve)
- After-backstay 桅後牽繩
- After-peak tank 船尾尖處水箱
- After-sails 後帆
- After-shift of planking 船板向後對節
- After waterballast tank 後水壓儎箱
- After-yards 後桅橫桿
- Ahead-motion 向前動
- Air-casing (of funnel) 烟通邊通風殼
Air cock
- Air-course 通氣路
- Air-extractor 抽氣器
- Air-pipe 通氣管
- Air-port 通氣孔

TABLE 63 CT, page 1 (*cont.*)

- After-balance-frame 後半船相稱肋骨[217]	- Air-pressure 空氣壓力
- After-beam 後橫樑	2 Air-pumps[218] 恒升車抽氣筩
- After-body 後半體	3 Air-pumps back link 恒升車後連桿
- After-capstan 後直立絞車	4 Air-pumps barrel 恒升車筩
After-crank-shaft	5 Air-pumps bucket 恒升車鞲鞴
- After-hatchway 後貨艙口	6 Air-pumps bucket-ring 恒升車鞲鞴管
- After-hold 後貨艙	7 Air-pump bucket-valve 恒升車鞲鞴舌門
- After-hoods (of planking) 船底最後之内外板	8 Air-pumps chamber 恒升車膛[219]
- After-leech 帆後邊帶	9 Air-pump connecting-rod 升搖桿
- After-leech-rope 帆後邊繩	10 Air-pump cover 恒升車盖
- After-mast 後桅	1 Air-pump crosshead 恒升橫擔
- Aftermost-bulkhead 最後橫隔[220] 墻板	2 Air-pump crosshead journals 恒升車橫扭樞
- Aftermost-floor 最後肋骨底	3 Air-pump cylinder 恒升車筩
- After-peak 帆之後上角	4 Air-pumps delivery or discharge-valve 恒升車進水門

TABLE 64 CT, page 2

Air-pump delivery-valve guard 恒升車進水門護條
5 Air-pump delivery-valve seating 恒升車進水門座
6 Air-pumps delivery-valve spindle 恒升車進水門軸
7 Air-pumps discharge-pipe 恒升車出水管
9 Air-pump foot-valve 恒升車底門

217 The cancelled original translation, which can still be distinguished below *hou ban chuan xiangchen leigu* 後半船相稱肋骨, was *hou xiangchen leigu* 後相稱肋骨.

218 In Paasch, *From Keel to Truck*, from which original terms are drawn, all entries including "Air-pump" are in the singular.

219 Beside the character *tang* 膛, *dang* 擋 can be read, even if cancelled.

220 Unless otherwise indicated, this and the following occurrences of the character *ge* 隔 listed in the other pages drawn from the CT, present in the right part, in place of *yi* 一 and *kou* 口, the two components united, as in the character *gao* 高.

TABLE 65 *QJ* and CT
Synoptic table including page 27, 28, 29, 30, 31, and 32 of the "Vocabulary of Terms in Naval Architecture," compared with page 23 of the *Vocabulary of Terms Relating to the Steam Engine*

Original term	*QJ*	CT
Flap (of a port)[221]		船邊門外蓋
Flap valves	舌門, 或扇門, 或鏈門	
Flat of bottom		艙底平面處
Flat of deck		艙平面
Flat bastard-file[222]	平[223] 面亂紋銼	平面雜紋銼
Flat-bottomed ship		平底船
Flat brasses	平面銅襯	
Flat brasses inner surface	平面銅襯內面	
Flat caps	平蓋	
Flat-ceiling		平艙底
Flat-file	平面銼	平銼
Flat-floor		肋骨平中節
Flat pieces of gun metal	平面礮銅塊	
Flat-plate keel		平板脊骨

221 All CT entries listed in table 65, excluding "Flue-boiler," "Following-edge (of propeller blades)," "Foot-valve," "Force the fires (to-)," "Forcing-pump," "Forcing-pump rod," "Fore top staysail sheet," "7 Fork-link," "8 Forward crank-shaft," "Foundation (of engines and boilers)," "9 Foundation-plate," "0 Foundation-plate bolt," "1 Four bladed propeller," "2 Four cylinder engine," "3 Four cylinder tandem engine," "4 Four-way cock," "Framing of an engine," "6 Free-expansion," "7 Freezing-point," "8 Fresh-water-pipe," "1 Friction-brake," "2 Friction-clutch," "3 Friction-coupling," "4 Friction-pulley," "Friction-roller," and "Friction-wheel," are preceded by a sign similar to a stroke of red marker. Other additions, for instance numbers written in pencil, are directly signalled in table 65 close to the corresponding entry. Finally, the first entry of each of the pages of the "Vocabulary of Terms in Naval Architecture" considered here is indicated by the author of this work directly on the table.

222 In the CT manuscript, numerous entries present a small dash between the single words composing the original term; this was possibly originally meant to provide indications for the printing process. In the *QJ* manuscript, in turn, this feature is not present. The entries listed in the CT, if coinciding with those of the *QJ*, are reported in table 65 in the same way they were annotated in the first glossary's manuscript.

223 In the *QJ*, and more generally in the *TVM*, in the character *ping* 平, the alternation between this and another graphical variant, in which the two stroke *dian* 丶 and *pie* 丿 are changed respectively in *pie* 丿 and *fu* ㇏, is constant. This alternation is not present in the CT.

TABLE 65 *QJ* and CT (*cont.*)

Original term	*QJ*	CT
Flat-plate keelson		平板副脊骨
Flat-scarph		平斜搭節
Flat-seizing		細繩平繞大繩
Flaws	蜂窩	
Flemish-eye		繩頭眼
Flemish-horse		桅横桿外踏繩
Flexible tube valves	韌性管舌門	
Flight of frames		船首尾肋骨底級
Float & feed apparatus	浮表添水器	
Float gauge	浮表	
Float for boiler	鍋爐浮表	
Float in boiler	鍋爐內浮表	
Float rods	明輪翼柄	
Floating-anchor		浮錨
Floating steam fire engine	浮水面滅火汽機水龍	
Floats	輪翼	
Floats of paddle wheel	明輪翼	
Floats outside of	明輪翼外端	
Flogging-hammer	向前打錘	向前打錘
Flood-anchor		漲潮錨
Floor		肋骨中節, 或脊骨連肋骨平處
Floor-ceiling		肋骨中節[224] 板面
Floor-dowel		肋骨中節通捎
Floor-head		肋骨中節頭
Floor-head chock		肋骨中節頭補塊
Floor-plate		肋骨中節上板
Floor-rider		肋骨中節彎膝
Floor-timber		肋骨中節
2 Flue boiler	空篃鍋爐	空篃鍋[225] 爐
Flue brush	火路刷子	
Flue cleaner	修淨火路器	
Flue hammer	火路錘	

224 Before the character *ban* 板, a cancelled *nei* 内 can be read.

225 The right part of the character corresponds to that of the simplified version *guo* 锅.

TABLE 65 *QJ* and CT (*cont.*)

Original term	*QJ*	CT
Flue heating surface	空筩火切面	
Flue plate	火路板	
Flue scraper	火路刮器	
Flue surface	火路面積	
Flues	鍋爐火管, 或空筩	
Fluid	流質	
Fluke of an anchor		錨鈎掌
Flush-deck		平艙面
Flush-decked vessels		平艙面船
Flush-deck pump		艙面中[226] 水筩
Flush flue	平砌火路	
Flush-head rivet	板面平齊帽釘	平頭帽釘
(Page 28) Flush-joint		平節
Fly governor	飛輪汽制	
Fly wheel	飛輪, 或外輪	
Fly-wheel (of a pump)		起水筩飛輪
Fly wheel rim	飛輪牙	
Flying-bridge		飛橋, 或連台橋
Flying-jib		三節前伸桿帆
Flying-jib boom		三節前伸桿, 或三喘頭弼標
Flying-jib boom foot-rope		三節前伸桿踏繩
Flying-jib boom guy		前伸桿三節[227] 中牽繩
Flying-jib boom martingale stay		三節前伸桿下牽繩
Flying-jib downhaul		三節前伸桿落帆繩

226 In the translation, different corrections can be read; it is not easy to ascertain whether Fryer meant to cancel the character *zhong* 中 or not.

227 The employment of these characters is presumably a mere oversight by Fryer. In the other entries in which these characters were included, *qian shen gan san jie* 前伸桿三節 was annotated on the manuscript and an arrow changed the order of the characters in *san jie qian shen gan* 三節前伸桿; in this entry the order *san jie qian shen gan* was instead correct. Nevertheless, plausibly by analogy with the other entries, Fryer has inverted the order anew, bringing it back to *qian shen gan san jie*. More specifically, the four characters within the Chinese rendering of the entries "Flying-jib boom martingale stay," "Flying-jib sheet," and "Flying-jib stay" are annotated in the correct order since the first writing and do not present corrections; the translation of "Flying-jib tack" shows instead the order *qian shen gan san jie*, without revisions.

TABLE 65 *QJ* and CT (*cont.*)

Original term	*QJ*	CT
Flying-jib halliard		三節前伸桿扯帆繩
Flying-jib sheet		三節前伸桿拽帆繩
Flying-jib stay		三節前伸桿牽繩
Flying-jib tack		前伸桿三節帆下底角
Flying-kites		四層以上各橫帆
Flying-royal		四層桅飛帆, 或四層桅鬆帆
Flying topgallant-sail		三層桅飛帆, 或三層桅鬆帆
Fog-horn		重霧號喇叭
Fog-signal		下務號令
Folding-chair		能收放椅[228]
Follower	從輪, 或鞲韛蓋, 或軟墊臼[229] 蓋	
Following edge	螺輪翼之後邊	
Following-edge (of propeller blades)		
Foot (of a sail)		帆底邊
Foot-band (of a sail)		帆底邊帶
Foot-brail		過滑車底繩
Foot-gore (of a sail)		帆底補塊
Foot-grating (in a boat)		艇底木柵
Foot-lining (of a sail)		帆底襯布
Foot of propeller frame		螺輪架底
Foot path	馬路兩旁之人路	
Foot-rope (of a yard)		桅橫桿踏繩
Foot-rope (of a sail)		帆底邊[230] 繩
3 Foot-valve	底門, 或底舌門	底門
Foot-waling		隔水旁
Force	力	力

228 The four characters of the Chinese translation in the manuscript are rather confusing; the four listed in table 65 are the author's deduction.

229 In the *QJ*, the character is incomprehensible; the manuscript is also not helpful in clarifying it. The character in the cell, *jiu* 臼, is the result of a comparison conducted by the author with entries listed in Fryer and Xu, *Qiji biyi*.

230 In the manuscript, the graphical variant of the character *bian* 邊 presents, above *chuo* 辶, the upper part of *gao* 高, without *kou* 口, connected with *jia* 家; this graphical variant is extremely rare.

TABLE 65 *QJ* and CT (*cont.*)

Original term	*QJ*	CT
Force accelerating	漸增速力	
Force bending	彎力	
Force constant	常力	
Force expansive	漲力	
Force moving	動力	
Force percussive	擊力	
Force pump rod	壓水箭桿	
Force shearing	剪力	
Force the fires	强加爐火力	
4 Force the fires (to-)		强爐火力
Force twisting	扭力	
Forced draught	進風强加火力	
5 Forcing-pump	壓水箭	壓水箭
6 Forcing-pump rod	壓水箭桿	壓水箭桿
Fore-and after (in a hatch-way)		艙口[231] 縱門
Fore-and aft sails		縱帆, 或後伸帆
Fore-and aft schooner		二杖桅縱帆船
Fore-balance frame		前半船相稱肋骨
Fore-body		前半體
Fore-boom		前桅後伸桿
Fore-boom topping-lift		前桅後伸桿弔繩
Fore-bowline		前下桅御帆總繩
Fore-brace		前下桅橫桿拉繩
Fore-buntline		前桅帆腹繩
Fore-gaff		前桅後伸斜桿[232]
Fore-castle		船首臺 或船首上艙
Forecastle-awning		首臺遮篷
Forecastle-beam		船首台橫樑
Forecastle bilts		首台絆繩柱
Forecastle-bulkhead		首臺隔艙

231 The character *kou* 口 has been added over another that was, presumably, *men* 門 or *shuan* 閂.

232 The order of *xie* 斜 and *gan* 桿 was originally the opposite; the correction has been realised at a later time.

TABLE 65 *QJ* and CT (*cont.*)

Original term	QJ	CT
Forecastle covering-board		首臺蓋板
Forecastle-deck		首臺艙面
Forecastle-deck-beam		首臺艙面横樑
Forecastle-deck-beam-clamp		首臺艙面横樑搭
Forecastle-deck-beam-shelf		首臺艙面横樑托板
Forecastle-deck-beam tie-plate		首臺艙面横樑連板
(Page 29) Forecastle-spirketting		首臺横樑上厚板
Forecastle-deck stringer		首臺艙面銕連條
Forecastle-ladder		首臺梯
Forecastle-rail		首臺欄杆
Forecastle-rails and stanchions		首臺欄杆并柱
Forecastle-skylight		首臺天牕
Fore chain-plate		前桅船边牽繩銕根
Fore-channel		前桅船边牽繩銕根輔板
Fore clew-garnet		前桅柱帆下角定滑車
Fore-course		前下桅帆
Fore-deck		前艙面
Fore foot		船前足
Fore front-rope; Fore-yard foot-rope		前桅横桿底踏繩
Fore futtock-rigging		前桅盤下梯
Fore-gaff		前桅斜後伸桿
Fore-gaff-topsail		前桅斜後伸上帆
Fore-hatchway or Fore-hatch		前艙口[233]
Fore-hold		前貨艙
Fore-hoods (of planking)		船底最前内外板
Fore-leech (of a sail)		帆前邊帶

233 Before the character *kou* 口, *men* 門 can be read, even if cancelled; also the word "or" within the original lemma has apparently been added at a later time.

TABLE 65 *QJ* and CT (*cont.*)

Original term	*QJ*	CT
Fore-leech-line		帆前邊連綫
Fore-leech-rope		帆前邊繩
Fore-lift		前桅橫桿弔繩
Fore-lock		錨橫桿捎
Fore lower-rigging		前下桅繩梯
Fore lower-shrouds		前下桅繩 梯直條
Fore lower-studding sail		前下桅伸桿帆
Fore mast partner plate		前桅孔旁板
Fore lower-studdingsail [sic] downhaul		前下桅伸桿收帆繩
Fore lower studding sail halliard		前下桅伸桿扯帆繩
Fore lower-studding sail sheet		前下桅伸桿拽帆繩
Fore-mast		前桅
Fore-mast cap		前桅節箍[234]
Fore-mast crosstrees		前桅盤[235] 橫木
Fore-mast hole		艙面容前桅孔
Fore-mast pole		前桅頂桿
Fore-mast trestle-trees		前桅盤縱木
Foremost-backstay		最前左右牽繩
Foremost-beam		最前橫樑
Foremost-bulkhead		最前隔墻
Foremost-floor		最前肋骨中節
Foremost-frame		最前肋骨
Foremost-mast		最前桅
Foremost-shroud		最前繩梯直條
Fore-peak		船首底
Fore-peak-halliard		前桅後伸斜桿扯帆繩
Fore-peak tank		船首夹處[236] 水箱

234 In the character *gu* 箍, *mu* 木 is used in place of *shou* 扌; the alternation between the two radicals is also constantly found in the other *TVM*'s glossaries.

235 In the character *pan* 盤, the top right part only presents the upper component of *shu* 殳, to wit *ji* 几.

236 Originally *di* 底 could be read in place of the two characters *jia* 夹 and *chu* 處.

TABLE 65 *QJ* and CT (*cont.*)

Original term	*QJ*	CT
Fore-reaf-tackle		前桅減帆全套轆轤
Fore-rigging		前下桅繩梯
Fore-royal		前四層桅帆
Fore-royal backstay		前四層桅左右牽繩
Fore-royal brace		前四層桅橫桿拉繩
Fore-royal buntline		前四層桅帆背繩
Fore-royal clew-line		前四層桅帆角繩
Fore-royal halliard		前四層桅扯帆繩
(Page 30) Fore-royal lift		前四層橫桿弔繩
Fore-royal mast		前四層桅
Fore-royal pole		前四層桅頂桿
Fore-royal sheet		前四層桅拽帆繩
Fore-royal stay		前四層桅前牽繩
Fore-royal studdingsail		前四層桅伸桿帆
Fore-royal studdingsail boom		前四層桅伸桿
Fore-royal studdingsail downhaul		前四層桅伸桿落帆繩
Fore-royal studdingsail sheet		前四層桅伸桿拽帆繩
Fore-royal studdingsail tack		前四層桅伸桿帆底角繩
Fore-royal-yard		前四層桅橫桿
Fore-sail		前下桅帆
Fore-sail-boom		前下桅伸桿
Fore-sheet		前下桅拽帆繩
Fore-shift of planking		外板向前參差縫
Fore-skysail		前頂桅帆
Fore-skysail-backstay		前頂桅左右牽繩
Fore-skysail-brace		前頂桅橫桿拉繩
Fore-skysail-clew-line		前頂帆下角繩
Fore-skysail-haillard		前頂桅扯帆繩
Fore-skysail-lift		前頂桅橫桿弔繩
Fore-skysail-mast		前頂桅
Fore-skysail-pole		前頂桅頂桿
Fore-skysail-sheet		前頂桅拽帆繩
Fore-skysail-stay		前頂桅前牽繩

TABLE 65 *QJ* and CT (*cont.*)

Original term	*QJ*	CT
Fore-skysail-yard		前頂桅横桿
Fore-spencer		前桅後伸帆
Fore-stay		前下桅前牽繩
Fore-staysail		前桅前牽 繩帆
Fore-staysail-halliard		前桅前牽 繩 帆之扯帆繩
Fore-staysail-downhaul		前桅前牽 繩 帆之落帆繩
Fore-staysail-sheet		前桅前牽 繩帆之拽帆繩
Fore-tack		前桅帆底角繩
Fore-top		前桅盤
Fore-top-bowline		前桅盤御帆總繩
Fore-topgallant backstay		前三層桅左右牽繩
Fore-topgallant-bowline		前三層桅御帆總繩
Fore-topgallant-buntline		前三層桅帆腹繩
Fore-topgallant-brace		前三層横桿拉繩
Fore-topgallant-clew-line		前三層桅帆下角繩
Fore-topgallant-crosstrees		前二節桅盤横木
Fore-topgallant-futtocks		前二節桅盤下梯
Fore-topgallant-halliard		前三層桅扯帆繩
Fore-topgallant-lift		前三層横桿弔繩
Fore-topgallant-mast		前三層桅
Fore-topgallant-mast-cap		前二節[237] 桅箍
Fore-topgallant-pole		前三層桅頂桿
Fore-topgallant-rigging		前三層桅繩梯
Fore-topgallant-sail		前三層帆
Fore-topgallant-sheet		前三層桅拽帆繩
Fore-topgallant-stay		前三層桅前牽繩
Fore-topgallant-studding-sail		前三層伸桿帆
Fore-topgallant-studding-sail boom		前三層伸桿

237 Under *er* 二 and *jie* 節, *san* 三 and *ceng* 層 can be read, even if cancelled. The subsequent character *gu* 箍 is a graphical variant with *mu* 木 in place of *shou* 扌; moreover, in the manuscript a cancelled *jie* 節 can be read before *gu* 箍.

TABLE 65 *QJ* and CT (*cont.*)

Original term	*QJ*	CT
(Page 31) Fore-topgallant-studding-sail downhaul		前三層伸桿落帆繩
Fore-topgallant-studding-sail halliard		前三層伸桿扯帆繩
Fore-topgallant-studding-sail sheet		前三層伸桿拽帆繩
Fore-topgallant-studding-sail tack		前三層伸桿帆底角繩
Fore-topgallant-tye		起前三層横桿鏈索
Fore-topgallant-trestle-trees		前三層桅盤縱木
Fore-topgallant-yard		前三層桅横桿
Fore-topmast		前二層桅
Fore-topmast-backstay		前二層桅左右牽繩
Fore-topmast-cap		前桅下節箍
Fore-topmast-cross-trees		前下桅盤横木
Fore-topmast-pole		前二層桅頂桿
Fore-topmast-rigging		前二層桅繩梯
Fore-topmast-stay		前二層桅前牽繩
Fore-topmast-staysail		前二層桅前牽繩帆
Fore-topmast-staysail downhaul		前二層前牽繩落帆繩
Fore-topmast-staysail halliard		前二層前牽繩扯帆繩
Fore-topmast-staysail sheet		前二層前牽繩拽帆繩
Fore-topmast-staysail tack		前二層前牽繩放帆繩
Fore-topmast-studding sail		前二層伸桿帆
Fore-topmast-studding-sail boom		前二層伸桿
Fore-topmast-studdingsail downhaul		前二層伸桿落帆繩
Fore-topmast-studding-sail halliard		前二層伸桿扯帆繩
Fore-topmast-studding-sail sheet		前二層伸桿拽帆繩

TABLE 65 *QJ* and CT (*cont.*)

Original term	*QJ*	CT
Fore-topmast-studding-sail tack		前二層伸桿帆底角繩
Fore-topmast-trestle trees		前下[238] 桅盤縱木
Fore-topsail		前二層帆
Fore-topsail-brace		前二層橫桿牽繩
Fore-topsail-buntline		前二層桅帆腹繩
Fore-topsail-clew-line		前二層帆下角繩
Fore-topsail-halliard		前二層扯帆繩
Fore-topsail lift		前二層橫桿弔繩
Fore-topsail reef-tackle		前二層減帆轆轤
Fore-topsail sheet		前二層桅牽繩帆拽帆繩
Fore-topsail tye		起前二層橫桿鏈索
Fore-topsail yard		前二層橫桿
Fore-top-staysail		前二層桅前牽繩帆
Fore-top-staysail downhaul		前二層桅前牽繩落帆繩
Fore-top-staysail-halliard		前二層桅前牽繩扯帆繩
Fore top staysail sheet		前二層桅前牽繩拽帆繩
Fore-trysail		前後伸帆
Fore-trysail-bail		前後伸桿帆繩
Fore-trysail-gaff		前下桅後斜伸桿
(Page 32) Fore-trysail-inhaul		收[239] 前桅後伸帆繩
Fore-trysail-outhaul		開前桅後伸帆繩
Fore-trysail-sheet		拽前桅後伸帆繩
Fore-trysail-vang		前桅後伸斜桿下[240] 牽繩
Fore-waterballast-tank		前壓儎水箱
Fore-yard		前桅下橫桿
Fore-yard foot-rope		前桅下橫桿踏繩
Fore-yards		前桅各層橫桿
Forge	打熟鐵爐	
Forge-tongs	打熟鐵爐鉗	銕匠鉗
Fork, connecting rod	搖桿叉形端	

238 A cancelled *ceng* 層 can be read after *xia* 下.

239 The character *shou* 收 in this translation and *kai* 開 in the following were originally inserted after *qian wei* 前桅, with an arrow placing them in the correct position.

240 The character *xia* 下 has been added at a later time; the same applies to the translation for "Fore-yard" further below.

TABLE 65 *QJ* and CT (*cont.*)

Original term	*QJ*	CT
Fork & lashing eye		弔斜伸桿之叉端與眼
Fork head	叉形頭	
7 Fork-link	叉形進退桿	叉形連桿
Fork of connecting rod	搖桿叉形端	
Fork of rod	桿之叉支端	
Fork-wrench	叉支起子	叉形起子
Forked connecting rod	有叉形之搖桿, 或雙端搖桿	
Formula	算式	
8 Forward crank-shaft	前曲拐軸	前曲拐軸
Foundation	根基, 或座	
Foundation bolt	根基螺釘	
Foundation (of engines and boilers)		
9 Foundation-plate	根基地板	根基地
0 Foundation-plate bolt	根基鐵板螺釘	根基铁板螺釘
1 Four bladed propeller	四翼螺輪	四翼螺輪
Four bladed screw propeller	四翼螺輪	
2 Four cylinder engine	四汽筩汽機	四汽筩机器
3 Four cylinder tandem engine	四汽筩魚貫汽機	四汽筩魚貫机器
Four fold purchase		全套四層轆轤
Four masted ship		四枝桅船
Four sheave block		四車轆轤
Four stranded rope		四股繩
4 Four-way cock	四路塞門	四路塞門
Foxes		手紋綁物繩[241]
5[242] Fracture	折	折
Frame		正肋骨
Frame angle-iron		肋骨角銕
Frame-bolt		肋骨螺釘
Frame for locomotive	汽車輿架	

241 The translation was originally annotated as *shou wen sheng* 手紋繩, with a cancelled character before *sheng* 繩.

242 In this entry, as well as in "Fresh-water-tank" and "Friction," the number written in pencil is covered by a sign in red marker. It is plausible to infer that Fryer wanted to cancel these numbers, since entries with numbers beside them never include red signs.

TABLE 65 *QJ* and CT (*cont.*)

Original term	*QJ*	CT
Frame for propeller	螺輪架	
Frame-liner		肋骨襯, 肋骨與鐵板之補料
Frame-rider		肋骨斜牽板
Frame-saw	有邊鋸	有邊鋸
Frame solid sides of	架樑實心邊	
Framing	架座	
Framing central line of	架座中線	
Framing of an engine		
Framing of a winch[243]	絞車之架	絞車架
Framing of body of railway cars	輿上架	
Framing sides of	架邊	
Framing two sides of	架座之兩邊	
Free-bord		從水面至[244] 欄杆底之船板
6 Free-expansion	任意自機	任意自機
Freeing-port		放漏水孔, 欄杆門
7 Freezing-point	冰界	冰界
Freight car	貨車	
Freight engine	運貨汽機	
French-sennit		法國鬆緶繩
French-shroud-knot		法國繩中接雙結
8 Fresh-water-pipe	通淡水管	通淡水筩
9 Fresh-water-tank	淡水池, 或淡水箱	淡水池
o Friction	磨阻力	面阻力
Frictional-resistance		磨阻力
1 Friction-brake[245]	磨阻力停器	磨阻力停器
2 Friction-clutch	磨阻力接脫之筍	磨阻力夾器
Friction cone	磨阻力圓錐形輪	
3 Friction-coupling	磨阻力接盤	磨阻力接節

243 In the *QJ*, "Framing of a winch" is listed as "Framing of winch," therefore inserted one entry below.

244 The character *zhi* 至 has been added at a later time.

245 In the *QJ*, "Friction-brake" is listed as "Friction break"; furthermore, in the characters *ting* 停 within the CT, *tou* 亠 and *kou* 口 are united, similarly to the upper part of *gao* 高.

TABLE 66 STFNC
"Syllabary for the transfer of foreign names into Chinese," page 5

"Syllabary for the transfer of foreign names into Chinese," page 5

			Dĭ-ăn. 典[246]
Chŏn. 真		Dăn. 旦	Dĭ-ĕ. 疊
Choo. 朱	D. 德	Däng. 當	Dĭ-ĕn. 典
Choon. 準	Dā. 德	Dăng. 當	Dĭm. 典
Chow. 召	Dä. 達	Dä-ō. 道	Dīn. 典 Dĭn. 典[247]
Chū. 朱	Da[248] 道	Dē. 第	Dĭ-ow. 調[249]
Chŭ. 遮[250]	Dă. 達	Dĕ. 德	Diow. 調[251]
Chŭm. 真	Dām. 典	Dĕm. 典	Ding. 定
Chŭn. 真	Däm. 旦	Dēn. 典	Dĭ-ō. 丟
Chŭng. 正	Dăm. 旦	Dĕn. 典	Dĭ-oo. 丟
Chwā. 追	Dān. 典	Dī. 代	Dĭ-ŏn 典
Chwä. 撾	Dän. 旦	Dĭ. 第	Dj. 支

TABLE 67 STFNC, second copy
"Syllabary for the transfer of foreign names into Chinese," second copy, page 1

"Syllabary for the transfer of foreign names into Chinese," second copy, page 1

Ā. 哀	Bä. 巴	Bĕng. 迸	Bŏm. 笨
Ä 阿	2 Ba. 保	Bī. 敗	Bōn. 笨
2 A 奧	Bă. 巴	Bĭ. 比 Bĭ-ăn 卞[252]	Bŏn. 笨

246 The first line has in reality been added here only to include this entry in table 66; in the manuscript, "Dĭ-ăn. 典" has indeed been annotated beyond the lines of the page considered here.

247 "Dĭn. 典," cancelled, is in reality annotated in the line between the precedent and the subsequent line. The STFNC page reported in this table, contrary to the next one, is indeed written leaving an empty line between every line containing the phonetic representation and its correspondent character.

248 A diacritic corresponding to two dots can be found below "A"; in the subsequent entries in table 66, the diacritic will be indicated with "2" before the entry.

249 The whole entry has been cancelled in the manuscript.

250 In place of the radical *huo* 灬, the character *zhe* 蔗 presents the graphical variant composed of two *ren* 人.

251 Beside the entry, a cancelled character, *dian* 典, can still be read.

252 "Bĭ-ăn 卞" has been added in the space between the precedent entry, "Bĭ. 比," and the following one, "Bŏn. 笨."

TABLE 67 STFNC, second copy (*cont.*)

"Syllabary for the transfer of foreign names into Chinese," second copy, page 1			
Ă. 阿	Bäm. 班	Bĭ-ĕ. 別	Bōng. 迸
Ām. 恩	Băm. 班	Bĭ-ĕn. 卞	Bŏng. 迸
Äm. 安	Bān. 笨	Bĭm. 賔	Boo. 布
Ăm. 安	Bän. 班	Bĭn. 賔	Boon. 笨
Ān. 恩	Băn. 班	Bĭng. 丙	Bow. 保
Än. 安	Bäng. 邦	Bĭ-ō. 表	Bū. 布
Ăn. 安	Băng. 邦	Bĭ-ŏn. 卞	Bŭ. 伯
Äng. 昂	Bē. 比	Bō. 波	Bü. 比
Ăng. 昂	Bĕ. 備	Bŏ. 波	Bŭn. 笨
	Bĕm. 笨	Bö. 伯	Bün. 賔
B. 伯	Bēn. 賔	Boi. 波伊 or 波	Bŭng 迸
Bā. 備	Bĕn. 笨	Bōm. 笨	

TABLE 68 BTWDCL
"Botanical Terms from William's Syllabic Dictionary of the Chinese language," page 3

"Botanical Terms from William's Syllabic Dictionary of the Chinese language," page 3			
501	Air-plants[253]	吊蘭 : 風 ![254]	
1071	Alder	赤楊	The alder, a species with red bark.
953	Algae	海藻	Algae; long leaved seaweed.
855	"	蕁	A kind of marine a. or delicate sea-weed likened to hair.
848	" (species)	苔菜	A specs of A used for food.
833	Alliaceous plants	蒜	A. plts with ligulate leaves, called 葷菜 or rank herbs by priestly people.

253 The first column includes numbers indicating the page of Williams' dictionary in which the characters of the third column are annotated.

254 The colon indicates that the two were different translations, while the sign similar to an exclamation point specifies that the second character is the same as the second of the precedent translation.

TABLE 68 BTWDCL (*cont.*)

"Botanical Terms from William's Syllabic Dictionary of the Chinese language," page 3

1024	" "	葱	A gen term for A. plts with fistular leaves; onions, garlies [sic]; a leek green.
535	Allium (species)	菱 (薤)	A spec of scallions or Allium.
414	" "	麥門冬	A spec of A like the bulb bearing tree onion, wh produces bulbs on the stems;
188	" (wild ")	山薤	A wild spec of A without bulbs.
188	" ascalonium	薤	A spec of onion, the shallot or scallions (a: as:) with fistular leaves, com in Hu-kwuang; it is forbidden to those who fast; (name of a mat).
1024	" "	胡葱	Scallions (a. as:)
1056	" porrum	薈	A spec of leek or squills (a. por:), called 鴻 ! used as a pot-herb
1024	" schœnoprasum	細香葱	Chives (a. sch:)
414	" setaceum or uliginosum	韭 (韮)	A plt wh grows a long time fr one root, perhaps denoting especially the A. set: or ulig:; scalliols [sic] or chives; a salad onion, with fistular ligulate leaves a[255] minute bulbs.
497	Allspice, Japan	蠟(蜡) 梅花	Japan Allspice or the Chimonanthes fragrans.
539	Aleurites (nuts of)	石栗	Nuts of the Aleurites.
209	Almonds	杏	杏兒 name includes the sorts of Prunus genlly, almonds a plums
209	"	杏仁	Almonds; also apricot pits
584	Almond (flowering)	梅	A gen name for plums, prunes, a the bullace; the flowering al:

255 The sign reported here is unrecognisable; it might be an "&" written in cursive to indicate "and."

TABLE 69 "Vegetable kingdom ect of Shanghai"

Page 2 Williams Dic:			
	Coarse hemp	苧蔴	
	Coriander	芫荽	
	Cotton	木棉	There are 2 spec of C. called the 木 arb-oraceous, a the 草 herbaceous
	Cucumber	王瓜	
	" (bitter)	苦瓜	
	" (northern)	北瓜	
766 233	Dendrobium ceraia	石蔛	one of the dry land herbs a flos wh are com here.
	Earthnail plant	地丁草	med plt, wh beaten out a spread over the pt affected is reported to cure nameless poisons.
	Edible flag (a sort of)	茭白	
	" rush (a species of)	蔞蒿	
	Egg-plant	茄	
	Figs	無花果	wh are also called 優鉢曇
	Flag (a sort of edible)	茭白	
	Foreign ginseng	洋參	com in this district
	Fragrant boletus	香蕈	or wood ears
	" tars	香芋	
	Garlic	蒜	
	Ginger	薑	

TABLE 70 "Sz-Ch'uan plants—List-no 1"

Page 1		
1	Abelia serrata, S a[256] Z.	蓪花梗
2	Abutilon Avicennae, Gaertn.	桐麻
3	Aleurites vernicia, Hassk.	桐子
4	Ammannia rotundifolia, Roxb.	水莧菜

256 Here "a" has the meaning "and."

TABLE 70 "Sz-Ch'uan plants—List-no 1" (*cont.*)

Page 1		
5	Arctium maius, Schk.	牛榜
6	Boltonia indica, Bth.	魚秋竄兒
7	Bromus japonicus, Thunb.	野梅簽
8	Canna indica, Linn.	水芭蕉
9	Cephalanthus naucleoïdes, D, C.	風箱 or 獺(pro tᶜa') 狗耳
10	Chrysanthemum Roxburghii, DC.	銅蒿
11	Cnicus segetum, Bth a Hook. f.	鵝居婆
12	Commelyna communis, Linn.	竹葉菜
13	Convolvulus acetosifolius, Turez.	面根籐兒 or C. sepium, Linn.
14	" (Calystegia) acetosifolius, Turez.	來江籐兒
15	" sepium, Linn.	面根籐 or 鷄屎條
16	Cornus paucineruis, Hance.	臭黃荊[257] or 南天種
17	Crepis japonica, Benth.	雀雀草 (or 菜) (or 包) (or 台) or 牛屎花 or 牛石花 or 鴨屎條
18	Daucus carota, Linn.	山羅卜[258]
19	Dichrocephalia latifolia, Dc.	山胡椒
20	Eclipta alba, Linn.	麥豆草

TABLE 71 "Sz-Ch'uan plants—List. No ...2"

Page 2		
13	Ficus	山地瓜 or 翦刀子兒
14	Frasinus	白蠟樹
15	Geranium	老灌草
16	Gnaphalium	清明菜
17	Hypericum	對葉草
18	Indigofera	苡威油
19	Iussiaea	對對草 or 過江藤

257 Between the two Chinese translations reported in table 70, another one, *ji shi tiao* 鷄屎條 can be read, even if cancelled; every character of *ji shi tiao* is followed, on the top right, by a small circle.

258 The character *bu* 卜 presents the radical *cao* 艹 on the top.

TABLE 71 "Sz-Ch'uan plants—List. No …2" (*cont.*)

Page 2		
20	Lactuca	萵笋
21	Lespedeza	串子草 or 野瓜門兒
22	Linnanthemum	午時 or 午心花
23	Lysimachia	野刷把
24	Panicum	水子米
25	Phaseolus	黃豆
26	"	四季豆
27	Phylanthus [sic]	算盤子兒
28	Pinus	松
29	Polygonum	青蓼 or 蓼子草 (or 菜)
30	Pteris	鳳尾 or 冷蕨鷄
31	Ranunculus	貓脚草
32	Rosa	金英子 or 塘古兒
34	Rumex	土大黃 or 牛耳 (黃 or) 朶
33	Rosa[259]	青皮子

TABLE 72 *RMDM* manuscript and its printed version, page 1
Synoptic table including page 1 of the *Vocabulary of Proper Names*, corresponding to page 1, first count, of the Vocabulary of proper names occurring in Royle & Headland's "Materia Medica & therapics"

Original term	***RMDM***	***RMDM* manuscript, first version**	***RMDM* manuscript, second version**
Abel (mustard)	亞比勒	阿里伯	阿伯里
Abernethy (dark sky blue)	阿伯內替	阿伯內替	
Abercrombie (dark sky blue)	阿白克浪比	阿白克浪比	
X[260] Abercrombie		阿白克浪比	

259 An arrow indicates the need to insert "Rosa" in the line above.

260 In table 72, the sign "x" precedes all entries that have been cancelled in the manuscript. In numerous cases, these entries were repetitions of ones already present, and therefore

TABLE 72 *RMDM* manuscript and its printed version, page 1 (*cont.*)

Original term	*RMDM*	*RMDM* manuscript, first version	*RMDM* manuscript, second version
Abich (olive green)	阿比克	阿比[261] 克	
Adans	阿但司	阿但司	
Adderley (purple)	阿特里	阿特里	
Addison (dark sky blue)	阿的生	阿的生	
X Addison (purple)		阿的生	
Aegineta Paulus	依幾尼他保羅司	依幾尼他保羅司	
Aetius	依替由司	依替由司	
Aettinger	依丁格	依丁格	
Agricola	阿格里哥賴	阿格里哥賴	
Ahmed Shah	阿米特沙	阿米特沙	
Aikin	愛根	愛根	
Ailsa, Marquis (purple)	愛勒薩侯	愛勒薩侯	
Ainslie	恩司里	恩司里	
Aitken (dark sky blue)	阿欽	阿欽[262]	
Alan (purple)	阿蘭	阿蘭	
Albertus Magnus	阿白佗司馬搿奴司	阿白佗司馬搿奴司	
Alderson	阿勒突生	阿勒突生	
X Alderson (purple)		阿勒突生	
Alexander	亞勒散得	亞勒散得	
Alison (purple)	阿利孫	阿利生	阿利孫
Allen	阿林	阿林	
X Allen (olive green)		阿林	
X Allen (dark sky blue)		阿林	
Almeida Dr.[263]	阿勒米達	阿勒米達	

were excluded from the *RMDM*, while others, such as “Aix,” have been included in the printed version.

261 Before the character *bi* 比, a correction was surely present, but it is unrecognisable.

262 Before the character *qin* 欽, another can be read, but it is unrecognisable.

263 In the *RMDM*, “Almeida Dr” will be listed as “Almeda.”

TABLE 72 *RMDM* manuscript and its printed version, page 1 (*cont.*)

Original term	*RMDM*	*RMDM* manuscript, first version	*RMDM* manuscript, second version
Alpine	亞勒品	亞卑斯山者	
Alsop	阿索白	阿索白	
Ambrose (purple)	安白路斯	安白路斯	

TABLE 73 *RMDM* manuscript and its printed version, page 14
Synoptic table including page 14 of the *Vocabulary of Proper Names*, "Names of places," corresponding to page 1, first count, of the Vocabulary of proper names occurring in Royle & Headland's "Materia Medica & therapics," Names of Places

Original term	*TVM*	*RMDM* manuscript, first version	*RMDM* manuscript, second version
Abai	阿背河	阿背河	
Abuarisch	阿婆阿里施	阿婆阿里施	
Abydos (purple)	阿比佗司	阿[264] 比佗司	
X Abyssinia (mustard)		阿皮西尼亞	
X Abysinnia (wine)		阿比西尼亞	
Abyssinnia[265]	阿比西尼	阿比西尼	
X Abyssinia (dark sky blue)		阿比西尼亞	
X Abysinnia (purple)		阿比西尼[266]	
Abysinnian	阿比西尼者	阿比西尼者	
Abyssinnians	阿比西尼人	阿比西尼人	
Acajutla	阿揩由他拉	阿揩由他拉	
Acapulco	亞加補可	亞加補可	
Adalia	阿大利亞	阿大利亞	
Adel	阿地立	阿地立	
Aden	亞丁	亞丁	

264 Between *a* 阿 and *fei* 非, *er ti ke* 尔替克 can be read; the three characters are cancelled with a small dot, perhaps indicating that this strip of paper had been previously used.

265 In the *RMDM*, "Abyssinnia" will be erroneously listed as "Abyinnia."

266 Before this translation, another one can be read, namely *oudaliya* 區大利亞; these characters are cancelled with a small dot.

TABLE 73 *RMDM* manuscript and its printed version, page 14 (*cont.*)

Original term	*TVM*	*RMDM* manuscript, first version	*RMDM* manuscript, second version
X Aden (dark sky blue)		亞丁	
Aden cape	亞丁角	亞丁角	
Afghanistan	阿富汗	阿富汗	
Afghans	阿富汗人	阿富汗人	
Africa (wine)	阿非利加	阿非利加	
X Africa		阿非利加	
X Africa (mustard)		阿非利加	
X Africa (dark sky blue)		亞非加利	亞非利加
X Africa (purple)		阿[267] 非利加	
African	阿非利加者	阿非利加者	
X African (dark sky blue)		阿非利加者	
Africans	阿非利加人	阿非利加人	
Agra (dark sky blue)	亞加拉	亞加拉	
Airthrey (dark sky blue)	愛阿得來	愛阿得來	
X Aix		愛斯	
Aix (dark sky blue)	愛克斯	愛克斯	

TABLE 74 LGN
"List of geographical names," page 1

"List of geographical names," page 1			
Aar or Aren	阿耳 或作 阿連	Acapulco	阿嘎布勒哥
Abaco	阿巴哥	Acaray	阿嘎來
Abbeokuta	阿表古達	Acheen	阿真
Abbitibe	阿比第比	Aconcagua	阿根嘎瓜
Aberdeen	阿伯典	Acre	阿革
Aberistwith	阿伯利綏	Adalia	阿達利亞
Abo	阿波	Adana	阿達那

267 The two characters *er* 尔 and *le* 勒, cancelled with a small dot, can be read after *a* 阿.

TABLE 74 LGN (*cont.*)

Abomey	阿波美	Adel	阿德勒
Abookir	阿布結	Aden	阿典
Abrolhos	阿波羅斯	Adelaide	阿第雷德
Abyssinia	阿比辛伊亞	Adige	阿第遮

TABLE 75 VGT, page 5
"Vocabulary of geological terms," page 5

Apatite (yellow)	Archegosaurus monor (red)
Apennines, origin of (brown)	Archeoniscus (yellow)
Aphanite (light green)	Archeopteryx (yellow)
1 Aphides[268] (blue)	Archimedes revers (brown)
Apiocrinite (yellow)	Arching (light violet)
Apiocrinites rotundus (red)	Arctic climate in coal area (brown)
Appalachian coal (brown)	Arctic climate in Cenozoic (brown)
Apparatus (purple)	Arctic climate in Carboniferous (brown)
Aptychus latus (red)	Arctic climate in Jurassic (brown)
Apus dubius (red)	Ardeche lava (red)
Aquapendente (red)	2 Arenaceous (blue) 沙石
1 Aqueous agency (blue) 水工	Arenaceous limestone (light green)
Aqueous[269] deposits (red)	2 Arenaceous rocks (blue) 沙石類
Aqueous erosion (red)	Arenig (red)
Aqueous rocks (blue) 經水石	2 Argillaceous (blue) 泥石
1 Aqueous works (blue) 經水石	Argillaceous flagstone (light green)
Aquitanien (red)	Argillaceous limestone (light green)
1 Arachnids (blue) 蜘蛛	1 Argillaceous Rock (blue) 泥石
Arachnoids (blue) 蜘蛛	2 Argillaceous rocks (blue) 泥石類
Aralo-Caspian formations (red)	Argillaceous sandstones (light green)

268 All the lemmata within the VGT have been typewritten, for the great majority in black ink. Numerous entries marked with blue, for which in the majority of cases translations are present, are instead annotated in ink more similar to blue. However, at times they are also annotated in black ink, or in other circumstances, the relevant Chinese translation is not present. In table 75, if such entries are written in blue ink, they are preceded by "1"; if in turn they are annotated on yellowed strips of paper, they are preceded by "2."

269 "Aqueous" is written in lower case, analogously to "Lepidurus." Further up, "Apparatus" was originally written in lower case. Nevertheless, it has been subsequently corrected, apparently with a typewriter.

TABLE 75 VGT, page 5 (*cont.*)

Araucaria (red)	Argillaceous schist (brown)
Araucarites (yellow)	Argile plastique (red)
Arbroth[270] paving stone (red)	Argillite (brown)
Archaeoniscus Brodiei (brown)	Arles or Earles (purple)
Archaeopterix macrura (red)	1 Arsenic (blue) 信石
2 Archean age (blue) 元生世	1 Artesian Wells (blue) 自流井
Archeocidaris (yellow)	Articulata (yellow)
Archegosaurus (yellow)	1 Articulates (blue) 節虫

TABLE 76 VGT, page 60

Laths (purple)	Lepidodendron primaevum (brown)
Lauberis slates (red)	Lepidodendron Sternbergu (red)
Launder (purple)	1 Lepidodendrons (blue) 鱗樹
Laurentian (yellow)	Lepidobostrobus ornatus (red)
1 Lava (blue) 噴流石	Lepidosteus (brown)
1 Lava beds (blue) 噴石床	Lepidostrobus (yellow)
Lava-cones (brown)	Lepidophyllum (yellow)
Lava streams (red)	Lepidotus gigas (red)
1 Laws of matter (blue) 質理	Lepidotus Mantelli (red)
Lay out (purple)	Lepidurus (brown)
1 Layer (blue) 一行	Leptaena depressa (red)
Layered (purple)	Leptaena transversalis (brown)
Laying out pit-eye (light green)	Leptena sericea (brown)
Lea-stone (purple)	Leptignite (red)
1 Lead (blue) 鉛	Leptina Moorei (red)
Leader (purple)	1 Level (blue) 平
Leading bank (purple)	Level-bearing (light green)
Leading Man (purple)	Level-free (purple)
Leaf-bed (red)	Level-tons (light violet [?])
Lean (purple)	Levels (purple[?])
Leap (purple)	Levels, thirlings (light green)
Leather-bed (purple)	Levels, timbering (light green)
Lecanocrinus elegans (brown)	Levels, walling (light green)

270 Between "o" and "t" within "Arbroth," a correction in pencil has been added, but the correction is cut and therefore incomprehensible.

TABLE 76 VGT, page 60 (*cont.*)

Led (purple)	Leys (purple)
1 Ledge (blue) 一墱石	1 Lias (blue) 縹牌
Leg (purple)	Lias, flower (light green)
Leguminosites (brown)	Lias, upper (light green)
Leperditia, Anna (brown[?])	Lias, white (light green)
Lepidodendra (brown)	
Lepidodendron (brown)	

TABLE 77 LBT, page 1
"List of botanical terms," page 1

Abelia serrata (red) 蒲花梗	Acacia (green and red) 聲息花
Abies (ultramarine blue) 樅[271]	Acacia (ultramarine blue) 銷寒花
Abies davidiana (indigo and green)[272] 馬尾松 牛尾松	Acacia concinna[273] (yellow ochre) 肥皂核
Abies firma 秾[274] 松	Acacia concinna (yellow ochre) 肥皂莢[275]
Abies homolepsis (ultramarine blue) 落葉松	Acacia julibrissin (white) 絨花樹
Abies homolepsis (ultramarine blue) 金錢松	Acacia nemu (ultramarine blue) 合歡
Abies leptolepsis (white) 杆 "[276]	Acacia nemu (yellow ochre) 夜合

271 The character *cong* 樅 presents the rare graphical variant without *chi* 彳 in the right part.

272 There are two translations reported for "Abies davidiana," as occasionally also happens in other pages of the LBT. The two renderings are at times listed one above the other, at times one beside the other. In the latter circumstance, when the space left is inadequate, it is frequently difficult to realise if they represent two different renderings or the same translation. In table 77, the colour of the legend close to the lemma is written in brackets, if present; other parentheses are those annotated by Fryer directly on the manuscript. A question mark beside the colour indicates those cases in which the colour itself has faded or is difficult to identify, particularly for the two groups of colours purple-ultramarine blue-ink, and red-purple and red.

273 Both this and the entry below are apparently annotated as "Concinnia"; the alternation and indecision among different spellings are rather common in the LBT. Mistakes are also frequent, for example in the two entries on page 40, "Coix lachryma" and "Coix lachrymalis," sepia, are originally annotated as "Coix lachyma" and "Coix lachymalis."

274 The character *nong* 秾 presents the radical *mu* 木 in place of *he* 禾.

275 The character is almost unrecognisable; *jia* 莢 is the author's deduction.

276 The sign is apparently the same used to indicate the reproduction of the same content of the line above; it is impossible to state it with the utmost certainty, as it is only attested in the LBT in this entry.

TABLE 77 LBT, page 1 (*cont.*)

Abrus precatorius (white) 相思豆	Acalypha gemina (indigo and green) 鐵莧菜
Abrus precatorius (yellow ochre) 紅豆	Acanthaceae (b) 艽 (277)
Abrus precatorius (ultramarine blue) 紅相思	Acanthaceae (purple[278]) 莨芳科
Abrus precatorius (yellow ochre) 想思子	Acanthopanax aculeatum (indigo and green) 三加皮
Abutilo Avicennae (red) 桐麻	Acanthopanax ricinifolium (indigo and green) 刺楸
Abutilon avicennae (indigo and green) 青麻	Acanthus ilicifolius (red) 老鼠力
Abutilon avicennae (purple) 茵麻	Acanthus ilicifolius (ultramarine blue) 莨花
Abutilon avicennae (indigo and green) 檾麻 茵麻	Acanthus ilicifolius (ink) 莨花芳
Abutilon hemp (white) 檾麻	Acer (white) 楓
Abutilon hemp (white) 黃麻	Acer japonicum (ultramarine blue) 明月
Abutilon hemp 蹟 (茼) 蔴	Acer oblungum (indigo and green) 飛蛾子樹
Abutilon hemp stalks 藒	Acer oblungum (indigo and green) 雞火樹
Abutilon indicum (indigo and green) 冬葵子	Acer trifidum (ultramarine blue) 楓雅
	Acer trifidum (ultramarine blue) 楓紅樹
	Achillea 芛 蓍芛

TABLE 78 LBT, page 40

Cocks' claw fruit (yellow ochre) 枳椇子	Coir palm[279] 椶樹
Cocks' claw fruit (yellow ochre) 鷄距子	Coir-palm (allied) (white) 栟
Cocksclaws (white) 鷄爪子	Coix (barley) (white) 苡薏米

277 In brackets, a variant to *jiao* 艽 is provided; it is represented by a character composed of the radical *cao* 艹 and what apparently is *dou* 斗, with only a stroke *dian* 丶; by analogy with the other following characters, though, it may be a cursive form of *niu* 牛.

278 The colour is very faded and seems to be ultramarine blue, but since the list corresponding to purple is characterised by the use of the character *ke* 科, it is more plausible that it is the latter colour.

279 In this and other entries, the interpretation of the colour is arduous; the colour is diffused and includes different tonalities, from ultramarine blue to purple.

TABLE 78 LBT, page 40 (*cont.*)

Cockscomb (ink) 雞冠花	Coix lachryma (indigo and green) 大豌子
Cockscomb ([280]) (white) 芊	Coix lachryma (purple and red?) 薏苡仁
Cockscomb flower (white) 鷄冠花	Coix lachrymalis (yellow ochre) 薏苡仁
Cockshead (white) 鷄頭	Coix lachrymalis (sepia) 薏仁米
Cock's head, seeds (sepia) 鷄頭子	Colldion (yellow ochre) 棉膏
Cocoa (yellow ochre) 哥哥	Colocasia antiquorum (purple) 芊
Cocoa (yellow ochre) 果膏茶	Colocasia antiquorum (purple and red) 芋頭
Cocoa-nut (yellow ochre) 椰子	Colocasia (species) (purple and red) 芋頭
Cocoa-nut (white) 椰(梛) 子	Coltsfoot (sepia) 欵[281] 冬花
Cocoa-nut dipper (white) 椰(梛) 子瓢	Comfrey (yellow ochre) 地黃
Cocoa-nutmilk (yellow ochre) 樹頭酒	Commelina (ultramarine blue) 鴨跖
Cocoa-nut pulp (white) 椰(梛) 肉	Commelina (purple?) 鴨跖草一種
Cocoa-nut shells (white) 椰(梛) 殼	Commelina (ink) 淡竹葉 鴨跖
Cocoa-nut toddy (white) 椰(梛) 花酒	Commelina Communis (purple?) 鴨[282] 跖草
Cocos nucifera (purple and red) 椰樹	Commelina medica (purple and red) 麥門冬
Codonopsis lanceolata (indigo and green) 奶樹	Commelina polygama (ultramarine blue?) 鴨跖草
Coffee (sepia) 架非	Commelinacea (purple?) 鴨跖草科
Coffee (yellow ochre) 咖啡茶	Commelinaceae (purple?) 鴨跖草科
Coir (sepia) 棕	Commelyna Bengalensis (yellow ochre) 火炭頭
Coir (sepia) 椶	Commelyna communis (indigo and green) 竹葉菜
Coir (white) 棕 (椶)	Commelyna communis (indigo and green) 水竹葉
Coir (sepia) 棕片	Commelyna medica (yellow ochre) 寸冬
Coir (sepia) 棕梠	Commelyna polygama (yellow ochre) 竹葉菜
Coir palm (white) 椶櫚	Commelyna polygama (yellow ochre) 鴨跖花

280 The term in brackets is scarcely recognisable; perhaps it is "Allied," by analogy with other entries within the same page.

281 The one reported here is the most common graphical variant of the character *kuan* 款 within the LBT.

282 The character *ya* 鴨 is written with a simplified *huo* 灬.

TABLE 79 VPN "Vocabulary of proper names," page 6

1 Bauer,[283] 波爾
1 Bauhin, C. 賽保希捺
1 Bauhin, J. 垂保希捺
2 Baumgartner 包麥加得捺
1 Baup, 波潑
2 Bavaria 不費利亞邦
1 Baxt, 巴格司德
2 Baxter 巴斯得
1 Bayard, 拜耶特
1 Bayle, 倍勒
1 Baynard, 貝那特
3 Bazaine 巴賽印
1 Beale, 比勒
2 Beardslee 比爾子里
2 Beathe 比阿蒂
2 Beatson 皮得生
2 Beaufort 波福的
2 Beaufort 布佛德
2 Beaufoy 步弗
1 Beaumont, 盤芒脫
2 Beauregard 波勒嘎得
3 Beauregard 波利嘎德
1 Beaujean 布正
2 Beccaria 白賈利亞
2 Beccaria 白加靈[284] 阿
2 Beccaria 白加利阿
1 Beccher, 比綽
2 Becher 白客
3 Belisarius 備力撒力牾
2 Bell 貝拉
1 Bell, 彼拉
1 Bellamy, 備拉美
2 Bellerophon 白勒勒分
1 Bellingham, 貝林哈末
1 Belloc, 比路克
1 Belon, 比倫
3 Belshazzar 伯沙撒
1 Beltz, 卑勒智
2 Benbow 奔薄
3 Benjamin 便雅閔
1 Bennach, 白那克
2 Bennett 彼尼得
1 Bennett, 白尼德
2 Benoit 丙奴華
1 Benoit, 貝奴埃
2 Benson 邊生
2 Bentham 本地麥
1 Bentham, 本他末
2 Bentick 變定格
2 Bentley 扁得利
1 Bentley, 本得里
2 Beortric 比奧的得
1 Beral, 畢賴勒
1 Bergen, 白爾根
1 Berger, 白爾格
1 Berghaus, 白高司

283 Entries within the *RMDM* manuscript, annotated on strips of paper, can be grouped into three categories: the first includes entries written on more yellowed strips of paper, with the original terms and their Chinese translations typewritten in black ink; the second comprises entries annotated on brighter strips of paper, with original terms and their Chinese translations handwritten in black ink; the third finally consists of entries annotated on more yellowed strips of paper, with original terms printed in blue ink and with Chinese renderings handwritten in black ink. In table 79, the three categories are represented respectively by numbers "1," "2," and "3" added before the entries. Finally, it should be pointed out that the entries of the first group usually include a comma, listed in table 79, where the original term ends.

284 The character *ling* 靈 presents the graphical variant with *wang* 罒 in place of three *kou* 口.

TABLE 79 VPN "Vocabulary of proper names," page 6 (*cont.*)

1 Becher, 倍叉
1 Beck, 白割
1 Becker, 白格
2 Becket 北革[285]
3 Becket, Thomas'a 卑克特[286] 多馬
1 Beclard, 貝克賴特
2 Becquerel 白格路
2 Becquerel 彼格路
1 Becquerel, 伯克來勒
2 Becquerel 白加路
2 Becquerel 白格路
2 Becquerel 執格路
1 Beddingfield, 貝丁非勒特
1 Beddoe, 皮度
2 Bede 比的
2 Beeby 畢比
2 Beechy 避器
1 Beissinhirtz, 倍性合爾茲
1 Bergmann, 罷格門
1 Berkeley 白克里
1 Berlin 栢林
2 Bernoulli 卜奴里
1 Bernt, 伯納脫
1 Berry, 白里
1 Bert, 白爾德
2 Bertha 伯爾他
2 Bertha 伯爾達
1 Berthelot, 白替路
1 Bertholet, 白拖來
1 Berthollet, 白土來
1 Bertrand, 白脫蘭特
1 Bernard, 白那德
2 Bernard 拔[287] 乃特
3 Bernard, (St)[288] 卑納德
2 Bernouilli 百納里
1 Bertrand, de Rois[289] 白脫蘭特得魯阿
1 Berzelius, 栢西里烏司
2 Berzelius 白雪利亞司
2 Berzilius 八西里烏司

285 In the graphical variant of the character *ge* 革, the final vertical stroke does not cut *kou* 口. Generally, the other handwritten characters which include *ge* as one of their components, such as *le* 勒, present the same peculiarity.

286 A space is left between the characters *te* 特 and *duo* 多, in order to separate the surname and name in the phonemic loan used for the Chinese rendering.

287 The character is scarcely recognisable; it might also be *ban* 扳.

288 "Bernard, (St)" is followed by the sign "x," annotated directly on the sheet of paper.

289 The entry presents a really peculiar feature: it is annotated on two lines, "Bertrand, de" and "Rois," with the new line within the Chinese translation corresponding to the new line also within the original term, plausibly to simplify the reading of the phonemic loan. Within other entries of the *RMDM* manuscript, Fryer also tries to help the person in charge of printing, with this and other modalities, to better understand the split words of the original term.

Bibliography

"Abstract of Minutes—School and Textbook Series Committee." *The Chinese Recorder and Missionary Journal, Volume XVIII*, no. 11, 441–42. Shanghai: Presbyterian Mission Press, 1887.

Ai, Suzhen 艾素珍. "Qingdai chuban de dizhixue yizhu ji tedian 清代出版的地质学译著及特点 [Translated works on geology published during the Qing dynasty and their peculiarities]." *Zhongguo keji shiliao* 中国科技史料—*China Historical Materials of Science and Technology* 19, no. 1 (1998): 11–25.

Allen, Young John, and Zheng Changyan 鄭昌棪, trans. *Gezhi qimeng gewuxue* 格致啟蒙格物學 [Science primer: physics]. Shanghai: Jiangnan zhizaoju 江南製造局, 1880.

Alleton, Viviane. "Chinese Terminologies: On Preconceptions." In *New Terms for New Ideas: Western Knowledge and Lexical Change in Late Imperial China*, edited by Michael Lackner, Iwo Amelung and Joachim Kurtz, 15–34. Leiden, Boston, Köln: Brill, 2001.

Alleton, Viviane, and Jean-Claude Alleton. *Terminologie de la chimie en chinois moderne*. Paris: Mouton, La Haye, 1966.

Amelung, Iwo. "Some Notes on Translations of the *Physics Primer* and Physical Terminology in Late Imperial China." *Wakumon* 或問, no. 8 (2004): 11–34.

Amelung, Iwo. "*The Complete Compilation of New Knowledge*, Xinxue beizuan 新學備纂 (1902): Its Classification Scheme and its Sources." In *Chinese Encyclopaedias of New Global Knowledge (1870–1930): Changing Ways of Thought*, edited by Milena Doleželová-Velingerová and Rudolf G. Wagner, 85–102. Heidelberg: Springer, 2013.

Anderson, Gerald H., ed. *Biographical Dictionary of Christian Missions*. New York: Simon & Schuster Macmillan, 1998.

Andres, Erwin. *A Practical Treatise on the Fabrication of Volatile and Fat Varnishes, Lacquers, Siccatives, and Sealing-waxes*. Philadelphia: Henry Carey Baird & Co.; London: Sampson Low, Marston, Searle & Rivington, 1882.

Andrews, Bridie. *The Making of Modern Chinese Medicine, 1850–1960*. Vancouver: University of British Columbia Press, 2014.

Andrews, Bridie. "Blood in the History of Modern Chinese Medicine." In *Historical Epistemology and the Making of Modern Chinese Medicine*, edited by Chiang Howard, 113–36. Manchester: Manchester University Press, 2015.

Arakawa, Kiyohide 荒川清秀. *Kindai nicchū gakujutsu yōgo no keisei to denpa: chirigaku yōgo wo chūshin ni* 近代日中学術用語の形成と伝播： 地理学用語を中心に [The formation and circulation of technical terminology in modern Japanese and Chinese: a focus on geographical terminology]. Tokyo: Hakuteisha 白帝社, 1997.

Arcodia, Giorgio Francesco. *Lexical Derivation in Mandarin Chinese*. Taiwan Journal of Linguistics—Book series in Chinese linguistics. Taipei: Crane Publishing Co., Ltd, 2012.

Bai, Limin. "Practicality in Curriculum Building: A Historical Perspective on the Mission of Chinese Education." *Frontiers of Education in China*, 8, no. 4 (2013): 518–39.

Beach, Frederick Converse, and George Edwin Rines, ed. *The Encyclopedia Americana*. Vol. 7. New York, Chicago: The Americana Company, 1904.

Bell, Jacob, ed. *Pharmaceutical Journal and Transactions: Volume I; 1841–2*. London: John Churchill; Edinburgh: Maclachlan & Stewart; Dublin: Fannin & Co., 1842.

Bennett, Adrian Arthur. *John Fryer: The Introduction of Western Science and Technology into Nineteenth-century China*. Cambridge, MA: Harvard University Press, 1967.

Bentham, George. *Flora Honkongensis: A Description of Flowering Plants and Ferns*. London: Lovell Reeve, 1861.

Biggerstaff, Knight. "The T'ung Wen Kuan." *The Chinese Social and Political Science Review* 18 (1934): 307–40.

Biggerstaff, Knight. *The Earliest Modern Government Schools in China*. Ithaca, NY: Cornell University Press, 1961.

Bloxam, Charles Loudon. *Chemistry Inorganic and Organic: With Experiments and a Comparison of Equivalent and Molecular Formulæ*. London: John Churchill and Sons, 1867.

Bloxam, Charles Loudon. *Chemistry, Inorganic and Organic: With Experiments*. Philadelphia: Henry C. Lea, 1873.

Boltz, William G. *The Origin and Early Development of the Chinese Writing System*. American Oriental Series, vol. 78. New Haven, CT: American Oriental Society, 1994.

Bottéro, Françoise. "La vision de l'écriture de Xu Shen à partir de sa présentation des liushu." *Cahiers de linguistique. Asie orientale* 27, no. 2 (1998): 161–91.

Bourne, John. *A Catechism of the Steam Engine in its Various Applications to Mines, Mills, Steam Navigation, Railways, and Agriculture; With Practical Instructions for the Manufacture and Management of Engines of Every Class*. Eleventh edition. London: Longman, Green, Longman, Roberts, & Green, 1865.

Bourne, John. *Recent Improvements in the Steam Engine: Containing Descriptions of the More Important Modern Engines, and among Them of the Steam, Air and Gas Engines Shown at the Paris Exhibition in 1867*. Philadelphia, J. B. Lippincott & Co.; London: Longmans & Co., 1868.

Bowman, John Eddowes. *An Introduction to Practical Chemistry, Including Analysis*. Second edition. London: John Churchill, 1854.

Bretschneider, Emil. *Early European Researches into the Flora of China*. Shanghai: American Presbyterian Mission Press, 1881.

Bretschneider, Emil. *Botanicon Sinicum: Notes on Chinese Botany from Native and Western Sources*. London, Trübner & Co., 1882. Originally published in *Journal of the North-China Branch of the Royal Asiatic Society* 16 (1881), article 3.

Bridgman, Elijah Coleman. *A Chinese Chrestomathy in the Canton Dialect*. Macao: Samuel Wells Williams, 1841.

Burgh, Nicholas Procter. *Practical Rules for the Proportions of Modern Engines and Boilers for Land and Marine Purposes*. Philadelphia: Henry Carey Baird, 1865.

Cadbury, William Warder, and Mary Hoxie Jones. *At the Point of a Lancet: One Hundred Years of the Canton Hospital, 1835–1935*. Shanghai: Kelly and Walsh, 1935.

Cai, Meian 蔡梅庵. *Cefu tongzong* 策府統宗 [A summary of the imperial library]. Shanghai: Hongwen shuju 鴻文書局, 1889.

Cai, Shangsi 蔡尚思, and Fang Xing 方行, ed. *Tan Sitong quanji* 谭嗣同全集 [Complete works of Tan Sitong]. Beijing: Zhonghua shuju 中华书局, 1981.

Cao, Yuanyu 曹元宇. *Zhongguo huaxue shi hua* 中国化学史话 [Historical anecdotes on Chinese chemistry]. Nanjing: Jiangsu kexue jishu chubanshe 江苏科学技术出版社, 1979.

Cao, Guangshun, and Yu Hsiao-jung. "Language Contact and its Influence on the Development of Chinese Syntax." In *The Oxford Handbook of Chinese Linguistics*, edited by William S-Y. Wang and Chaofen Sun: 203–214. Oxford: Oxford University Press, 2015.

Catalogue of the Chinese Imperial Maritime Customs Collection, at the United States International Exhibition, Philadelphia, 1876: Published by Order of the Inspector General of Chinese Maritime Customs. Shanghai: Statistical Department of the Inspectorate General of Customs, 1876.

Chemical Terminology: Serial No. I—Elements: Approved by the Ministry of Education—Huaxue mingci (yi)—Yuanzhi—Jiaoyubu shending 化学名词（一）— 原质—教育部审定. General Committee on Scientific Terminology, 1922.

*Chemical Terminology: Serial No. II—Chemical Terms: Approved by the Ministry of Education—Huaxuemingci(er)—Huaxueshuyu—Jiaoyubushending*化學名詞（二）— 化學術語—教育部審定. General Committee on Scientific Terminology, 1922.

Chen, Duxiu 陳獨秀. "Xiwen yiyin siyi 西文譯音私議 [Personal considerations on the transliteration of Western languages]." *Xin qingnian* 新青年 2, no. 4 (1916): 1–6.

Chen, Liwei 陈力卫. "*Jiangnan zhizaoju de chubanwu zai Riben de chuanbo he liyong* 江南制造局的出版物在日本的传播和利用 [The circulation and use in Japan of the publications of the Jiangnan Arsenal]." In *Lishi shang de Zhongguo chuban yu Dongya wenhua jiaoliu* 历史上的中国出版与东亚文化交流 [The historical Chinese publishing and cultural exchanges in East Asia], edited by Fudan daxue lishi xi chuban bowuguan 复旦大学历史系出版博物馆, 44–60. Shanghai: Shanghai baijia chubanshe 上海百家出版社, 2009.

Chen, Longchang 陳龍昌. *Zhong-Xi binglüe zhizhang* 中西兵略指掌 [An easy explanation of the military strategies of China and the West]. Dongshan caotang 東山草堂, 1897.

Chen, Pingyuan 陈平原, and Xia Xiaohong 夏晓虹, ed. *Tuxiang wan Qing* 图像晚清 [Images in the late Qing dynasty]. Tianjin: Baihua wenyi chubanshe 百花文艺出版社, 2001.

Chen, Yuanlong 陳元龍. *Gezhi jingyuan* 格致鏡原 [Mirroring the origin for the investigation of things and extension of knowledge], 1735.

Chen, Zhongyi 陳忠倚, ed. *Huangchao jingshi wen sanbian* 皇朝經世文三編 [Third collection of Qing dynasty writings on statecraft]. Shanghai: Baowen shuju 寶文書局, 1898. Republished in Idem. *Huangchao jingshi wen sanbian* 皇朝經世文三編. Taipei: Guofeng chubanshe 國風出版社, 1965.

Chen, Zhu 陳洙, ed. *Jiangnan zhizaoju yishu tiyao* 江南製造局譯書提要 [A summary of the books translated at the Jiangnan Arsenal]. Shanghai: Jiangnan zhizaoju 江南製造局, 1909.

Chen, Jiang. "Western Learning and Social Transmutation in the Late Qing." In *China's Quest for Modernization: A Historical Perspective*, edited by Frederic Wakeman Jr. and Wang Xi, 150–74. Berkeley: Institute of East Asian Studies, University of California, 1997.

Chun, Doris Sze. "John Fryer, the First Agassiz Professor of Oriental Languages and Literature, Berkeley." *Chronicle of the University of California* (Fall 2005): 1–18.

Cohen, Paul A. *Between Tradition and Modernity: Wang T'ao and Reform in Late Ch'ing China*. Cambridge, MA: Harvard University Press, 1974.

Colby, Frank Moore, and Talcott Williams. *The New International Encyclopædia: Volume IX*. Second edition. New York: Dodd, Mead and Company, 1917.

Corbett, Charles Hodge. *Shantung Christian University (Cheeloo)*. New York: United Board for Christian Colleges in China, 1955.

Cordier, Henry. *Bibliotheca sinica: Dictionnaire bibliographique des ouvrages relatifs à l'Empire chinois. Supplément*. Paris: Leroux, 1885.

"Correspondence—School and Textbook Series Committee." *The Chinese Recorder and Missionary Journal, Volume XVIII*, no. 4: 162. Shanghai: Presbyterian Mission Press, 1887.

"Correspondence—School and Text Book Series Committee." *The Chinese Recorder and Missionary Journal, Volume XX*, no. 9: 430. Shanghai: Presbyterian Mission Press, 1889.

Cousland, Philip Brunelleschi. *An English-Chinese Lexicon of Medical Terms Compiled for the Terminology Committee*. Shanghai: Medical Missionary Society of China, 1908.

Covell, Ralph. *W.A.P. Martin: Pioneer of Progress in China*. Washington: Christian University Press, 1978.

Crawford, Tarleton Perry. "A System of Phonetic Symbols for Writing the Dialects of China." *The Chinese Recorder and Missionary Journal, Volume XIX*, no. 3, 101–10. Shanghai: Presbyterian Mission Press, 1888.

Crosland, Maurice Pierre. *Historical Studies in the Language of Chemistry*. London: Heinemann, 1962. Reprinted Mineola, NY: Dover Publications, 2004.

Culp, Robert J. "'Weak and Small Peoples' in a 'Europeanizing World': World History Textbooks and Chinese Intellectuals' Perspectives on Global Modernity." In *The Politics of Historical Production in Late Qing and Republican China*, edited by Tze-ki Hon and Robert J. Culp, 211–46. Leiden, Boston: Brill, 2007.

Dagenais, Ferdinand. *The John Fryer Papers*. Three volumes. Guilin: Guangxi shifan daxue chubanshe 广西师范大学出版社, 2010.

Darlington, William. *Agricultural Botany: An Enumeration and Description of Useful Plants and Weeds, which Merit the Notice, or Require the Attention, of American Agriculturists*. Philadelphia: J. W. Moore; New York: Mark. H. Newman, 1847.

Davidson-Houston, James Vivian, and Raymond Varley Dewar-Durie. *Han-Ying junshi cidian 漢英軍事辭典—Chinese and English Modern Military Dictionary*. Peiping: Henri Vetch, The French Bookstore, 1935.

Davies, D. I., D. C. Lyon, and R. J. Spring. "Charles Loudon Bloxam—A Victorian University and Military Academy Chemistry Teacher." *Ambix* 33, 1 (1986): 11–32.

De Candolle, Augustin Pyrame, Alphonse Louis Pierre Pyrame de Candolle, and Heinrich Wilhelm Buek. *Prodromus systematis naturalis regni vegetabilis, sive enumeratio contracta ordinum, generum, specierumque plantarum hucusque cognitarum, juxta methodi naturalis normas digesta*. Paris (and others): Treuttel et Würtz (and others), 1824–1873.

De Crespigny, Rafe. *A Biographical Dictionary of Later Han to the Three Kingdoms (23–220 AD)*. Leiden, Boston: Brill, 2007.

De Loureiro, João. *Flora Cochinchinensis: sistens plantas in regno Cochinchina nascentes; quibus accedunt aliæ observatæ in Sinensi imperio, Africa Orientali, Indiæque locis variis; omnes dispositæ secundum systema sexuale Linnæanum*. Lisbon: Real Academia de Ciências, 1790.

Deng, Liang 邓亮, and Han Qi 韩琦. "'Zhongxue' banben liuchuan ji qi yingxiang '重学' 版本流传及其影响 [The diffusion of the editions of Mechanics and their influence]." *Wenxian* 文献, no. 3 (2009): 151–57.

Descriptive Catalogue and Price List of the Books, Wall Charts, Maps, Etc. Published or Adopted by the Educational Association of China. Shanghai: Presbyterian Mission Press, 1907.

Devan, Thomas T. *The Beginner's First Book in the Chinese Language, (Canton Vernacular): Prepared for the Use of the House-Keeper, Merchant, Physician, and Missionary*. Hong Kong: China Mail, 1847.

Doolittle, Justus. *Ying-Hua cuilin yunfu 英華萃林韻府—A Vocabulary and Hand-Book of the Chinese Language, Romanized in the Mandarin Dialect*. Fuzhou: Rozario, Marcal, and Company, 1872–1873.

Du, Shiran 杜石然, ed. *Zhongguo gudai kexuejia zhuanji* 中国古代科学家传记 [Biographies of ancient Chinese scientists]. Two volumes. Beijing: Kexue chubanshe 科学出版社, 1993.

Duan, Fang 端方, ed. *Da Qing Guangxu xin faling* 大清光緒新法令 [New laws and regulations of the Guangxu era of the Great Qing dynasty]. Shanghai: Shangwu yinshuguan 商務印書館, Xuantong period.

Dudgeon, John. "Review of a New Medical Vocabulary." *The Chinese Recorder and Missionary Journal* 13, no. 1: 30–43; no. 3, 177–82; no. 4, 259–65. Shanghai: American Presbyterian Mission Press, 1882.

Dunne, George H. *Generation of Giants: The Story of the Jesuits in China in the Last Decades of the Ming Dynasty*. Notre Dame: University of Notre Dame Press, 1962.

"Editorial Notes and Missionary News." *The Chinese Recorder and Missionary Journal, Volume XVIII*, no. 6: 243. Shanghai: Presbyterian Mission Press, 1887.

Edkins, Joseph, and Li Shanlan 李善蘭, trans. *Zhongxue* 重學 [Mechanics]. Shanghai: Mohai shuguan 墨海書館, 1859.

"Educational Works for the Chinese—Minutes of Meeting of the Committee for the Publication of 'A Series of School and Text-books', Held at the London Mission, Shanghai, 15th, 16th and 17th March, 1880." *The Chinese Recorder and Missionary Journal, Volume XI*, no. 1: 138–42. Shanghai: Shanghai, American Presbyterian Mission Press, 1880.

"Educational Notes—The Educational Association of China." *The Chinese Recorder and Missionary Journal, Volume XXIII*, no. 1: 30–36. Shanghai: American Presbyterian Mission Press, 1892.

Elman, Benjamin A. *From Philosophy to Philology: Intellectual and Social Aspects of Change in Late Imperial China*. Harvard East Asian Monographs 110. Cambridge, MA: Council on East Asian Studies, Harvard University, 1984.

Elman, Benjamin A. "The Relevance of Sung Learning in the Late Ch'ing: Wei Yuan and the *Huang-ch'ao ching-shih wen-pien*." *Late Imperial China* 9, no. 2 (1988): 56–85.

Elman, Benjamin A. *On Their Own Terms: Science in China, 1550–1900*. Cambridge, MA: Harvard University Press, 2005.

Elman, Benjamin A. "Toward a History of Modern Science in Republican China." In *Science and Technology in Modern China, 1880s–1940s*, edited by Jing Tsu and Benjamin A. Elman, 15–38. Leiden, Boston: Brill, 2014.

Elwick, James. *Styles of Reasoning in the British Life Sciences: Shared Assumptions, 1820–1858*. London: Pickering and Chatto, 2007.

Encyclopaedia Britannica; Or a Dictionary of Arts, Sciences, and Miscellaneous Literature. Vol. 4. Third edition. Edinburgh: Printed for A. Bell and C. Macfarquhar, 1797.

Encyclopaedia Britannica: Or a Dictionary of Arts, Sciences, and Miscellaneous Literature; Enlarged and Improved. Vol. 12. Fifth edition. Edinburgh: Printed for Archibald Constable and Company, 1817.

Endō, Shōji 遠藤正治. "*Sōmoku zusetsu* no kōzō—Shokubutsu yōgo no seikaku '草木図説' の構造 – 植物用語の性格 [The structure of An illustrated book of plants and trees—The nature of botanical terminology]." *Yokusai kenkyūkai dayori* 慾斎研究会だより, no. 118 (2008): 1–8.

Endō, Shōji 遠藤正治, Katō Nobushige 加藤僖重, Kōda Masataka 幸田正孝, and Matsuda Kiyoshi 松田清, ed. *Udagawa Yōan shokubutsugaku shiryō no kenkyū: Kyōu shooku shozō* 宇田川榕菴植物学資料の研究. 杏雨書屋所蔵 [Research on botanical materials by Udagawa Yōan stored in the *Kyōu shooku*]. Osaka: Takeda kagaku shinkō zaidan 武田科学振興財団, 2014.

Fairbank, John King, and Liu Kwang-Ching, ed. *The Cambridge History of China, Vol. 11, Late Ch'ing 1800–1911, Part II*. Cambridge: Cambridge University Press, 1980.

Fan, Fa-ti. *British Naturalists in Qing China: Science, Empire, and Cultural Encounter*. Cambridge, MA: Harvard University Press, 2004.

Fang, Yizhi 方以智. *Tong ya* 通雅 [The general guide]. Fang shi fushan ci cang xuan keben 方氏浮山此藏軒刻本, 1666.

Fisher, Daniel Webster. *Calvin Wilson Mateer: Forty-Five Years a Missionary in Shantung, China; A Biography*. Philadelphia: The Westminster Press, 1911.

Fogel, Joshua A., ed., trans. *The Emergence of the Modern Sino-Japanese Lexicon: Seven Studies*. Leiden, Boston: Brill, 2015.

Fong, Foo Sec 鄺富灼. "The Co-operation of Chinese and Foreign Educationists in the Work of the Association." In *Records of the Sixth Triennial Meeting of the Educational Association of China, Held at Shanghai, May 19–22, 1909*, 66–74. Shanghai: The Educational Association of China, 1909.

Forsyth, Robert Coventry. *The China Martyrs of 1900: A Complete Roll of the Christian Heroes Martyred in China in 1900 with Narratives of Survivors*. New York, Chicago, Toronto: Fleming H. Revell Company, 1904.

Fryer, John. *John Fryer Papers*. BANC MSS C-B 968. Courtesy of the Bancroft Library, University of California, Berkeley.

Fryer, John. "Lun xinyi *Xiyao lüeshi* 論新譯'西藥略釋' [On the recently translated Brief explanations of Western medicine]." *Gezhi huibian* 格致彙編 8 (1876): 8–9.

Fryer, John. *An account of the Department for the Translation of Foreign Books at the Kiangnan Arsenal, Shanghai: With Various Lists of Publications in the Chinese Language*. Shanghai: American Presbyterian Press, 1880. Originally published in *North China Herald* (1880, January 29): 77–81. Republished in "Science in China." *Nature* (1881, May 5): 9–11; (May 19): 54–57.

Fryer, John. "Jiangnan zhizao zongju fanyi xishu shilüe 江南製造總局翻譯西書事略 [An Account of the Department for the Translation of Foreign Books at the Kiangnan Arsenal]." *Gezhi huibian* 格致彙編 3, no. 5 (June 1880): 10A–12B; no. 6 (July 1880): 9A–11B; no. 7 (August 1880): 9A–11B; no. 8 (September 1880): 9A–10B. Republished in *Zhongguo jindai chuban shiliao chubian* 中国近代出版史料初编 [First collection of historical materials on publishing in modern China], edited by Zhang Jinglu 张静庐, 1–23. Shanghai: Shanghai chubanshe 上海出版社, 1953.

Fryer, John, ed. *Gezhi xuzhi* 格致須知 [Science Outline Series—Essentials of the investigation of things and extension of knowledge]. Shanghai: Gezhi shushi 格致書室, 1882–1883.

Fryer, John, ed. *Jinshi Zhong-Xi mingmu biao* 金石中西名目表 [Vocabulary of Mineralogical Terms]. Shanghai: Jiangnan zhizaoju 江南製造局, 1883.

Fryer, John. *Fourth Report of the Chinese Polytechnic Institution and Reading Rooms, Shanghai: From March, 1883, to June, 1885*. North China Herald (July 10): 44–45. Shanghai: North-China Herald Office, 1885.

Fryer, John, ed. *Huaxue cailiao Zhong-Xi mingmu biao* 化學材料中西名目表 [Vocabulary of Names of Chemical Substances]. Shanghai: Jiangnan zhizaoju 江南製造局, 1885.

Fryer, John, ed. *Xiyao dacheng yaopin Zhong-Xi mingmu biao* 西藥大成藥品中西名目表 [Vocabulary of Names of Materia Medica]. Shanghai: Jiangnan zhizaoju 江南製造局, 1887.

Fryer, John, ed. *Gezhi xuzhi yi jiben* 格致須知一集本 [First volume of Science Outline Series]. Shanghai: Gezhi shushi 格致書室, 1887.

Fryer, John, ed. *The Translator's Vade-mecum, A Collection of Vocabularies of Chinese Terms Used in the Translation of Scientific Books at the Kiangnan Imperial Government Arsenal, Shanghai, China*. Shanghai: Presbyterian Mission Press, 1888.

Fryer, John, ed. *Qiji Zhong-Xi mingmu biao* 汽機中西名目表 [Vocabulary of Terms Relating to the Steam Engine]. Shanghai: Jiangnan zhizaoju 江南製造局, 1889.

Fryer, John. "Scientific Terminology: Present Discrepancies and Means of Securing Uniformity." In *Records of the General Conference of the Protestant Missionaries of China, Held at Shanghai, May 7–20, 1890*, edited by W. J. Lewis, William Theodore Aquila Barber and John R. Hykes, 531–51. Shanghai: American Presbyterian Mission Press, 1890.

Fryer, John. *Catalogue of Educational Books, Works of General Knowledge, Scientific and Technical Treatises &c., &c., &c., in the Chinese Language: Translated or Written by John Fryer*. Shanghai: Presbyterian Mission Press, 1894.

Fryer, John. *Descriptive Catalogue and Price List of the Books, Wall Charts, Maps, &c., Published or Adopted by the Educational Association of China (Formerly the School and Text-Book Series Committee)*. Shanghai: American Presbyterian Mission Press, 1894.

Fryer, John. "Chemical Terminology." *The Chinese Recorder and Missionary Journal, Volume XXVI*, no. 6: 288–89. Shanghai: American Presbyterian Mission Press, 1895.

Fryer, John. *Educational Directory for China*. Shanghai: Educational Association of China, 1895.

Fryer, John, ed. *Gezhi shuyuan xixue kecheng* 格致書院西學課程 [Curriculum of Western learning at the Polytechnic institution]: Shanghai: 1895.

Fryer, John. "Qiuzhu shixin xiaoshuo qi 求著時新小說啟 [A plea for the writing of new fiction]." *Wanguo gongbao* 萬國公報 7, no. 77 (1895): 31.

Fryer, John. "Shixin xiaoshuo chu'an 時新小說出案 [A proposal for new fiction]." *Wanguo gongbao* 萬國公報 8, no. 86 (1896): 30–31.

Fryer, John. "The Present Outlook for Chinese Scientific Nomenclature." In *Records of the Second Triennial Meeting of the Educational Association of China, Held at Shanghai, May 6–9, 1896*, 155–67. Shanghai: American Presbyterian Mission Press, 1896.

Fryer, John, ed. *Gezhi xuzhi er jiben* 格致須知二集本 [Second volume of Science Outline Series]. Shanghai: Gezhi shushi 格致書室, 1899.

Fryer, John, and Hua Beiyu 華備鈺, trans. *Bingchuan qiji* 兵船汽機 [The warship steam engine—The Marine Steam Engine]. Shanghai: Jiangnan zhizaoju 江南製造局, 1894.

Fryer, John, and Pan Song 潘松, trans. *Qiukuang zhinan* 求礦指南 [The guide of prospecting for minerals—The Prospector's Handbook]. Shanghai: Jiangnan zhizaoju 江南製造局, 1899.

Fryer, John, and Wang Jilie 王季烈, trans. *Tongwu dianguang* 通物電光 [The electric light piercing through all things—The X-ray; Or, Photography of the Invisible and Its Value in Surgery]. Shanghai: Jiangnan zhizaoju 江南製造局, 1899.

Fryer, John, and Xu Jianyin 徐建寅, trans. *Qiji biyi* 汽機必以 [Necessary instruments for the steam engine—A Catechism of the Steam Engine]. Shanghai: Jiangnan zhizaoju 江南製造局, 1872.

Fryer, John, and Xu Jianyin 徐建寅, trans. *Qiji xinzhi* 汽機新製 [New patterns in the steam engine—Pocket-book of Practical Rules for the Proportions of Modern Engines and Boilers for Land and Marine Purposes]. Shanghai: Jiangnan zhizaoju 江南製造局, 1872.

Fryer, John, and Xu Jianyin 徐建寅, trans. *Shengxue* 聲學 [Acoustics—Sound: A Course of Eight Lectures Delivered at the Royal Institution of Great Britain]. Shanghai: Jiangnan zhizaoju 江南製造局, 1874.

Fryer, John, and Xu Jianyin 徐建寅, trans. *Dianxue yuanliu* 電學源流 [The origin of electricity—The Student's Textbook of Electricity]. Shanghai: Datong shuju 大同書局, 1888.

Fryer, John, and Xu Shou 徐壽, trans. *Huaxue jianyuan* 化學鑑原 [Mirror of chemistry—Principles and Applications of Chemistry]. Shanghai: Jiangnan zhizaoju 江南製造局, 1871.

Fryer, John, and Xu Shou 徐壽, trans. *Huaxue jianyuan xubian* 化學鑑原續編 [Sequel to the Mirror of chemistry]. Shanghai: Jiangnan zhizaoju 江南製造局, 1875.

Fryer, John, and Xu Shou 徐壽, trans. *Huaxue jianyuan bubian* 化學鑑原補編 [Supplement to the Mirror of chemistry]. Shanghai: Jiangnan zhizaoju 江南製造局, 1882.

Fryer, John, and Zhao Yuanyi 趙元益, trans. *Yejin lu* 冶金錄 [Records of metallurgy—Moulder's and Founder's Pocket Guide]. Shanghai: Jiangnan zhizaoju 江南製造局, 1873.

Fryer, John, and Zhou Xun 周郇, trans. *Dianxue gangmu* 電學綱目 [Compendium of electricity—Notes of a Course of Seven Lectures on Electrical Phenomena and Theories]. Shanghai: Jiangnan zhizaoju 江南製造局, 1881.

Fu, Yunlong 傅雲龍. *Youli Riben tujing yuji* 遊歷日本圖經餘記 [Additional records of a chronicle of Japan with maps]. 1889.

Fulton, Mary Hannah, and Zhou Zhongyi 周仲彝, trans. *Hao Ci er kexue* (*chuji*) 豪慈兒科學(初集) [Paediatrics by Holt: first collection]. Shanghai: Zhongguo Boyihui 中國博醫會, 1915.

Gaide, Laurent. "La médecine européenne en Annam, autrefois et de nos jours." *Bulletin des Amis du Vieux Hué*, no. 4 (1921): 189–214.

Gao, Mingkai 高名凯, and Liu Zhengtan 刘正埮. *Xiandai hanyu wailaici yanjiu* 现代汉语外来词研究 [Research on modern Chinese loanwords]. Beijing: Wenzi gaige chubanshe 文字改革出版社, 1958.

Gao, Xi, 高晞. "Jingshi Tongwenguan de yixue jiangzuo 京师同文馆的医学讲座 [Lectures on medicine at Beijing School of combined learning]." *Zhongguo keji shiliao* 中国科技史料—*China Historical Materials of Science and Technology* 11, no. 4 (1990): 42–47.

Gao, Xi, 高晞. "'Jiepouxue' zhongwen yiming de youlai yu queding—Yi De Zhen *Quanti tongkao* wei zhongxin '解剖学'中文译名的由来与确定—以德贞'全体通考'为中心 [The origin and establishing of the translation of 'anatomy' in Chinese—A focus on Dudgeon's Comprehensive investigation of the whole body (Grey's Anatomy)]." *Lishi yanjiu* 历史研究, no. 6 (2008): 80–104.

Gao, Xi, 高晞. *De Zhen zhuan: yi ge Yingguo chuanjiaoshi yu wan Qing yixue jindaihua* 德贞传：一个英国传教士与晚清医学近代化 [Biography of John Dudgeon: an English missionary and the modernisation of medicine in the late Qing dynasty]. Shanghai: Fudan daxue chubanshe 复旦大学出版社, 2009.

General Committee on Medical Terminology—*Yixue mingci shenchahui* 醫學名詞審查會: *Second Report of Chemical Terms—Huaxue mingci shenchazu di er ci shenchaben—Huaxue shuyu* 化學名詞審查組第二次審查本—化學術語. Publication place and year unknown.

General Committee on Scientific Terminology. *Fifth Report of General Terms in Organic Chemistry—Huaxue mingci shenchazu di wu ci shenchaben—Youji huaxue putong mingci*—化學名詞審查組第五次審查本—有機化學普通名詞. Publication place and year unknown.

Gibbs, Hill Michael. *Lin Shu, Inc.: Translation and the Making of Modern Chinese Culture*. New York: Oxford University Press, 2013.

Gregory, Thomas G., and Gao Xian 高銛. *Huaxue yaopin cidian* 化學藥品辭典 [Pharmaceutical dictionary]. Shanghai: Xinya shudian 新亞書店, 1946.

Godwin, Christopher D. "Writing Foreign Terms in Chinese." *Journal of Chinese Linguistics* 7, no. 2 (1979): 246–267.

Gong, Yi 龚益. "Gui qu lai xi 硅去来矽 [The replacement of the term *gui* with *xi*]." *Zhongguo keji shuyu* 中国科技术语, no. 5 (2008): 62–64.

Gore, James Howard. *American Members of Foreign Orders*. Washington: Press of W. F. Roberts Co., 1910.

Guoli bianyiguan 國立編譯館. *Huaxue mingming yuanze* 化學命名原則 [Principles of chemical nomenclature]. Chongqing: Guoli bianyiguan 國立編譯館, 1933.

Guoli bianyiguan 國立編譯館. *Tianwenxue mingci* 天文學名詞 [Terms in astronomy]. Shanghai: Shangwu yinshuguan 商務印書館, 1934.

Guoli bianyiguan 國立編譯館. *Wulixue mingci* 物理學名詞 [Terms in physics]. Shanghai: Shangwu yinshuguan 商務印書館, 1934.

Guoli bianyiguan 國立編譯館. *Kuangwuxue mingci* 礦物學名詞 [Terms in mineralogy]. Shanghai: Shangwu yinshuguan 商務印書館, 1936.

Guoli bianyiguan 國立編譯館. *Xijunxue mianyixue mingci* 細菌學免疫學名詞 [Terms in bacteriology and immunology]. Shanghai: Shangwu yinshuguan 商務印書館, 1937.

Guoli bianyiguan 國立編譯館. *Dianji gongcheng mingci—Dianhua bu* 電機工程名詞—電化部 [Terms in electrical engineering—Chemistry]. Shanghai: Zhengzhong shuju 正中書局, 1946.

Guoli bianyiguan 國立編譯館. *Huaxue gongcheng mingci* 化學工程名詞 [Terms in chemical engineering]. Shanghai: Zhengzhong shuju 正中書局, 1946.

Guoli bianyiguan 國立編譯館. *Jixie gongcheng mingci (putong bu)* 機械工程名詞(普通部) [Terms in mechanical engineering—General part]. Chongqing: Zhengzhong shuju 正中書局, 1946.

Guoli bianyiguan 國立編譯館. *Dianji gongcheng mingci—Dianli bu* 電機工程名詞—電力部 [Terms in electrical engineering—Power]. Shanghai: Zhengzhong shuju 正中書局, 1947.

Guoli bianyiguan 國立編譯館. *Huaxue mingming yuanze (zengding ben)* 化學命名原則（增訂本）[Principles of chemical nomenclature (Revised and enlarged edition)]. Nanjing: Zhengzhong shuju 正中書局, 1947.

Han, Qi 韩琦. "*Shuli gezhi* de faxian—Jian lun 18 shiji Niudun xiangguan zhuzuo zai Zhongguo de chuanbo '数理格致'的发现—兼论18世纪牛顿相关著作在中国的传播 [The discovery of Principles of number: investigation of things and extension of knowledge (Principia) and the circulation of Newton's relevant works in China in the eighteenth century]." *Zhongguo keji shiliao* 中国科技史料—*China Historical Materials of Science and Technology* 19, no. 2 (1998): 78–85.

Han, Qi 韩琦. “Le début de la diffusion des techniques d'imprimerie occidentales en Chine à la fin des Qing: l'exemple de l'introduction du procédé lithographique.” *Histoire et civilisation du livre: Revue internationale III*: 135–52. Geneva: Librairie Droz S. A., 2007.

Hanan, Patrick. *Chinese Fiction of the Nineteenth and Early Twentieth Centuries*. New York: Columbia University Press, 2004.

Hao, Bingjian 郝秉键. “Shanghai Gezhi shuyuan ji qi jiaoyu chuangxin 上海格致书院及其教育创新 [The Shanghai Polytechnic institution and its innovation in education].” *Qing shi yanjiu* 清史研究—*Studies in Qing History*, no. 3 (2003): 85–96.

He, Songling 何崧齡. *Biaozhun hanyi waiguo renming diming biao: fu hanwen suoyin, xifang yiming zongfen ge biao* 標準漢譯外國人名地名表： 附漢文索引， 西方譯名總分各表 [Tables of names of places and persons with standard Chinese translations: with a Chinese index and Western translated names appended]. Shanghai: Shangwu yinshuguan 商務印書館, 1924.

Hemeling, Karl Ernst Georg. *English-Chinese Dictionary of the Standard Chinese Spoken Language* (官話) *and Handbook for Translators, Including Scientific, Technical, Modern and Documentary Terms*. Shanghai: Statistical Department of the Inspectorate General of Customs, 1916.

Henry, Augustine. *Notes on Economic Botany of China*. Shanghai: Presbyterian Mission Press, 1893.

Ho, Yi Kai, ed. *Science in China, 1600–1900: Essays by Benjamin A. Elman*. Singapore: World Century, World Scientific, 2015.

Hobson, Benjamin. *A Medical Vocabulary in English and Chinese*. Shanghai: Shanghae Mission Press, 1858.

Hooker, Joseph Dalton, and George Bentham. *Genera plantarum ad exemplaria imprimis in herbariis kewensibus servata definita*. London: A. Black, 1862–1883.

Horng, Wann-sheng. “Li Shanlan: The Impact of Western Mathematics in China During the Late 19th Century.” PhD diss., City University of New York, 1991.

Hua, Linfu, Paul D. Buell, and Paul U. Unschuld, ed. *Dictionary of the Ben cao gang mu, Volume II: Geographical and Administrative Designations*. Oakland, CA: University of California Press, 2017.

Huanqiushe bianjibu 环球社编辑部. *Tuhua ribao* 图画日报 [Illustrated daily]. Shanghai: Shanghai guji chubanshe 上海古籍出版社, 1999.

Huang, Heqing 黄河清. *Jinxiandai ciyuan* 近现代辞源 [Modern etymology]. Shanghai: Shanghai cishu chubanshe 上海辞书出版社, 2010.

Huang, Max Ko-wu. “The War of Neologisms: The Competition between the Newly Translated Terms Invented by Yan Fu and by the Japanese in the Late Qing.” In *China and Its Others: Knowledge Transfer through Translation, 1829–2010*, edited by James St. André and Peng Hsiao-yen, 51–70. Amsterdam, New York: Rodopi, 2012.

Huang, Moxi 黃摩西. *Putong baike xin da cidian* 普通百科新大詞典 [New general encyclopaedic dictionary]. Shanghai: Zhongguo cidian gongsi 中國詞典公司, 1911.

Huang, Qingcheng 黄庆澄. *Zhong-Xi putong shumu biao* 中西普通書目表 [A list of Chinese and Western general books]. Shanghai: Suanxue baoguan 算學報館, 1898.

Huang, Shuanfan. "Chinese Morphology: Anatomy of a Headless Language." In *Zhongyang yanjiuyuan di er jie guoji hanxue huiyi lunwenji: yuyan yu wenzi zu* 中央研究院第二屆國際漢學會議論文集： 語言與文字組 [Proceedings of the second international conference on sinology at Academia Sinica: language and writing], edited by Zhongyang yanjiuyuan 中央研究院, 571–587. Two volumes. Taipei: Academia Sinica, 1989.

Huang, Zhending 黄振定. "Zhong-Xi keji jiaoliu ji qi fanyi de chuangzaoxing 中西科技交流及其翻译的创造性 [The scientific and technical exchanges between China and the West and the creativeness of their translations]." *Zhongguo fanyi* 中国翻译—*Chinese Translators Journal* 24, no. 2 (2003): 66–68.

Huang, Zunxian 黃遵憲. *Riben guozhi* 日本國志 [A description of Japan]. Taipei: Wenhai chubanshe 文海出版社, 1968.

Huaxue mingci shencha weiyuanhui gongzuo xiaozu 化學名詞審查委員會工作小組. *Huaxue mingci cao'an: A-L* 化學名詞草案: A-L [A draft of terms in chemistry: A-L]. Zhongguo huaxuehui 中國化學會.

Huaxue mingci shencha weiyuanhui gongzuo xiaozu 化學名詞審查委員會工作小組. *Huaxue mingci cao'an: M-Z* 化學名詞草案: M-Z [A draft of terms in chemistry: M-Z]. Zhongguo huaxuehui 中國化學會.

Hummel, Arthur William, ed. *Eminent Chinese of the Ch'ing Period, 1644–1912*. Washington: United States Government Printing Office, 1943. Reprinted in Taipei: Ch'eng Wen Publishing Company, 1970.

Hunter, Stephen Alexander, trans. *Wanguo yaofang* 萬國藥方 [Prescriptions from the ten thousand nations]. Shanghai: Meihua shuguan 美華書館, 1886.

Iinuma, Yokusai 飯沼慾斎. *Sōmoku zusetsu* 草木圖説 [An illustrated book of plants and trees]. Edo: Suharaya Mohei 須原屋茂兵衛, 1856.

Iinuma, Yokusai 飯沼慾斎, Tanaka Yoshio 田中芳男, Ono Motoyoshi 小野職愨, Kubo Hiromichi 久保弘道, and Yokokawa Masatoshi 横川政利. *Sōmoku zusetsu* 草木図説 [An illustrated book of plants and trees]. Gifu: Hirabayashisō 平林荘, 1874.

Inoue, Tetsujirō 井上哲次郎, and Motora Yūjirō 元郎勇次郎. *Tetsugaku jii* 哲學字彙 [Dictionary of philosophy]. Third edition. Tokyo: Maruzen kabushiki kaisha 丸善株式會社, 1912.

Irwin, Richard Gregg, and Ferdinand Dagenais. *John Fryer's Legacy of Chinese Writings—Edited, with Supplementary Material Provided by Ferdinand Dagenais*. Berkeley: Center for Chinese Studies, University of California, 1999.

Itō, Keisuke 伊藤圭介. *Taisei honzō meiso* 泰西本草名疏 [Notes on Western names of herbs]. Owari: Kajō Shooku zōhan 花繞書屋藏版, 1829.

James, Francis Huberty, and Fan Xiyong 范熙庸, trans. *Nongxue chuji* 農學初級 [A primary of agronomy]. Shanghai: Jiangnan zhizaoju 江南製造局, 1898.

Jami, Catherine, Peter Engelfriet, and Gregory Blue, ed. *Statecraft and Intellectual Renewal in Late Ming China: The Cross-Cultural Synthesis of Xu Guangqi (1562–1633)*. Leiden, Boston, Köln: Brill, 2001.

Janku, Andrea. "Translating Genre: How the 'Leading Article' Became the *Shelun*." In *Mapping Meanings: The Field of New Learning in Late Qing China*, edited by Michael Lackner and Natascha Vittinghoff, 329–53. Leiden, Boston: Brill, 2004.

Jevons, William Stanley. *Elementary Lessons in Logic: Deductive and Inductive; With Copious Questions and Examples, and a Vocabulary of Logical Terms*. London: Macmillan and Co., 1870.

Ji, Meng, and Ukai Atsuko 鵜飼敦子, ed. *Translation, History and Arts: New Horizons in Asian Interdisciplinary Humanities Research*. Cambridge: Cambridge Scholars Publishing House, 2013.

Jiangnan zhizaoju 江南製造局. *Jiangnan jiqi zhizao zongju shumu* 江南機器製造總局書目 [A book list of the Jiangnan Arsenal]. Shanghai: 1902.

Jiangnan zhizaoju 江南製造局. *Shanghai zhizaoju yiyin tushu mulu* 上海製造局譯印圖書目錄 [An index of books translated and printed at the Shanghai Arsenal]. Shanghai: 1911.

Kaske, Elisabeth. *The Politics of Language in Chinese Education, 1895–1919*. Leiden, Boston: Brill, 2008.

Kerr, John Glasgow. *A Vocabulary of Diseases, Based on Thomson's Vocabulary and Whitney's Anatomical Terms, Prepared by J. G. Kerr, M. D., for the Committee on Nomenclature of the Medical Missionary Association of China*. Shanghai: American Presbyterian Mission Press, 1894.

Kerr, John Glasgow, and He Liaoran 何瞭然, trans. *Huaxue chujie* 化學初階 [First steps in chemistry]. Boji yiju 博濟醫局: Guangzhou, 1870.

Kerr, John Glasgow, and Kong Jiliang 孔繼良, trans. *Xiyao lüeshi* 西藥略釋 [Brief explanations of Western medicine]. Third edition. Guangzhou: Boji yiju 博濟醫局, 1886.

Kexue mingci shenchahui 科學名詞審查會 (General Committee on Scientific Terminology), ed. *Yixue mingci huibian: La-Ying-De-Han duizhao* 醫學名詞彙編: 拉英德漢對照—*A Latin-English-German-Chinese Medical Terminology*. Shanghai: Kexue mingci shenchahui 科學名詞審查會, 1931. Republished in 1940.

Kexue mingci shenchahui 科學名詞審查會 (General Committee on Scientific Terminology), ed. *Dong zhiwu mingci huibian: kuangwu mingci fu* 動植物名詞彙編. 礦物名詞坿 [Collection of zoological and botanical terms: with an appendix of mineralogical terms]. Shanghai: Kexue mingci shenchahui 科學名詞審查會, 1935.

Kimura, Shunkichi 木村駿吉. "Guangshuo 光說 [On light]." In *Xinxue da congshu* 新學大叢書 [Great collectanea of new learning], translated by Gong Chai 龔柴 and Xu Bin 許彬. Shanghai: Jishan qiaoji shuju 積山喬記書局, 1903.

King, Robert C., and William D. Stansfield. *A Dictionary of Genetics*. Fifth edition. New York, Oxford: Oxford University Press, 1997.

Kokusai Kirisutokyō daigaku Ajia bunka kenkyū iinkai 国際基督教大学アジア文化研究委員会, ed. *Nihon Kirisutokyō bunken mokuroku: Meiji ki.* 日本キリスト教文献目録：明治期 [An index of Japanese Christian texts: Meiji era]. Tokyo: Kokusai Kirisutokyō daigaku 国際基督教大学, 1965.

Kokusho kankōkai 国書刊行会, ed. *Bunmei genryū sōsho* 文明源流叢書 [Collectanea of the source of civilisation]. Tokyo: Kokusho kankōkai 国書刊行会, 1914.

Koselleck, Reinhart. *The Practice of Conceptual History: Timing History, Spacing Concepts*. Translated by Todd Samuel Presner and others. Stanford, California: Stanford University Press, 2002.

Kuang, Qizhao 鄺其照. *Hua-Ying zidian jicheng* 華英字典集成—*English and Chinese Dictionary*. Hong Kong, 1882.

Kubitzki, Klaus, ed. *The Families and Genera of Vascular Plants—Volume X: Flowering Plants; Eudicots; Sapindales, Cucurbitales, Myrtaceae*. Berlin, Heidelberg: Springer-Verlag, 2011.

Kuiper, Koos. "Dutch Loan-words and Loan-translations in Modern China: An Example of Successful Sinification by Way of Japan." In *Words from the West: Western Texts in Chinese Library Context; Essay to Honor Erik Zürcher on his Sixty-Fifth Birthday*, edited by Lloyd Haft, 116–44. Leiden: Centre of Non-Western Studies, 1993.

Kuiper, Koos. *The Early Dutch Sinologists (1854–1900): Training in Holland and China, Functions in the Netherlands Indies*. Two volumes. Leiden, Boston: Brill, 2017.

Kunth, Karl Sigismund. *Enumeratio plantarum omnium hucusque cognitarum, secundum familias naturales disposita, adjectis characteribus, differentiis et synonymis*. Stuttgart and Tübingen: J. G. Cotta, 1833–1850.

Kurtz, Joachim. *The Discovery of Chinese Logic*. Leiden, Boston: Brill, 2011.

Kwong, Luke S. K. "Tan Sitong he Fu Lanya de yi ci huijian 譚嗣同和傅兰雅的一次会见 [An encounter between Tan Sitong and John Fryer]." *Jindaishi yanjiu* 近代史研究—*Modern Chinese History Studies*, no. 6 (1994): 194–99.

Kwong, Luke S. K. *T'an Ssu-t'ung, 1865–1898: Life and Thought of a Reformer:* Leiden, New York, Köln: E. J. Brill, 1996.

Lackner, Michael, Iwo Amelung, and Joachim Kurtz. *MCST Databases: An Electronic Repository of Chinese Scientific, Philosophical and Political Terms Coined in the Nineteenth and Early Twentieth Century*, 2001. Last consultation: August 2018, http://mcst.uni-hd.de/search/searchMCST_short.lasso

Lai, Xinxia 来新夏, ed. *Qingdai jingshi wen quanbian* 清代经世文全编 [Complete collection of Qing dynasty writings on statecraft]. Vol. 129. Beijing: Xueyuan chubanshe 学苑出版社, 2010.

Lehmann, Ulrich, and Gero Hillmer. *Fossil Invertebrates*. Translated by Janine Lettau. Cambridge: Cambridge University Press, 1983.

Lemaire, Gabriel, and Prosper Marie Giquel. *Dictionnaire de poche français-chinois: suivi d'un dictionnaire technique des mots usités à l'arsenal de Fou-Tcheou*. Shanghai: American Presbyterian Mission Press, 1874.

Leonard, John William, ed. *Who's Who in America: A Biographical Dictionary of Notable Living Men and Women of the United States; 1901–1902*. Chicago: A. N. Marquis & Company, 1901.

Levenson, Rosemary, and Elizabeth Huff. *Elizabeth Huff: Teacher and Founding Curator of the East Asiatic Library—From Urbana to Berkeley by Way of Peking*. Berkeley: Bancroft Library, University of California, Regional Oral History Office, 1977.

Lewis, W. J., William Theodore Aquila Barber, and John R. Hykes, ed. *Records of the General Conference of the Protestant Missionaries of China, Held at Shanghai, May 7–20, 1890*. Shanghai: American Presbyterian Mission Press, 1890.

Li, Nanqiu 黎难秋. "Qing mo keji ziliao fanyi chutan 清末科技资料翻译初探 [A preliminary exploration on the translation of scientific and technical materials in the late Qing dynasty]." *Zhongguo fanyi* 中国翻译, no. 2 (1981): 30–33.

Li, Nanqiu 黎难秋. "Shuxue wenxian fanyijia Hua Hengfang 数学文献翻译家华蘅芳 [Hua Hengfang, translator of mathematical texts]." *Zhongguo fanyi* 中国翻译, no. 6 (1982): 54–56.

Li, Nanqiu 黎难秋. "Shuangyu keji cidian shi lüeshu (shang) 双语科技辞典史略述（上）[A brief outline of the history of bilingual scientific and technical dictionaries (first part)]." *Shanghai keji fanyi* 上海科技翻译—*Shanghai Journal of Translators for Science and Technology*, no. 2 (1993): 41, 46–48.

Li, Nanqiu 黎难秋. "Shuangyu keji cidian shi lüeshu (xia) 双语科技辞典史略述（下）[A brief outline of the history of bilingual scientific and technical dictionaries (second part)]." *Shanghai keji fanyi* 上海科技翻译—*Shanghai Journal of Translators for Science and Technology*, no. 3 (1993): 45, 46–48.

Li, Nanqiu 黎难秋. *Zhongguo kexue fanyi shiliao* 中国科学翻译史料 [Historical materials on Chinese scientific translation]. Hefei: Zhongguo kexue jishu daxue chubanshe 中国科学技术大学出版社, 2006.

Li, Zipeng 黎子鹏. *Fuyin yanyi: wan Qing hanyu jidujiao xiaoshuo de shuxie* 福音演義：晚清漢語基督教小說的書寫 [Romance of the Gospel: the writing of Christian novels in Chinese in the late Qing dynasty]. Taipei: Taida chuban zhongxin 臺大出版中心, 2017.

Li, Li 李丽. *Jindai huaxue yizhu zhong de huaxue yuansuci yanjiu* 近代化学译著中的化学元素词研究 [Research on chemical elements nomenclature in translated works on modern chemistry]. Beijing: Zhongyang minzu daxue chubanshe 中央民族大学出版社, 2012.

Li, San-pao 李三寶. "Letters to the Editor in John Fryer's *Chinese Scientific Magazine*, 1876–1892: An Analysis." *Bulletin of the Institute of Modern History* 4 (1974): 729–77. Taipei: Academia Sinica.

Li, Wen 李文, and Dai Wusan 戴吾三. "*Qiji Zhong-Xi mingmu biao* yu Zhongguo jindai jixie gongcheng shuyu de fanyi '汽机中西名目表'与中国近代机械工程术语的翻译 [The Vocabulary of Terms Relating to the Steam Engine and the translation of Chinese modern technical terminology of mechanical engineering]." In *Di san jie Zhong-Ri jixie jishu shi guoji xueshu huiyi lunwenji* 第三届中日机械技术史国际学术会议论文集 [Proceedings of the third international academic conference on the history of Chinese and Japanese mechanical technique], edited by CJICHMT—2002 bianji weiyuanhui CJICHMT—2002 编辑委员会, 454–59. Zhongguo jixie gongcheng xuehui jixie shi fenhui 中国机械工程学会机械史分会, 2002.

Li, Xiaoti 李孝悌. "Jianli xin shiye: wan Qing de baike quanshu jia 建立新事业: 晚清的百科全书家 [Establishing a new cause: the encyclopædists of the late Qing dynasty]." *Dongwu xueshu* 东吴学术, no. 2 (2012): 45–55.

Li, Xiaoti 李孝悌. "Jianli xin shiye: wan Qing de baike quanshu jia (xu) 建立新事业: 晚清的百科全书家（续） [Establishing a new cause: the encyclopædists of the late Qing dynasty (second part)]." *Dongwu xueshu* 东吴学术, no. 3 (2012): 46–57, 89.

Li, Yadong 李亚东. "Xu Shou suo yi huaxue zhuzuo de yuanben 徐寿所译化学著作的原本 [The original texts of translations on chemistry by Xu Shou]." *Huaxue tongbao* 化学通报, no. 3 (1985): 52–55.

Li, Zhijun 李志军. "Gezhi shuyuan yu shixue jiaoyu 格致书院与实学教育 [The Polytechnic institution and the education of practical learning]." *Qing shi yanjiu* 清史研究, no. 3 (1999): 45–54.

Liang, Qichao 梁启超. "Xixue shumu biao 西學書目表 [Bibliography of Western learning]." In *Zhixue congshu chuji* 質學叢書初集 [Collectanea of substantive learning: first collection], edited by Zhixuehui 質學會, vol. 9–10. Wuchang: Zhixuehui 質學會, 1896.

Liao, Wei-wen Roger. "Morphology." In *The Handbook of Chinese Linguistics*, edited by C.-T. James Huang, Y.-H. Audrey Li and Andrew Simpson, 3–25. Malden: Wiley Blackwell, 2014.

Lin, Yutang 林語堂. "Guanyu yiming tongyi de tiyi 關於譯名統一的提議 [Suggestions on the standardisation of translated terms]." In *Yuyanxue luncong* 語言學論叢. Shanghai: Kaiming shudian 開明書店, 1933.

Liu, Guangding 劉廣定. *Zhongguo kexue shi lunji* 中國科學史論集 [Collected essays on the history of Chinese science]. Taipei: Guoli Taiwan daxue chuban zhongxin 國立台灣大學出版中心, 2002.

Liu, Zhengtan 刘正埮, Gao Mingkai 高名凯, Mai Yongqian 麦永乾, and Shi Youwei 史有为. *Hanyu wailaici cidian* 汉语外来词词典 [A dictionary of Chinese loanwords]. Shanghai: Shanghai cishu chubanshe 上海辞书出版社, 1984.

Liu, Lydia He, ed. *Tokens of Exchange: The Problem of Translation in Global Circulations*. Durham: Duke University Press, 1999.

Liu, Lydia He. *The Clash of Empires: The Invention of China in Modern World Making*. Cambridge, MA: Harvard University Press, 2004.

Lobscheid, Wilhelm. *Ying-Hua zidian* 英華字典*—English and Chinese Dictionary, with the Punti and Mandarin Pronunciation*. Hong Kong: Daily Press Office, 1866–1869.

Lodwick, Kathleen Lorraine. *The Chinese Recorder Index: A Guide to Christian Missions in Asia, 1867–1941*. Two volumes. Wilmington: Scholarly Resources Inc., 1986.

Luan, Xueqian 欒學謙. "Gezhi shuyuan jiaoyan huaxue ji 格致書院教演化學記 [A memoir on the teaching and demonstrations of chemistry at the Polytechnic institution]." *Xinxue bao* 新學報, no. 1 (1897).

Luo, Changpei 羅常培. *Guoyin zimu yanjin shi* 國音字母演進史 [A history of the development of the national phonetic alphabet]. Shanghai: Shangwu yinshuguan 商務印書館, 1934.

Luo, Jialun 羅家倫. "Zhongguo ruo yao you kexue, kexue yingdang xian shuo zhongguohua 中國若要有科學， 科學應當先說中國話 [If China wants to have science, science needs in the first place to speak Chinese]." *Tushu pinglun* 圖書評論 1, no. 3 (1932): 1–5.

Macgowan, Daniel Jerome, and Hua Hengfang 華蘅芳, trans. *Jinshi shibie* 金石識別 [Identification of minerals—Manual of Mineralogy]. Shanghai: Jiangnan zhizaoju 江南製造局, 1871.

Mai, Zhonghua 麥仲華, ed. *Huangchao jingshi wen xinbian* 皇朝經世文新編 [New collection of Qing dynasty writings on statecraft]. Publication place and year unknown.

Main, Thomas John, and Thomas Brown. *The Marine Steam-Engine: Designed Chiefly for the Use of the Officers of Her Majesty's Navy*. Fourth edition. London: Longman Brown, Green and Longmans, 1860.

Maisch, John Michael, ed. *The American Journal of Pharmacy: Published by Authority of the Philadelphia College of Pharmacy; Volume XLIII; Fourth Series; Vol. I*. Philadelphia: Merriehew & Son, 1871.

Marcon, Federico. *The Knowledge of Nature and the Nature of Knowledge in Early Modern Japan*. Chicago: University of Chicago Press, 2015.

Marquis, Albert Nelson, ed. *Who's Who in America: A Biographical Dictionary of Notable Living Men and Women of the United States; Vol. VI, 1910–1911*. Chicago: A. N. Marquis & Company; London: Kegan Paul, Trench, Trubner & Co., 1910.

Martin, William Alexander Parsons. *Gewu rumen* 格物入門 [Introduction to the investigation of things]. Beijing: Tongwenguan 同文館, 1868.

Martin, William Alexander Parsons. *The Awakening of China*. New York: Doubleday, Page & Company, 1907.

Masini, Federico. *The Formation of Modern Chinese Lexicon and Its Evolution Toward a National Language: The Period from 1840 to 1898*. Berkeley: Journal of Chinese Linguistics, Monograph series number 6, 1993.

Mateer, Calvin Wilson. "School Books for China." *The Chinese Recorder and Missionary Journal* 8, no. 5 (1877): 427–32. Shanghai.

Mateer, Calvin Wilson. *Technical Terms, English and Chinese Prepared by the Committee of the Educational Association of China*. Shanghai: Presbyterian Mission Press, 1904.

Matsumura, Jinzō 松村任三. *Shokubutsu mei-i: Enumeration of Selected Scientific Names of Both Native and Foreign Plants, with Romanized Japanese Names, and in Many Cases Chinese Characters*. Tokyo: Maruzen kabushiki kaisha 丸善株式會社, 1897.

Matsumura, Jinzō 松村任三 *Shokubutsu mei-i* 植物名彙 [A collection of names of plants]. Revised and enlarged edition. Tokyo, Osaka, Kyoto, Fukuoka: Maruzen kabushiki kaisha 丸善株式會社, 1916.

McIntosh, Gilbert. *The Mission Press in China, Being a Jubilee Retrospect of the American Presbyterian Mission Press, with Sketches of Other Mission Presses in China, as well as Accounts of the Bible and Tract Societies at Work in China*. Shanghai: American Presbyterian Mission Press, 1895.

Médard, Jules. *Fa-Han zhuanmen cidian* 法漢專門詞典—*Vocabulaire Français-Chinois des sciences morales et politiques*. Tianjin: Société française de librairie et d'édition, 1927.

Medhurst, Walter Henry. *English and Chinese Dictionary: In Two Volumes*. Shanghai: Mission Press, 1847–1848.

Medhurst, Walter Henry. *General Description of Shanghae and its Environs, Extracted from Native Authorities*. Shanghai: Mission Press, 1850.

Métailié, Georges. "La création lexicale dans le premier traité de botanique occidentale publié en chinois (1858)." *Documents pour l'histoire du vocabulaire scientifique*, no. 2 (1981): 65–73.

Métailié, Georges. "Botanical Terminology of Li Shizhen in *Bencao gangmu*." In *Essays on Science*, edited by Hakim Mohammed Said, 140–53. Karachi: Hamdard Foundation Pakistan, 1990.

Métailié, Georges. "Plantes et noms, plantes sans nom dans le *Zhiwu mingshi tukao*." *Extrême-Orient Extrême-Occident—Le Juste Nom*, no. 15 (1993): 138–48.

Métailié, Georges. "À propos du sexe des fleurs: le cas des 'rui.'" *Cahiers de linguistique—Asie Orientale* 23, no. 1 (1994): 223–30.

Métailié, Georges. "The Formation of Botanical Terminology: A Model or a Case Study?" In *New Terms for New Ideas: Western Knowledge and Lexical Change in Late Imperial China*, edited by Michael Lackner, Iwo Amelung and Joachim Kurtz, 327–38. Leiden, Boston, Köln: Brill, 2001.

Métailié, Georges. "Terminologie et approche interculturelle: terminologies scientifiques du chinois." *Langages*, no. 157 (2005): 117–27.

Miao, Qiyu 缪启愉, and Miao Guilong 缪桂龙, ed. *Qimin yaoshu yizhu* 齐民要术译注 [Essential techniques for common people: translated and annotated]. Jinan: Qilu shushe 齐鲁书社, 2009.

Milne, Colin. *A Botanical Dictionary: Or, Elements of Systematic and Philosophical Botany*. Third edition. London: Printed for H. D. Symonds by Bye and Law, 1805.

Milne-Edwards, Henri, and Pierre Dominique Vavasseur. *A Manual of Materia Medica and Pharmacy, Comprising a Concise Description of the Articles Used in Medicine; Their Physical and Chemical Properties; The Botanical Characters of the Medicinal Plants; And the Formulæ for the Principal Officinal Preparations of the American, Parisian, London, Dublin, Edinburgh, &c. Pharmacopœiæ; With Observations on the Proper Mode of Combining and Administering Remedies*. Philadelphia: Carey, Lea & Carey, 1829.

Mitchell, Leslie. *Maurice Bowra: A Life*. Oxford: Oxford University Press, 2009.

Mittler, Barbara. "China's 'New' Encyclopaedias and Their Readers." In *Chinese Encyclopaedias of New Global Knowledge (1870–1930): Changing Ways of Thought*, edited by Milena Doleželová-Velingerová and Rudolf G. Wagner, 399–424. Heidelberg: Springer, 2013.

Miyazaki, Ryūjō 宮崎柳条, ed. *Hyakka kōgyō shinsho* 百科工業新書 [A new encyclopaedic book for the industry]. Tokyo: Seifūkaku 清風閣, 1880.

Morgan, Evan. *Chinese New Terms and Expressions, with English Translations, Introduction and Notes*. Shanghai: Kelly & Walsh, 1913.

Morgan, Evan. *Chinese New Terms: Revised and Enlarged with English Translations, Classifications, Introduction and Index*. Shanghai: Kelly & Walsh, 1932.

Morohashi, Tetsuji 諸橋轍次. *Dai kanwa jiten* 大漢和辞典 [Great Sino-Japanese dictionary]. Tokyo: Taishūkan shoten 大修館書店, 1971.

Morrison, Robert. *A Dictionary of the Chinese Language, in Three Parts*. Macao: East India Company's Press, 1815–1823.

Morrison, John Robert. *A Chinese Commercial Guide, Consisting of a Collection of Details Respecting Foreign Trade in China*. Canton: Albion Press, 1834.

Morrison, John Robert. *A Chinese Commercial Guide, Consisting of a Collection of Details and Regulations Respecting Foreign with China*. Second edition. Macao: Samuel Wells Williams, 1844.

Morrison, John Robert. *A Chinese Commercial Guide, Consisting of a Collection of Details and Regulations Respecting Foreign with China*. Third edition. Canton: Office of the Chinese Repository, 1848.

Mouat, Melissa. "The Establishment of the Tongwen guan and the Fragile Sino-British Peace of the 1860s." *Journal of World History* 26, no. 4 (2015): 733–55.

Mullaney, Thomas S. *The Chinese Typewriter: A History*. Cambridge, MA: The MIT Press, 2017.

Nagoya shi Hōka bunko 名古屋市蓬左文庫, ed. *Nagoya sōsho sanpen* 名古屋叢書三編 [Third collection of Nagoya's collectanea]. Nagoya: Nagoya shi kyōiku iinkai 名古屋市教育委員会, 1982–1990.

Nakamura, Seiji 中村清二. *Jinshi wulixue jiaokeshu* 近世物理學教科書 [A textbook of modern physics]. Translated by Xuebu bianyi tushuju 學部編譯圖書局. Beijing: Jingshi guanshuju 京師官書局, 1906.

Nakamura, Tamekuni 中村為邦. *Jiangsu shifan jiangyi—Wuli* 江蘇師範講義—物理 [Teaching notes for normal schools in Jiangsu—Physics]. Jiangsu xuewuchu 江蘇學務處, 1906.

Nakayama, Shigeru 中山茂. "Translation of Modern Scientific Terms into Chinese Characters—The Chinese and Japanese Behavior in Comparison." In *Jindai Zhongguo keji shi lunji* 近代中國科技史論集 [Collected essays on the history of modern Chinese science and technology], edited by Yang Cuihua 楊翠華 and Huang Yinong 黃一農, 295–305. Taipei: Zhongyang yanjiuyuan jindaishi yanjiusuo 中央研究院近代史研究所, Guoli Qinghua daxue lishi yanjiusuo 國立清華大學歷史研究所, 1990.

Nakayama, Shigeru 中山茂, and Nathan Sivin, ed. *Chinese Science: Explorations of an Ancient Tradition*. Cambridge, MA: The MIT Press, 1973.

Needham, Joseph, Ho Ping-yü, and Lu Gwei-djen. *Science and Civilization in China, vol. 5, part 3—Spagyrical Discovery and Invention: Historical Survey, from Cinnabar Elixirs to Synthetic Insulin*. Cambridge: Cambridge University Press, 1976.

Norman, Jerry. *Chinese*. Cambridge: Cambridge University Press, 1988.

Novotná, Zdenka. "Linguistic Factors of the Low Adaptability of Loan-words to the Lexical System of Modern Chinese." *Monumenta Serica* 26 (1967): 103–18.

Novotná, Zdenka. "Contribution to the Study of Loan-words and Hybrid Words in Modern Chinese." *Archiv Orientální* 35 (1967): 613–48; 36 (1968): 295–325; 37 (1969): 48–75.

(Hermanova-) Novotná, Zdenka. *Affix-Like Word-Formation Patterns in Modern Chinese*. Dissertationes Orientales, vol. 21. Prague: Oriental Institute in Academia, 1969.

(Hermanova-) Novotná, Zdenka. "Morphemic Reproduction of Foreign Lexical Models in Modern Chinese." *Archiv Orientální* 43 (1975): 146–71.

Paasch, Heinrich. *From Keel to Truck—A Marine Dictionary in English, French and German Amply Illustrated by Explanatory Diagrams of the Most Important Details—For the Use of Ship-Owners, Builders, Brokers, Insurance-Societies, Average-Staters, Barristers, Solicitors, Translators, Surveyors, Masters, Mates, Engineers, Naval-schools, Nautical Journalists, etc.* Antwerp: Ratinckx Frères, 1885.

Packard, Jerome Lee, ed. *New Approaches to Chinese Word-Formation: Morphology, Phonology and the Lexicon in Modern and Ancient Chinese*. Trends in Linguistics, Studies and Monographs 105. Berlin, New York: Mouton de Gruyter, 1998.

Pan, Jixing 潘吉星. "Tan 'zhiwuxue' yi ci zai Zhongguo he Riben de youlai 谈'植物学'一词在中国和日本的由来 [Discussing the origin of the word 'botany' in China and Japan]." *Da ziran tansuo* 大自然探索, no. 3 (1984): 167–72.

Pan, Jixing 潘吉星. *Zhong wai kexue zhi jiaoliu* 中外科學之交流 [The scientific exchanges between China and abroad]. Hong Kong: Xianggang zhongwen daxue 中文大學出版社, 1993.

Parker, Edward Harper. *Up the Yang-tse*. Hong Kong: China Mail Office, 1891.

Parker, Alvin Pierson, and Xie Honglai 謝洪賚, trans. *Gewu zhixue* 格物質學 [Investigation of things and their properties—Popular Physics]. Shanghai: Meihua shuguan 美華書館, 1898.

Perny, Paul. *Dictionnaire français-latin-chinois de la langue mandarine parlée*. Paris: Librairie de Firmin Didot frères, fils et cie., 1869.

Perny, Paul. *Appendice du dictionnaire français-latin-chinois de la langue mandarine parlée*. Paris: Maisonneuve et cie., Ernest Leroux, 1872.

Playfair, George Macdonald Home. *The Cities and Towns of China: A Geographical Dictionary*. Hong Kong: Noronha & Co., 1879.

Qian, Yimin 钱益民. "Zheng Zhenwen yu wo guo huaxue mingci tongyi gongzuo 郑贞文与我国化学名词统一工作 [Zheng Zhenwen and the standardisation work for chemistry in China]." *Keji shuyu yanjiu* 科技术语研究 4, no. 3 (2002): 40–43.

Qilu zhuren 杞盧主人, ed. *Shiwu tongkao* 時務通攷 [Comprehensive examination of current affairs]. Shanghai: Dianshizhai 点石齋, 1897.

Qiu ziqiangzhai zhuren 求自彊齋主人. *Xizheng congshu* 西政叢書 [Collectanea on Western governance]. Shanghai: Shenji shuzhuang 慎記書莊, 1897.

Quanguo jingji weiyuanhui shuili chu 全國經濟委員會水利處, ed. *Shuili gongcheng mingci cao'an* 水利工程名詞草案 [A draft for terms in hydraulic engineering]. Nanjing: Quanguo jingji weiyuanhui 全國經濟委員會, 1935.

Rahman, Abdur, ed. *History of Science, Philosophy, and Culture in Indian Civilization—Volume III Part 2: India's Interaction with China, Central and West Asia*. New Delhi, New York: Oxford University Press, 2002.

Reardon-Anderson, James. *The Study of Change: Chemistry in China, 1840–1949*. Cambridge: Cambridge University Press, 1991.

Records of the Triennial Meeting of the Educational Association of China, Held at Shanghai, May 2–4, 1893. Shanghai: American Presbyterian Mission Press, 1893.

Records of the Fifth Triennial Meeting of the Educational Association of China, Held at Shanghai, May 17–20, 1905. Shanghai: American Presbyterian Mission Press, 1906.

Records of the Sixth Triennial Meeting of the Educational Association of China, Held at Shanghai, May 19–22, 1909. Shanghai: The Educational Association of China, 1909.

Reddy, Danda Venkata Subba. "John Forbes Royle, Botanist, Medical Historian, Teacher, and Benefactor of British Empire." *Bulletin of the Institute of Medicine Hyderabad* 3, no. 2 (1973): 79–87.

Ren, Xueliang 任学良. *Hanyu zaocifa* 汉语造词法 [Methods of word formation in Chinese]. Beijing: Zhongguo shehui kexue chubanshe 中国社会科学出版社, 1981.

Richard, Timothy, and Donald MacGillivray, ed. *A Dictionary of Philosophical Terms: Chiefly from the Japanese*. Shanghai: Christian Literature Society for China, 1913.

Royle, John Forbes. *Essay on the Productive Resources of India*. London: Wm. H. Allen and Co., 1840.

Royle, John Forbes, and Frederick William Headland. *A Manual of Materia Medica and Therapeutics; Including the Preparations of the British Pharmacopœia, and Many Other Approved Medicines*. Fifth edition. London: John Churchill & Sons, 1867.

Ruoff, Henry Woldmar. *Masters of Achievement: The World's Greatest Leaders in Literature, Art, Religion, Philosophy, Science, Politics and Industry*. Buffalo, NY: Frontier Press Company, 1910.

Sa, Bendong 薩本棟, ed. *Wulixue mingcihui* 物理學名詞彙*—English-Chinese Physics Terminology*. Beiping: Zhonghua jiaoyu wenhua jijin dongshihui bianji weiyuanhui 中華教育文化基金董事會編輯委員會, 1932.

Sakade, Yoshinobu 坂出祥伸. *Chūgoku kindai no shisō to kagaku* 中国近代の思想と科学 [Modern Chinese thought and science]. Kyoto: Dōhōsha shuppan 同朋舎出版, 1983.

Schlegel, Gustave. *He-Hua wenyu leican* 荷華文語類參*—Nederlandsch-Chineesch Woordenboek met de Transcriptie der Chineesche Karakters in het Tsiang-Tsiu Dialekt*. Leiden: Brill, 1886.

Sennett, Richard. *The Marine Steam Engine: A Treatise for the Use of Engineering Students and Officers of the Royal Navy*. Second edition. London: Longmans, Green, and Co., 1885.

Shang, Zhicong 尚智丛. "1886–1894 nian jian jindai kexue zai wan Qing zhishi fenzi zhong de yingxiang—Shanghai Gezhi shuyuan gezhi lei keyi fenxi 1886–1894年间近代科学在晚清知识分子中的影响—上海格致书院格致类课艺分析 [The influence of modern science among the intellectuals of the late Qing dynasty from 1886 to 1894—An analysis of the classified prize essay competition at the Shanghai Polytechnic institution]." *Qing shi yanjiu* 清史研究*—Studies in Qing History*, no. 3 (2001): 72–82.

Shao, Jingyu 邵靖宇. "Gui zi de laili he bianqian 硅字的来历和变迁 [The origin and vicissitudes of the character *gui*]." *Zhongguo keji shuyu* 中国科技术语, no. 1 (2008): 46–48.

Shen, Guowei 沈国威. "*Xiandai hanyu zhong de riyu jieci zhi yanjiu—Xushuo* 现代汉语中的日语借词之研究—序说 [Research on Japanese loans in modern Chinese—Preliminary remarks]." *Riyu xuexi yu yanjiu* 日语学习与研究, no. 5 (1988): 14–19.

Shen, Guowei 沈国威. *Kindai nicchū goi kōryūshi: shin kango no seisei to juyō* 近代日中語彙交流史：新漢語の生成と受容 [A history of modern lexical exchanges between China and Japan: creation and acceptance of new Japanese words of Chinese origin]. Tokyo: Kasama shoin 笠間書院, 1994.

Shen, Guowei 沈国威. *Shin Jiga to sono goi—Kenkyū, sakuyin, eiinbon fu* '新爾雅' とその語彙一研究， 索引， 影印本附 [The New guide and its lexicon: research, index, and reproduction appended]. Tokyo: Hakuteisha 白帝社, 1995.

Shen, Guowei 沈国威. *Shokugaku keigen to Shokubutsugaku no goi: kindai nicchū shokubutsugaku yōgo no keisei to kōryū* 植学啓原と植物学の語彙： 近代日中植物学用語の形成と交流 [Principles of botany and the botanical lexicon: the formation and exchanges of modern botanical terminology between Japan and China]. Osaka: Kansai daigaku shuppanbu 関西大学出版部, 2000.

Shen, Guowei 沈国威. "The Creation of Technical Terms in English-Chinese Dictionaries from the Nineteenth Century." In *New Terms for New Ideas: Western Knowledge and Lexical Change in Late Imperial China*, edited by Michael Lackner, Iwo Amelung and Joachim Kurtz, 287–304. Leiden, Boston, Köln: Brill, 2001.

Shen, Guowei 沈国威. "Yici yu jieci—Chongdu Hu Yilu 'Lun yiming' (1914) 譯詞與借詞一重讀胡以魯'論譯名' (1914). [Translated words and loans—A rereading of On translated terms by Hu Yilu]." *Wakumon* 或問, no. 9 (2005): 103–12.

Shen, Guowei 沈国威. *Jindai Zhong-Ri cihui jiaoliu yanjiu: hanzi xinci de chuangzhi, rongshou yu gongxiang* 近代中日词汇交流研究： 汉字新词的创制、 容受与共享 [Research on modern lexical exchanges between China and Japan: creation, acceptance, and sharing of characters and new words]. Beijing: Zhonghua shuju 中华书局, 2010.

Shen, Guowei 沈国威. "Xifang xin gainian de rongshou yu zao xinzi wei yici—Yi Riben lanxuejia yu lai Hua chuanjiaoshi wei li 西方新概念的容受与造新字为译词一以日本兰学家与来华传教士为例 [The acceptance of new Western concepts and the creation of new characters for translating words—The examples of Japanese *Rangaku* and missionaries in China]." *Zhejiang daxue xuebao* (*Renwen shehui kexue ban*) 浙江大学学报（人文社会科学版）—*Journal of Zhejiang University* (*Humanities and Social Sciences*) 40, no. 1 (2010): 121–34.

Shen, Guowei 沈国威. "Chinese Language and the Modern: Contemplating Chinese in the East Asian Linguistic Environment." In: *A Selection of Essays on Oriental Studies of ICIS* 1, edited by Kansai daigaku 関西大学, 93–102. Osaka: Institute for Cultural Interaction Studies, Kansai University, 2011.

Shen, Guowei 沈国威. "Kindai kanji yakugo no kenkyū ni tsuite: chūgokugo kara no shiten 近代漢字訳語の研究について：中国語からの視点 [A Chinese perspective on the research of modern translated words composed of characters]." In *Higashi Ajia gengo sesshoku no kenkyū* 東アジア言語接触の研究 [Research on linguistic contact in East Asia], edited by Shen Guowei 沈国威 and Uchida Keiichi 内田慶市, 19–51. Osaka: Kansai daigaku shuppanbu 関西大学出版部, 2016.

Shen, Guowei 沈国威. *Yan Fu yu kexue* 严复与科学 [Yan Fu and science]. Nanjing: Fenghuang chubanshe 凤凰出版社, 2017.

Shen, Guowei 沈国威. "Women weishenme xuyao erzi ci? Yuyan jiechu yu hanyu de jindai yanhua: xushuo 我们为什么需要二字词？ 语言接触与汉语的近代演化：序说. [Why do we need disyllabic words? Language contact and the modern evolution of Chinese: preliminary remarks]." *Higashi Ajia bunka kōshō kenkyū* 東アジア文化交渉研究—*Journal of East Asian Cultural Interaction Studies* 10 (2017): 101–18.

Shu, Xincheng 舒新城, ed. *Zhonghua baike cidian* 中華百科辭典 [Chinese encyclopaedic dictionary]. Shanghai: Zhonghua shuju 中華書局, 1930.

Smith, Frederick Porter. *Contributions Towards the Materia Medica and Natural History of China: For the Use of Medical Missionaries & Native Medical Students*. Shanghai: American Presbyterian Mission Press; London: Trübner, 1871.

Sprengel, Kurt Polycarp Joachim. *Anleitung zur Kenntniss der Gewächse*. Halle: C. A. Kümmel, 1802–1804.

Stauffer, Milton Theobald, Tsinforn C. Wong, Malcolm Gardner Tewksbury, and China Continuation Committee—Special Committee on Survey and Occupation. *The Christian Occupation of China: A General Survey of the Numerical Strength and Geographical Distribution of the Christian Forces in China Made by the Special Committee on Survey and Occupation, China Continuation Committee, 1918–1921*. Shanghai: China Continuation Committee, 1922.

Stuart, George A. "Report on the Committee on Scientific Terminology." In *Records of the Sixth Triennial Meeting of the Educational Association of China, Held at Shanghai, May 19–22, 1909*. Part II, Minutes, Reports, Etc., 39–40. Shanghai: The Educational Association of China, 1909.

Stuart, George A. *Technical Terms, English and Chinese, Prepared by the Committee of the Educational Association of China*. Shanghai: Methodist Publishing House, 1910.

Sun, Banghua 孙邦华. "Yu Hua yingren Fu Lanya de zongjiao kexueguan chuyi 寓华英人傅兰雅的宗教科学观刍议 [A tentative suggestion on the scientific and religious view of John Fryer, an Englishman living in China]." *Xueshu yuekan* 学术月刊, no. 8 (1993): 65–73.

Sun, Lei 孙磊, and Lü Lingfeng 吕凌峰. "Jiangnan zhizaoju zheng qiji yizhu diben kao 江南制造局蒸汽机译著底本考 [An investigation on the original texts of books on the steam engine translated at the Jiangnan Arsenal]." *Wakumon* 或問, no. 20 (2011): 33–48.

Sun, Zengde 孙增德. "Yan Huiqing yu *Ying-Hua da cidian* 颜惠庆与'英华大辞典' [Yan Huiqing and An English and Chinese Standard Dictionary]." *Shizhi xuekan* 史志学刊, no. 3 (2015): 49–52.

Sun, Zuo 孫佐. *Shengli weisheng xin jiaokeshu* 生理衛生新教科書 [A new textbook on physical hygiene]. Shanghai: Shangwu yinshuguan 商務印書館, 1907.

Svarverud, Rune. "The Formation of a Chinese Lexicon of International Law 1847–1903." In *Mapping Meanings: The Field of New Learning in Late Qing China*, edited by Michael Lackner and Natascha Vittinghoff, 507–36. Leiden, Boston: Brill, 2004.

Svarverud, Rune. "Zhongguo de kexue yiming zhi zheng—Tongwenguan yu Jiangnan zhizaoju fanyiguan guanyu Zhongguo kexue shuyu fanyi de zhengyi 中国的科学译名之争—同文馆与江南制造局翻译馆关于中国科学术语翻译的争议 [The Chinese controversy on scientific translated terms—The dispute between the Tongwenguan and the Translation Department of the Jiangnan Arsenal on the translation of Chinese scientific and technical terminology]." In *Yinshua chuban yu zhishi huanliu—Shiliu shiji yihou de Dongya* 印刷出版与知识环流—十六世纪以后的东亚 [Printing and publishing and the circulation of knowledge—East Asia after the sixteenth century], edited by Guanxi daxue wenhua jiaoshexue jiaoyu yanjiu zhongxin 关西大学文化交涉学教育研究中心, Chuban bowuguan 出版博物馆, 402–22. Shanghai: Shanghai renmin chubanshe 上海人民出版社, 2011.

Tan, Chung, and Geng Yinzeng. *History of Science, Philosophy, and Culture in Indian Civilization—Volume III Part 6: India and China; Twenty Centuries of Civilization Interaction and Vibrations*. New Delhi: Project of History of Indian Science, Philosophy and Culture, Centre for Studies in Civilizations, 2005.

Tanabe, Keiya 田邊慶彌, and Wang Wozang 王我臧. *Han yi Riben falü jingji cidian* 漢譯日本法律經濟辭典 [A Japanese dictionary of law and economics translated into Chinese]. Shanghai: Shangwu yinshuguan 商務印書館, 1913.

Taranzano, Charles. *Vocabulaire français-chinois des sciences mathématiques, physiques et naturelles suivi d'un index anglais-français*. Sien-hsien: Imprimerie de la Mission Catholique, 1914.

Taranzano, Charles. *Supplément au vocabulaire français-chinois des sciences: classifications—Tableaux synoptiques*. Sien-hsien: Imprimerie de la Mission Catholique, 1923.

Teng, Ssu-yü, and John King Fairbank. *China's Response to the West: A Documentary Survey, 1839–1923*. Cambridge: Harvard University Press, 1954.

"The Advisability, or the Reverse, of Endeavouring to Convey Western Knowledge to the Chinese through the Medium of Their Own Language." *Journal of the North China Branch of the Royal Asiatic Society*, New series, vol. 21, no. 1 (1886): 1–21.

The Asiatic Journal and Monthly Register for British and Foreign India, China and Australasia: Vol. XXXVII—New Series. London: Wm. H. Allen and Co., 1842.

The Blue and Gold 1904. Junior Class, University of California, 1903.

Thomson, Joseph Clarke. "Medical Publications in Chinese." *The China Medical Missionary Journal* 1, no. 3 (1887): 115–21. Shanghai: Kelly & Walsh.

Thomson, Joseph Clarke. *Vocabulary of Medicines, in English and Chinese*. China Medical Association, 1889.

Thunberg, Carl Peter. *Flora Japonica: sistens plantas insularum Japonicarum secundum systema sexuale emendatum redactas ad XX classes, ordines, genera et species cum differentiis specificis, synonymis paucis, descriptionibus concinnis et XXXIX iconibus adiectis*. Leipzig: Müller, 1784.

Tiedemann, Gary. *Handbook of Christianity in China—Volume Two: 1800–present.* Leiden, Boston: Brill, 2010.

Tola, Gabriele. "Between Phonemic Loans and Descriptive Labelling: Principles, Examples and Lexical Innovation in the *Huaxue cailiao Zhong-Xi mingmu biao.*" *Rivista degli Studi Orientali* 90, no. 1–4, New series (2018): 251–74.

Tola, Gabriele. "John Fryer and *The Translator's Vade-mecum*: The Identification of the 'Vocabulary of Terms in Naval Architecture'." *Monumenta Serica* 66, no. 1 (2018): 91–114.

Tola, Gabriele. "The John Fryer Papers: State of the Art and New Perspectives." In *Italian Association for Chinese Studies. Selected Papers 2*, edited by Tommaso Pellin and Giorgio Trentin, 138–48. Venice: Cafoscarina, 2018.

Tsien, Tsuen-hsuin 钱存训. "Western Impact on China through Translation." *The Far Eastern Quarterly* 13, no. 3 (1954): 305–27.

T'sou, Benjamin Ka-Yin. "Language Contact and Lexical Innovation." In *New Terms for New Ideas: Western Knowledge and Lexical Change in Late Imperial China*, edited by Michael Lackner, Iwo Amelung and Joachim Kurtz, 35–53. Leiden, Boston, Köln: Brill, 2001.

Tsu Jing "Chinese Scripts, Codes, and Typewriting Machines." In *Science and Technology in Modern China, 1880s–1940s*, edited by Jing Tsu and Benjamin A. Elman, 115–51. Leiden, Boston: Brill, 2014.

Uchida, Keiichi 内田慶市. *Kindai ni okeru tōzai gengo bunka sesshoku no kenkyū* 近代における東西言語文化接触の研究 [Research on the modern linguistic and cultural contacts between East and West]. Osaka: Kansai daigaku shuppanbu 関西大学出版部, 2001.

Uchida, Keiichi 内田慶市. *Bunka kōshōgaku to gengo sesshoku: Chūgoku gengogaku ni okeru shūen kara no apurōchi* 文化交渉学と言語接触：中国言語学における周縁からのアプローチ [Cultural interaction and linguistic contact: approaching Chinese linguistics from the periphery]. Osaka: Kansai daigaku shuppanbu 関西大学出版部, 2010.

Uchida, Keiichi 内田慶市. *A Study of Cultural Interaction and Linguistic Contact: Approaching Chinese Linguistics from the Periphery*. Göttingen: V&R unipress GmbH, 2017.

Udagawa, Yōan 宇田川榕菴. *Shokugaku keigen* 植学啓原 [Principles of botany]. Edo: Suharaya Ihachi 須原屋伊八, 1833.

Van Campenhoudt, Marc. "Un apport du monde maritime à la terminologie notionnelle multilingue: étude du dictionnaire du capitaine Heinrich Paasch *De la quille à la pomme de mât* (1885–1901)." Two volumes. PhD diss., Paris XIII, 1994.

Von Möllendorff, Paul Georg. *Praktische Anleitung zur Erlernung der Hochchinesischen Sprache.* Shanghai: Kelly and Walsh, 1900.

Von Möllendorff, Paul Georg. *Catalogue of P. G. von Möllendorff's Library*. Shanghai: American Presbyterian Mission Press, 1905.

Von Möllendorff, Paul Georg, and Otto Franz von Möllendorff. *Manual of Chinese Bibliography, Being a List of Works and Essays Relating to China*. Shanghai, Kelly & Walsh; London: Trübner & Co., Görlitz, H. Tzschaschel, 1876.

Wagner, Rudolf G. "The Formation of Encyclopaedic Commonplaces during the Late Qing: Entries on the Newspaper." In *Chinese Encyclopaedias of New Global Knowledge (1870–1930): Changing Ways of Thought*, edited by Milena Doleželová-Velingerová and Rudolf G. Wagner, 103–35. Heidelberg: Springer, 2013.

Wang, Deyi 王德毅, ed. *Congshu jicheng xubian* 叢書集成續編 [Sequel to the collectaneous compendium]. Vol. 81, *Ziran kexue lei* 自然科學類. Taipei: Xinwenfeng chuban gongsi 新文豐出版公司, 1989.

Wang, Ermin 王爾敏. *Shanghai gezhi shuyuan zhilüe* 上海格致書院志略 [A short account of the Shanghai Polytechnic institution]. Hong Kong: Xianggang zhongwen daxue chubanshe 香港中文大學出版社, 1980.

Wang, Genyuan 王根元, and Cui Yunhao 崔云昊. "Guanyu *Jinshi shibie* de fanyi, chuban he diben 关于 '金石识别' 的翻译、 出版和底本 [On the translation, publication, and original text of Identification of minerals]." *Zhongguo keji shiliao* 中国科技史料—*China Historical Materials of Science and Technology* 11, no. 1 (1990): 89–96.

Wang, Gengtang 王赓唐, and Wang Xi'nan 王锡南. *Xu Shou, Xu Jianyin zhuanlun* 徐寿、 徐建寅传论 [A biography of Xu Shou and Xu Jianyin]. Beijing: Zuojia chubanshe 作家出版社, 2001.

Wang, Guowei 王國維, trans. *Bianxue* 辨學 [Logic—Elementary Lessons in Logic]. Beijing: Jingshi wudaomiao shoushuchu 京師五道廟售書處, 1908.

Wang, Hongxia 王红霞. "Fu Lanya de Xishu zhongyi shiye 傅兰雅的西书中译事业 [John Fryer's cause of translating Western books]." PhD diss., Fudan daxue 复旦大学, 2006.

Wang, Shi 王栻, Wang Zuoliang 王佐良, He Lin 贺麟, Feng Youlan 冯友兰, Hou Wailu 侯外庐, Li Zehou 李泽厚, Zhang Qizhi 张岂之, Yang Chao 杨超, and Wang Rufeng 王汝丰. *Lun Yan Fu yu Yan yi mingzhu* 论严复与严译名著 [A discussion on Yan Fu and his famous translated books]. Beijing: Shangwu yinshuguan 商务印书馆, 1982.

Wang, Tao 王韬, ed. *Gezhi shuyuan keyi* 格致書院課藝 [Polytechnic institution prize essay competition]. Fifteen volumes. Shanghai: Shanghai tushu jicheng yinshuju 上海圖書集成印書局, 1887–1894.

Wang, Xiqing 王西清, and Lu Tiqing 盧梯青, ed. *Xixue dacheng* 西學大成 [Great anthology of Western knowledge]. Shanghai: Zui liutang shufang 醉六堂書坊, 1885.

Wang, Yangzong 王扬宗. "Guanyu '*Huaxue jianyuan*' he '*Huaxue chujie*' 关于'化学鉴原'和'化学初阶' [On Mirror of chemistry and First steps in chemistry]." *Zhongguo keji shiliao* 中国科技史料—*China Historical Materials of Science and Technology* 11, no. 1 (1990): 84–88.

Wang, Yangzong 王扬宗. "Qing mo Yizhi shuhui tongyi keji shuyu gongzuo shuping 清末益智书会统一科技术语工作述评 [A review of the standardisation of scientific and technical terms by the Educational Association of China in the late Qing dynasty]." *Zhongguo keji shiliao* 中国科技史料—*China Historical Materials of Science and Technology* 12, no. 2 (1991): 9–19.

Wang, Yangzong 王扬宗. "Jiangnan zhizaoju fanyi shumu xinkao 江南制造局翻译书目新考 [A new investigation on the works translated at the Jiangnan Arsenal]." *Zhongguo keji shiliao* 中国科技史料—*China Historical Materials of Science and Technology* 16, no. 2 (1995): 3–18.

Wang, Yangzong 王扬宗. "*Gezhi huibian* yu xifang jindai keji zhishi zai Qing mo de chuanbo '格致汇编'与西方近代科技知识在清末的传播 [The Chinese Scientific Magazine and the circulation of Western modern scientific and technical knowledge in the late Qing dynasty]." *Zhongguo keji shiliao* 中国科技史料—*China Historical Materials of Science and Technology* 17, no. 1 (1996): 36–47.

Wang, Yangzong 王扬宗. *Fu Lanya yu jindai Zhongguo de kexue qimeng* 傅兰雅与近代中国的科学启蒙 [John Fryer and the scientific enlightenment of modern China]. Beijing: Kexue chubanshe 科学出版社, 2000.

Wang, Yangzong 王扬宗. "A New Inquiry into the Translation of Chemical Terms by John Fryer and Xu Shou." In *New Terms for New Ideas: Western Knowledge and Lexical Change in Late Imperial China*, edited by Michael Lackner, Iwo Amelung and Joachim Kurtz, 271–83. Leiden, Boston, Köln: Brill, 2001.

Wang, Yangzong 王扬宗. "Shanghai gezhi shuyuan de yi fen yishu qingdan 上海格致书院的一份译书清单 [A list of books translated at the Shanghai Polytechnic institution]." *Zhongguo keji shi zazhi* 中国科技史杂志—*The Chinese Journal for the History of Science and Technology* 27, no. 1 (2006): 54–60.

Wang, Yangzong 王扬宗. *Guodu shidai de qiren. Xu Shou de gushi* 过渡时代的奇人. 徐寿的故事 [An outstanding person in a transitional era: the story of Xu Shou]. Changchun: Jilin kexue jishu chubanshe 吉林科学技术出版社, 2012.

Wang, Yusheng 王渝生. "Hua Hengfang: Zhongguo jindai kexue de xianxingzhe he chuanbozhe 华蘅芳：中国近代科学的先行者和传播者 [Hua Hengfang: the forerunner and circulator of Chinese modern science]." *Ziran bianzhengfa tongxun* 自然辩证法通讯—*Journal of Dialectics of Nature*, no. 2 (1985): 60–74.

Wang, Yusheng 王渝生. "Hua Hengfang: Forerunner and Disseminator of Modern Science in China." In *Chinese Studies in the History and Philosophy of Science and Technology*, edited by Fan Dainian 范岱年 and Robert S. Cohen, 369–93. Dordrecht, Boston, London: Kluwer Academic Publishers, 1996.

Wang, Zhongchen 王中忱. "Xinshi yinshua, zujie dushi yu jindai chuban ziben de xingcheng—Shangwu yinshuguan chuangli de qianqian houhou 新式印刷、租界都市与近代出版资本的形成—商务印书馆创立的前前后后 [New-style printing, concessions, cities, and the formation of modern publishing capitals—Ins and outs of the foundation of the Commercial Press]." *Zhongguo xiandai wenxue yanjiu congkan* 中国现代文学研究丛刊, no. 1 (1999): 162–74.

Wang, Fengyun 汪丰云, and Gu Jiashan 顾家山. "*Gezhi xinbao* ji qi dui huaxue zhishi de chuanbo '格致新报'及其对化学知识的传播 [Revue Scientifique and its diffusion of chemical knowledge]." *Huaxue jiaoyu* 化学教育, no. 10 (2012): 91–94.

Wang, Guangren 汪广仁. "Xu Jianyin nianpu 徐建寅年谱 [Chronological life of Xu Jianyin]." *Qinghua daxue xuebao (Zhexue shehui kexue ban)* 清华大学学报 （哲学社会科学版）—*Journal of Tsinghua University* 12, no. 3 (1997): 39–50.

Wang, Guangren 汪广仁, ed. *Zhongguo jindai kexue xianqu Xu Shou fuzi yanjiu* 中国近代科学先驱徐寿父子研究 [Research on Xu Shou and his sons, the pioneers of China's modern science]. Beijing: Qinghua daxue chubanshe 清华大学出版社, 1998.

Wang, Guangren 汪广仁, and Xu Zhenya 徐振亚. *Haiguo xiezhu de Xu Shou fuzi* 海国撷珠的徐寿父子 [Xu Shou and his sons, who selected the quintessence of the West]. Beijing: Kexue chubanshe 科学出版社, 2000.

Wang, Rongbao 汪榮寶, and Ye Lan 葉瀾. *Xin erya* 新爾雅 [The new guide]. Shanghai: Mingquanshe 明權社, 1903.

Watson, Ernest. *The Principal Articles of Chinese Commerce (Import and Export)*. Shanghai: Inspectorate of General Customs, 1923.

Wei, Yungong 魏允恭, ed. *Jiangnan zhizaoju ji* 江南製造局記 [Records of the Jiangnan Arsenal]. Ten juan. Shanghai: Wenbao shuju 文寶書局, 1905. Reprinted in *Jindai Zhongguo shiliao congkan* 41 近代中國史料叢刊 41 [Collection of historical materials on modern China, 41], edited by 沈雲龍. Taipei: Wenhai chubanshe 文海出版社, 1969.

Wells, David Ames. *Principles and Applications of Chemistry; For the Use of Academies, High-Schools, and Colleges: Introducing the Latest Results of Scientific Discovery and Research, and Arranged with Special Reference to the Practical Application of Chemistry to the Arts and Employments of Common Life*. Fifth edition. New York: Ivison, Phinney and Co., 1862.

Wen, Changbin 温昌斌. "Minguo shiqi Guoli bianyiguan ji qi kexue mingci shending gongzuo 民国时期国立编译馆及其科学名词审定工作 [The National Institute for Compilation and Translation during the Republican era and its work for the examination and approval of scientific terms]." *Keji shuyu yanjiu* 科技术语研究 6, no. 3 (2004): 46–48.

Wen, Shufang 问书芳, Cui Ruifeng 崔瑞锋, and Qiao Yaming 乔亚铭. "Jiangnan zhizaoju yishu xuba wenxian jiazhi kaoshu 江南制造局译书序跋文献价值考述 [An examination and exposition of the documentary value of the prefaces and postscripts in the books translated at the Jiangnan Arsenal]." *Tushuguan lilun yu shijian* 图书馆理论与实践—*Library theory and practice*, 3 (2018): 69–74.

Wiebusch, Thekla, and Uri Tadmor. "Loanwords in Mandarin Chinese." In *Loanwords in the World's Languages: A Comparative Handbook*, edited by Martin Haspelmath and Uri Tadmor, 575–98. Berlin: De Gruyter Mouton, 2009.

Wieger, Léon. *Chinese Characters: Their Origin, Etymology, History, Classification and Signification; A Thorough Study from Chinese Documents*. Translated by L. Davrout. Ho-kien-fu: Catholic Mission Press, 1915.

Wilhelm, Richard. *De-Ying-Han wen kexue zidian* 德英華文科學字典*—Deutsch-Englisch-Chinesisches Fachwörterbuch*. Qingdao: Deutsch-Chinesische Hochschule, 1911.

Williams, Charles Alfred Speed. *Hai kuan yü yen pi hsü* 海關語言必須*—An Anglo-Chinese Glossary for Customs and Commercial Use*. Shanghai: Commercial Press Limited, 1914.

Williams, R. John. *The Buddha in the Machine: Art, Technology, and the Meeting of East and West*. New Haven: Yale University Press, 2014.

Williams, Samuel Wells. *A Chinese Commercial Guide, Consisting of a Collection of Details and Regulations Respecting Foreign Trade with China, Sailing Directions, Tables, &c*. Fourth edition, revised and enlarged. Canton: Office of the Chinese Repository, 1856.

Williams, Samuel Wells. *A Syllabic Dictionary of the Chinese Language; Arranged According to the Wu-Fang Yuen Yin, With the Pronunciation of the Characters as Heard in Peking, Canton, Amoy and Shanghai*. Shanghai: American Presbyterian Mission Press, 1874.

Williamson, Alexander. "Uniformity of Terms in the School and Text Book Series." *The Chinese Recorder and Missionary Journal* 10, no. 4 (1879): 303–304.

Williamson, Alexander. "Editorial Notes and Missionary News—School and Text Book Series Committee." In *The Chinese Recorder and Missionary Journal, Volume XXI*, no. 3, 138. Shanghai: Presbyterian Mission Press, 1890.

Williamson, Alexander, Joseph Edkins, and Li Shanlan 李善蘭, ed., trans. *Zhiwuxue* 植物學 [Botany]. Shanghai: Mohai shuguan 墨海書館, 1858.

Wirtz, Hans. *Deutsch-Chinesisches Verzeichnis von Fachausdrücken aus dem Gebiete der Physik und Verwandten Gebieten*. Qingdao: Deutsch-Chinesische Hochschule, Übersetzungsanstalt, 1910.

Witek, John W., ed. *Dicionário Português-Chinês*. Lisbon: Biblioteca Nacional Portugal; Macao: Instituto Português do Oriente; San Francisco: Ricci Institute for Chinese-Western Cultural History, University of San Francisco, 2001.

Wright, Arthur Frederick. "The Chinese Language and Foreign Ideas." In *Studies in Chinese Thought*, edited by Arthur Frederick Wright, 286–303. Chicago: University of Chicago Press, 1953.

Wright, David. "Careers in Western Science in Nineteenth-century China: Xu Shou and Xu Jianyin." *Journal of the Royal Asiatic Society*, third series, 5, no. 1 (1995): 49–90.

Wright, David. *Translating Science: The Transmission of Western Chemistry into Late Imperial China, 1840–1890*. Leiden, Boston, Köln: Brill, 2000.

Wu, Guoqing 吴国庆. "Gui zuowei yuansu ming yi kaishi jiu duzuo gui er bu duzuo xi 硅作为元素名一开始就读作gui而不读作xi [*Gui*, when indicating the chemical element, at first was read as *gui*, not as *xi*]." *Zhongguo keji shuyu* 中国科技术语, no. 4 (2010): 59–62.

Wu, Qiu 吳峸, trans. *Zhiwu shenglixue* 植物生理學—*Nongxue congshu* 農學叢書 [Plant physiology—Collectanea on agriculture]. Third edition. Shanghai: Xinxue shehui 新學社會, 1913.

Wu, Qiu 吳峸, trans. *Zhiwu shenglixue* 植物生理學—*Gaoji nongxue xiaoyong* 高級農學校用 [Plant physiology—For the use of agricultural high schools]. Sixth edition. Shanghai: Xinxue shehui 新學社會, 1926.

Wu, Xiaoxin, ed. *Christianity in China: A Scholars' Guide to Resources in the Libraries and Archives of the United States*. Second edition. London and New York: Routledge, 2015.

Wu, Shellen Xiao. *Empires of Coal: Fueling China's Entry into the Modern World Order, 1860–1920*. Stanford: Stanford University Press, 2015.

Wylie, Alexander. *Memorials of Protestant Missionaries to the Chinese: Giving a List of Their Publications, and Obituary Notices of the Deceased*. Shanghai: American Presbyterian Mission Press, 1867.

Wylie, Alexander, and Xu Shou 徐壽, trans. *Qiji faren* 汽機發軔 [A primer on the steam engine—The Marine Steam-Engine]. Shanghai: Jiangnan zhizaoju 江南製造局, 1871.

Xia, Jing 夏晶. "Wan Qing keji shuyu de fanyi—Yi Fu Lanya wei zhongxin 晚清科技术语的翻译—以傅兰雅为中心 [The translation of scientific and technical terms in the late Qing dynasty—A focus on John Fryer]." PhD diss., Wuhan daxue 武汉大学, 2011.

Xianggang Zhongguo yuwen xuehui 香港中国语文学会. *Jinxiandai hanyu xinci ciyuan cidian* 近现代汉语新词词源词典 [An etymological dictionary of modern Chinese neologisms]. Shanghai: Hanyu da cidian chubanshe 汉语大词典出版社, 2001.

Xie, Jiarong 謝家榮, ed. *Dizhixue* 地質學 [Geology]. Shanghai: Shangwu yinshuguan 商務印書館, 1926.

Xiong, Yuezhi 熊月之. "Shanghai Guang fangyanguan shilüe 上海广方言馆史略 [A brief history of the Shanghai Guangfang language school]." In *Shanghai shi yanjiu* 上海史研究 [Research on Shanghai], edited by Tang Zhenchang 唐振常 and Shen Hengchun 沈恒春, 176–211. Shanghai: Xuelin chubanshe 学林出版社, 1989.

Xiong, Yuezhi 熊月之. *Xixue Dongjian yu wan Qing shehui* 西学东渐与晚清社会 [The transfer of Western knowledge to the East and late Qing society]. Revised edition. Beijing: Zhongguo Renmin daxue chubanshe 中国人民大学出版社, 2011.

Xixue gezhi daquan 西學格致大全 [Comprehensive collection of the investigation of things and extension of knowledge of Western learning]. Hong Kong: Xianggang shuju 香港書局, 1897.

Xu, Hong 徐泓. "Xu Shou, Xu Jianyin yu Zhongguo jindai kexue yinjin 徐寿、 徐建寅与中国近代科学引进 [Xu Shou, Xu Jianyin, and the introduction of science in modern China]." *Chuban shiliao* 出版史料, no. 1 (2008): 109–13.

Xu, Weize 徐維則. *Dong Xi xue shulu* 東西學書錄 [Bibliography of Eastern and Western knowledge], 1899.

Xu, Zhenya 徐振亚. "Xu Shou fuzi dui Zhongguo jindai huaxue de gongxian 徐寿父子对中国近代化学的贡献 [The contribution of Xu Shou and his sons to modern Chinese chemistry]." *Daxue huaxue* 大学化学 15, no. 1 (2000): 58–62.

Xu, Zhenya 徐振亚. "Fu Lanya yu Zhongguo jindai huaxue 傅兰雅与中国近代化学 [John Fryer and Chinese modern chemistry]." *Beijing huagong daxue xuebao (Shehui kexue ban)* 北京化工大学学报 (社会科学版)—*Journal of Beijing University of Chemical Technology*, no. 2 (2001): 26, 55–64.

Xu, Zhenya 徐振亚, and Ruan Shenkang 阮慎康. "Xu Shou fuzi, zusun yizhu jianjie 徐寿父子、 祖孙译著简介 [A brief presentation of the translated works by Xu Shou, his sons, and his grandchildren]." *Zhongguo keji shiliao* 中国科技史料—*China Historical Materials of Science and Technology* 7, no. 1 (1986): 48–56.

Xu, Guangqiu 许光秋. *American Doctors in Canton: Modernization in China, 1835–1935*. New Brunswick, London: Transaction Publishers, 2011.

Xuebu shendingke 學部審定科. *Huaxue yuhui* 化學語彙 [A glossary of chemistry]. Shanghai: Shangwu yinshuguan 商務印書館, 1908.

Xuebu shendingke 學部審定科. *Wulixue yuhui* 物理學語彙 [A glossary of physics]. Shanghai: Shangwu yinshuguan 商務印書館, 1908.

Yan, Huiqing 顔惠慶. *An English and Chinese Standard Dictionary, Comprising 120,000 Words and Phrases, with Translations, Pronunciations, Definitions, Illustrations, Etc., Etc. with a Copious Appendix: In Two Volumes*. Shanghai: Shangwu yinshuguan 商務印書館, 1908.

Yan, Huiqing 顔惠慶. *Ying-Hua da cidian* 英華大辭典. *An English and Chinese Standard Dictionary, Comprising 120,000 Words and Phrases, with Translations, Pronunciations, Definitions, Illustrations etc., etc. with a Copious Appendix: In Two Volumes*. Third edition. Shanghai: Shangwu yinshuguan 商務印書館, 1910.

Yan, Huiqing 顔惠慶. *Ying-Hua da cidian* 英華大辭典. *An English and Chinese Standard Dictionary, Comprising 120,000 Words and Phrases, with Translations, Pronunciations, Definitions, Illustrations etc., etc. with a Copious Appendix. In Two Volumes*. Fourth edition. Shanghai: Shangwu yinshuguan 商務印書館, 1912.

Yan, Huiqing 顔惠慶. *An English and Chinese Standard Dictionary. Comprising 120,000 Words and Phrases, with Translations, Pronunciations, Definitions, Illustrations Etc., Etc. with a Copious Appendix: In Two Volumes*. Fourth edition. Shanghai: Shangwu yinshuguan 商務印書館, 1916.

Yan, Huiqing 顔惠慶. *Ying-Hua da cidian (xiaozi ben)* 英華大辭典(小字本)—*English and Chinese Standard Dictionary: Small Type Edition; Comprising 120,000 Words and*

Phrases with Translations, Pronunciations, Definitions, Illustrations, Etc., Etc. with a Copious Appendix; One volume. First edition. Shanghai: Shangwu yinshuguan 商務印書館, 1920.

Yan, Huiqing 顏惠慶. *Ying-Hua da cidian (xiaozi ben)* 英華大辭典(小字本)—*English and Chinese Standard Dictionary: Small Type Edition; Comprising 120,000 Words and Phrases with Translations, Pronunciations, Definitions, Illustrations, Etc., Etc. with a Copious Appendix; One volume*. Fifth edition. Shanghai: Shangwu yinshuguan 商務印書館, 1925.

Yan, Huiqing 顏惠慶. *East-West Kaleidoscope, 1877–1946: An Autobiography*. Jamaica, NY: St. John's University Press, 1974.

Yang, Gen 杨根. "Wo guo jindai huaxue xianquzhe Xu Shou de shengping ji zhuyao gongxian 我国近代化学先驱者徐寿的生平及主要贡献 [The life and main contributions of Xu Shou, the forerunner of Chinese modern chemistry]." *Huaxue tongbao* 化学通报, no. 4 (1984): 61, 71–76.

Yang, Gen 杨根, ed. *Xu Shou he Zhongguo jindai huaxue shi* 徐寿和中国近代化学史 [Xu Shou and the history of modern Chinese chemistry]. Beijing: Kexue jishu wenxian chubanshe 科学技术文献出版社, 1986.

Yang, Yi 样逸, et al. *Haishang molin—Guang fangyanguan quan'an—Fen mo congtan* 海上墨林—广方言馆全案—粉墨丛谈 [Shanghai's forest of ink—Complete documents of the Guang fangyanguan—Collected writings on powder and ink]. Shanghai: Shanghai guji chubanshe 上海古籍出版社, 1989.

Yates, Matthew Tyson, R. Nelson, and E. R. Barrett, ed. *Records of the General Conference of the Protestant Missionaries of China, Held at Shanghai, May 10–24, 1877*. Shanghai: Presbyterian Mission Press, 1878.

Yatsumimi, Toshifumi 八耳俊文. "Shin makki seijin choyaku kagaku kankei chūgokusho oyobi wakokubon shozai mokuroku 清末期西人著訳科学関係中国書および和刻本所在目録 [Chinese books related to science translated by Westerners in the late Qing dynasty, with an index of their Japanese editions]." *Kagakushi kenkyū* 化学史研究 22 (1995): 312–358.

Yip, Po-Ching. *The Chinese Lexicon: A Comprehensive Survey*. London, New York: Routledge, 2007.

Yong, Heming, and Jing Peng. *Chinese Lexicography: A History from 1046 BC to AD 1911*. Oxford: Oxford University Press, 2008.

Youji huaxue putong mingci cao'an 有機化學普通名詞草案 [A draft of common names for organic chemistry]. Year and place of publication unknown.

Yu, Wuheng 俞物恆, and Du Ruocheng 杜若城, ed. *Dizhixue* 地質學 [Geology]. Shanghai: Shangwu yinshuguan 商務印書館, 1932.

Yu, Yating 余雅婷. "*Guxin shengjing* no kenkyū '古新聖経'の研究 [Research on *Guxin shengjing*]." PhD diss., Kansai daigaku 関西大学, 2018.

Yuan, Hanqing 袁翰青. "Xu Shou—Wo guo jindai huaxue de qimengzhe 徐寿一我国近代化学的启蒙者 [Xu Shou, the enlightener of Chinese modern chemistry]." *Huaxue tongbao* 化学通报 (February 1955): 124–28.

Yuan, Hanqing 袁翰青. *Zhongguo huaxue shi lunwenji* 中国化学史论文集 [Collected essays on the history of Chinese chemistry]. Beijing: Sanlian shudian 三联书店, 1956.

Yuan, Junde 袁俊德, ed. *Fuqiangzhai congshu* 富強齋叢書 [Collectanea from the studio of wealth and power]. Shanghai: Xiaocang shanfang 小倉山房, 1897.

Yuan, Junde 袁俊德, ed. *Fuqiangzhai congshu xuji* 富強齋叢書續集 [A sequel to the Collectanea of the prosperous and strong studio]. Xiaocang shanfang 小倉山房, 1901.

Yuan, Yuan 袁媛. "Zhongguo zaoqi bufen shenglixue mingci de fanyi ji yanbian de chubu tantao 中国早期部分生理学名词的翻译及演变的初步探讨 [A preliminary discussion on the early translation of some Chinese words in physiology and their evolution]." *Ziran kexue shi yanjiu* 自然科学史研究—*Studies in the History of Natural Sciences* 25, no. 2 (2006): 170–81.

Yuan, Zonglian 袁宗濂, and Yan Zhiqing 晏志清, ed. *Xixue santong* 西學三通 [Three compilations on Western knowledge]. Shanghai: Cuixin shuguan 萃新書館, 1902.

Zeng, Pu 曾樸, and Xu Nianci 徐念慈. *Bowu da cidian* 博物大辭典 [The great dictionary of natural history]. Shanghai: Hongwenguan 宏文館, 1907.

Zeng, Zhaolun 曾昭掄. "Ershi nian lai Zhongguo huaxue zhi jinzhan 二十年來中國化學之進展 [The progress of Chinese chemistry in the last twenty years]." *Kexue* 科學 19, no. 10 (1935): 1514–554.

Zetzsche, Jost Oliver. *The Bible in China: The History of the Union Version or the Culmination of Protestant Missionary Bible Translation in China.* Monumenta Serica Monograph Series 45. Sankt Augustin: Institut Monumenta Serica, 1999.

Zhang, Daqing 张大庆. "Zaoqi yixue mingci tongyi gongzuo: Boyihui de nuli yu yingxiang 早期医学名词统一工作： 博医会的努力和影响 [The early standardisation work of medical terms: the commitment and influence of the Medical Missionary Society]." *Zhonghua yi shi zazhi* 中华医史杂志 24, no. 1 (1994): 15–19.

Zhang, Daqing 张大庆. "Gao Silan: yixue mingci fanyi biaozhunhua de tuidongzhe 高似兰：医学名词翻译标准化的推动者 [Philip Brunelleschi Cousland: the promoter of the standardisation of translations of medical terms]." *Zhongguo keji shiliao* 中国科技史料—*China Historical Materials of Science and Technology*, no. 4 (2001): 324–30.

Zhang, Hao 张澔. "Yang qing dan de fanyi: 1896–1944 nian 氧氢氮的翻译: 1896–1944 年 [The translation of oxygen, hydrogen, and nitrogen from 1896 to 1944]." *Ziran kexue shi yanjiu* 自然科学史研究—*Studies in the History of Natural Sciences* 21, no. 2 (2002): 123–34.

Zhang, Jinglu 张静庐, ed. *Zhongguo jindai chuban shiliao chubian* 中国近代出版史料初编 [First collection of historical materials on modern Chinese publishing]. Shanghai: Shanghai chubanshe 上海出版社, 1953.

Zhang, Qinglian 张青莲. "Xu Shou yu *Huaxue jianyuan* 徐寿与'化学鉴原' [Xu Shou and the Mirror of chemistry]." *Zhongguo keji shiliao* 中国科技史料—*China Historical Materials of Science and Technology* 6, no. 4 (1985): 54–56.

Zhang, Yinhuan 張蔭桓, ed. *Xixue fuqiang congshu* 西學富強叢書 [Collectanea of Western learning, wealth, and power]. Shanghai: Hongwen shuju 鴻文書局, 1896.

Zhang, Zigao 张子高. "He Liaoran de *Huaxue chujie* zai huaxue yuansu yiming shang de lishi yiyi 何了然的'化学初阶'在化学元素译名上的历史意义 [The historical significance of First steps in chemistry by He Liaoran in the translated terms for chemical elements]." *Qinghua daxue xuebao* 清华大学学报 9, no. 6 (1962): 41–47.

Zhang, Zigao 张子高. *Zhongguo huaxue shigao (gudai zhi bu)* 中国化学史稿（古代之部）[A draft of the history of Chinese chemistry]. Beijing: Kexue chubanshe 科学出版社, 1964.

Zhang, Zigao 张子高, and Yang Gen 杨根. "Cong *Huaxue chujie* he *Huaxue jianyuan* kan wo guo zaoqi fanyi de huaxue shuji he huaxue mingci 从化学初阶和化学鉴原看我国早期翻译的化学书籍和化学名词 [Looking at early Chinese translated books on chemistry and chemical terms through First steps in chemistry and Mirror of chemistry]." *Ziran kexue shi yanjiu* 自然科学史研究—*Studies in the History of Natural Sciences* 1, no. 4 (1982): 349–55.

Zhang, Zigao 张子高, and Yang Gen 杨根. "Xu Shou fuzi nianpu 徐寿父子年谱 [Chronological life of Xu Shou and his sons]." *Zhongguo keji shiliao* 中国科技史料—*China Historical Materials of Science and Technology*, no. 4 (1981): 55–62.

Zhang, Qing 章清. "'采西学'：学科次第之论辩及其意义—略论晚清对'西学门径'的探讨 'Cai xixue': xueke cidi zhi lunbian ji qi yiyi—Lüe lun wan Qing dui 'Xixue menjing' de tantao ['The adoption of Western learning': the debate on the hierarchy of branches of learning and its significance—A brief discussion of the exploration of 'approaches of Western learning' in the late Qing dynasty]." *Lishi yanjiu* 历史研究, no. 3 (2007): 107–28.

Zhang, Xiantao. *The Origins of the Modern Chinese Press; The Influence of the Protestant Missionary Press in Late Qing China*. London: Routledge, 2007.

Zhao, Pushan 赵璞珊. "Zhao Yuanyi he ta de bishu yishu 赵元益和他的笔述医书 [Zhao Yuanyi and the medical books he annotated]." *Zhongguo keji shiliao* 中国科技史料—*China Historical Materials of Science and Technology* 12, no. 1 (1991): 69–74.

Zhao, Zhongya 赵中亚. "Huaren bianji Luan Xueqian yu *Gezhi huibian* 华人编辑栾学谦与'格致汇编' [The Chinese editor Luan Xueqian and The Chinese Scientific Magazine]." *Shilin* 史林—*Historical Review*, no. 2 (2011): 101–108.

Zhao, Ziqin 趙子欽. "Zhongguo yiduo pin xiren cha kuang shuo lüe 中國宜多聘西人查礦說略 [A short explanation of why China should hire more Westerners to explore mines]." *Gezhi huibian* 格致彙編 8 (September 1877).

Zheng, Zhenwen 鄭貞文. *Huaxue mingming fa cao'an chugao* 化學命名法草案初稿 [A preliminary draft of the methods for the nomenclature of chemistry]. Shanghai: Zhonghua xueyi she 中華學藝社, 1933.

Zheng, Zhenwen 鄭貞文. *Ziran kexue cidian* 自然科學辭典 [A dictionary of natural science]. Shanghai: Huatong shuju 華通書局, 1934.

Zheng, Jinsheng, Nalini Kirk, and Paul D. Buell. *Dictionary of the Ben cao gang mu, Volume III: Persons and Literary Sources*. Oakland, CA: University of California Press, 2018.

Zhong, Shaohua 钟少华. *Renlei zhishi de xin gongju—Zhong-Ri jindai baike quanshu yanjiu* 人类知识的新工具—中日近代百科全书研究 [A new instrument for human knowledge—Research on modern Chinese and Japanese encyclopaedias]. Beijing: Beijing tushuguan chubanshe 北京图书馆出版社, 1996.

Zhong, Shaohua 钟少华. "Studies on the Characteristics of Late Qing Encyclopaedia Entries." In *Chinese Encyclopaedias of New Global Knowledge (1870–1930): Changing Ways of Thought*, edited by Milena Doleželová-Velingerová and Rudolf G. Wagner, 425–51. Heidelberg: Springer, 2013.

Zhongguo haiguan zong shuiwu sishu 中國海關總稅務司署. *Documents Illustrative of the Origin, Development and Activities of the Chinese Customs Service*. Vol. 3. Shanghai: Statistical Department of the Inspectorate General of Customs, 1938.

Zhongguo jixie gongcheng xuehui 中國機械工程學會, ed. *Ying-Han duizhao jixie gongcheng mingci* 英漢對照機械工程名詞 [Terms in mechanical engineering, with English-Chinese parallel text]. Shanghai: Shangwu yinshuguan 商務印書館, 1949.

Zhongguo shehui kexueyuan jindaishi suo fanyi shi 中国社会科学院近代史所翻译室. *Jindai lai Hua waiguo renming cidian* 近代来华外国人名辞典 [A dictionary of foreign persons arriving in China in modern times]. Beijing: Zhongguo shehui kexue chubanshe 中国社会科学出版社, 1981.

Zhonghua da zidian 中华大字典 [The great Chinese character dictionary]. Beijing: Zhonghua shuju 中华书局, 1978.

Zhou, Changshou 周昌寿. "Yikan kexue shuji kaolüe 譯刊科學書籍考略 [A short examination of the translations and publications of scientific books]." In *Zhang Jusheng xiansheng qishi shengri jinian lunwenji* 張菊生先生七十生日紀念論文集 [Commemorative essays for the seventieth birthday of Zhang Jusheng], edited by Hu Shi 胡適, Cai Yuanpei 蔡元培, and Wang Yunwu 王雲五, 409–70. Shanghai: Shangwu yinshuguan 商務印書館, 1937.

Zhou, Dingyi 周定一. "Yinyici he yiyici de xiaozhang 音译词和意译词的消长 [Growth and decline of phonemic loans and semantic loans)]." *Zhongguo yuwen* 中国语文, no. 119 (1962): 459–65, 476.

Zhou, Xinping 周欣平, ed. *Qing mo shixin xiaoshuo ji* 清末时新小说集 [A collection of new fiction of the late Qing dynasty]. Shanghai: Shanghai guji chubanshe 上海古籍出版社, 2011.

Zhou, Peter X. "At the Crossroads Between East and West: The C.V. Starr East Asian Library at the University of California, Berkeley." In *Collecting Asia: East Asian Libraries in North America, 1868–2008*, edited by Peter X Zhou, 67–79. Ann Arbor: Association for Asian Studies, 2010.

Zhu, Dawen 朱大文, and Ling Gengyang 淩賡颺. *Wanguo zhengzhi yixue quanshu* 萬國政治藝學全書 [A compendium of the politics and technical learning of the ten thousand nations]. Shanghai: Hongwen shuju 鴻文書局, 1902.

Zhu, Mouhan 朱谋埠. *Pian ya* 骈雅 [The rhythmical guide]. Beijing: Zhonghua shuju 中华书局, 1985.

Zhu, Youhuan 朱有瓛, ed. *Zhongguo jindai xuezhi shiliao* 中国近代学制史料 [Historical materials on modern Chinese school system]. Two volumes. Shanghai: Huadong shifan daxue chubanshe 华东师范大学出版社, 1983.

Zou Lingyuan 鄒淩沅. *Tongxuezhai congshu* 通學齋叢書 [Collectanea of the profound knowledge studio]. Publication place and year unknown.

Zou, Zhenhuan 邹振环. *Wan Qing xifang dilixue zai Zhongguo: Yi 1815 nian zhi 1911 nian xifang dilixue yizhu de chuanbo yu yingxiang wei zhongxin* 晚清西方地理学在中国：以1815至1911年西方地理学译著的传播与影响为中心 [Western geography in late Qing China: a focus on the circulation and influence of the translations of Western works on geography between 1815 and 1911]. Shanghai: Shanghai guji chubanshe 上海古籍出版社, 2000.

Zou, Zhenhuan 邹振环. "Guangxu huangdi de yingyu xuexi yu jinru Qing mo gongting de yingyu duben 光绪皇帝的英语学习与进入清末宫廷的英语读本 [The study of English during the Emperor Guangxu and the English readers which entered the late Qing court]." *Qing shi yanjiu* 清史研究—*The Qing History Journal*, no. 3 (2009): 107–15.

Zürcher, Erik. *The Buddhist Conquest of China: The Spread and Adaptation of Buddhism in Early Medieval China*. Third edition. Leiden: Brill, 2007.

Periodicals

Dongfang zazhi 東方雜誌 [The Eastern Miscellany]
Gezhi huibian 格致彙編 [The Chinese Scientific Magazine]
Gezhi xinbao 格致新報 [Revue Scientifique]
Iji shinbun 醫事新聞 [Medical journal]
Shenbao 申報 [Shanghai News]

Index

1) Names of people, institutions, and topics are in Roman; general topics are listed in lowercase letters, names of institutions, databases, and other proper nouns are in uppercase letters. For Chinese and Japanese names, characters are always provided.
2) Texts are indexed only when discussed in this volume; a full list of the works cited can be found in the bibliography. Chinese and Japanese texts are indexed under their original titles, in italics and with characters; their English translations are also listed, in Roman, with the first word written with an uppercase letter, and a reference to the corresponding relevant entry is provided.
3) Chinese names of non-Chinese authors, whenever present, are listed, with a reference to the corresponding relevant entry.
4) Most common names used throughout the book, such as names of cities and historical dynasties and eras, as well as frequent subjects or names, have not been indexed.
5) Chinese and Japanese terms in common use, such as *Rangaku* 蘭學 or *Tongwenguan* 同文館, are indexed as main entries; their English translations are also listed, with a reference to the corresponding relevant entry.
6) If alternative spellings for an entry or a sub-entry are listed in the text, they are provided in brackets; in the entry "terminology of the *TVM*", the alternative spellings or the alternative lemmata listed in the text are provided in brackets, without considering alternative spellings in lowercase and uppercase letters, or Latin translations.
7) The most frequently quoted works are listed in the book, after the first appearance, with abbreviations that are not indexed. For the abbreviations, see pages xvi–xvii.
8) Terms and relevant footnotes of the tables in the appendix are not indexed. For the list of tables, see pages xiii–xv.